Lecture Notes in Artificial Intelligence 8554

Subseries of Lecture Notes in Com~~puter Science~~

LNAI Series Editors

Randy Goebel
  University of Alberta, Edmonton, Canada
Yuzuru Tanaka
  Hokkaido University, Sapporo, Japan
Wolfgang Wahlster
  DFKI and Saarland University, Saarbrücken, Germany

LNAI Founding Series Editor

Joerg Siekmann
  DFKI and Saarland University, Saarbrücken, Germany

T0236331

Fabrizio Cariani   Davide Grossi
Joke Meheus   Xavier Parent (Eds.)

Deontic Logic and Normative Systems

12th International Conference, DEON 2014
Ghent, Belgium, July 12-15, 2014
Proceedings

 Springer

Volume Editors

Fabrizio Cariani
Northwestern University, Department of Philosophy
Evanston, IL, USA
E-mail: f-cariani@northwestern.edu

Davide Grossi
University of Liverpool, Department of Computer Science
Liverpool, UK
E-mail: d.grossi@liverpool.ac.uk

Joke Meheus
Ghent University, Centre for Logic and Philosophy of Science
Ghent, Belgium
E-mail: joke.meheus@ugent.be

Xavier Parent
University of Luxembourg, Luxembourg
E-mail: x.parent.xavier@gmail.com

ISSN 0302-9743 e-ISSN 1611-3349
ISBN 978-3-319-08614-9 e-ISBN 978-3-319-08615-6
DOI 10.1007/978-3-319-08615-6
Springer Cham Heidelberg New York Dordrecht London

Library of Congress Control Number: 2014941869

LNCS Sublibrary: SL 7 – Artificial Intelligence

© Springer International Publishing Switzerland 2014

Typesetting: Camera-ready by author, data conversion by Scientific Publishing Services, Chennai, India

Printed on acid-free paper

Springer is part of Springer Science+Business Media (www.springer.com)

Preface

This book constitutes the proceedings of ΔEON 2014, the 12th International Conference on Deontic Logic and Normative Systems held during July 12–15, 2014, at Ghent University, Belgium. The biennial ΔEON conferences are intended to promote interdisciplinary cooperation amongst scholars interested in linking the formal-logical study of normative concepts and normative systems with computer science, artificial intelligence, linguistics, philosophy, organization theory, and law. There have been eleven previous ΔEON conferences: Amsterdam, December 1991; Oslo, January 1994; Sesimbra, January 1996; Bologna, January 1998; Toulouse, January 2000; London, May 2002; Madeira, May 2004; Utrecht, July 2006; Luxembourg, July 2008; Fiesole, July 2010; Bergen, July 2012.

The conference has been renamed from "International Conference on Deontic Logic in Computer Science" to "International Conference on Deontic logic and Normative Systems", the acronym ΔEON being kept. This name change was decided by the ΔEON Steering Committee (http://www.deonticlogic.org), in order to broaden the scope of our conference series, which originated from within computer science.

The topics solicited for ΔEON 2014 included the following general themes:

- the logical study of normative reasoning, including formal systems of deontic logic, defeasible normative reasoning, logics of action, logics of time, and other related areas of logic
- the formal analysis of normative concepts and normative systems
- the formal specification of aspects of norm-governed multi-agent systems and autonomous agents, including (but not limited to) the representation of rights, authorization, delegation, power, responsibility, and liability
- the normative aspects of protocols for communication, negotiation, and multi-agent decision making
- the formal representation of legal knowledge
- the formal specification of normative systems for the management of bureaucratic processes in public or private administration
- the applications of normative logic to the specification of database integrity constraints

In addition to the above general themes, ΔEON 2014 had the following special theme: Deontic modalities in natural language. Deontic or normative modality is a subject of common interest for researchers in several fields, including moral philosophy, meta-ethics, linguistic semantics, and deontic logic. In the past, the deontic modalities were extensively studied on the logic side. Comparatively, much less attention was paid to them from a natural language perspective, at least in ΔEON. At the same time, there was a growing interest from linguists and philosophers in the study of the deontic modalities, mostly in the U.S.,

under the influence of Angelika Kratzer's work. A Deontic Modality Workshop
was held in Los Angeles during May 20–22, 2013. The aim of ΔEON 2014 was to
bring together these two communities. Topics of interest for this special theme
included:

- challenges from natural language for deontic logic
- the relationship between deontic and other types of modality: epistemic
  modality, imperatives, supererogatory, etc.
- the deontic paradoxes
- the modeling of normative concepts other than obligation and permission,
  e.g., values
- the game-theoretical aspects of deontic reasoning
- the emergence of norms
- norms from a conversational and pragmatic point of view
- norms and argumentation

Our call for papers attracted a variety of submissions, from both research communities. We received 43 abstracts, and reviewed 37 papers. This is slightly more
than usual. Of these, 21 were accepted for presentation at the conference and 17
are published in this volume. The titles themselves demonstrate commitment to
the themes of the conference. We believe that our attempt at bringing the two
communities together was a success.

The five keynote speakers were chosen in line with the special theme of the
conference. The first keynote speaker was Sven Ove Hansson from the Royal
Institute of Technology in Stockholm. His talk was titled "Deontic Diversity"
and the content of his presentation is displayed by the following abstract:

It is commonly assumed that deontic logic concerns "the" logic of normative concepts. However, a close look at actual usage shows that the
structural patterns of deontic notions differ between different usages.
Some of these differences are difficult to discern in natural language,
but may be easier to keep apart with the more precise tools of a formal
language. We should use the resources of deontic logic to discover and
distinguish between different meanings of the deontic terms in natural
language. Some of the ingrained disagreements on postulates in deontic logic may be resolvable if we recognize that the different viewpoints
correspond to different meanings of the normative terms of ordinary language.

Sven Ove Hansson's paper is included in the proceedings. The second keynote
speaker was Magdalena Kaufmann from the University of Connecticut. Her talk
was titled "Fine-Tuning Natural Language Modality", and is described by the
following abstract:

Over the past couple of years, the traditional binary classification of
modal expressions in natural languages into necessity operators and possibility operators has been called into question by a variety of distinctions

like descriptive vs. performative, information-sensitive vs. information-insensitive ("objective"), strong necessity vs. weak necessity, and subjective vs. objective. Not all of these distinctions are equally well-defined, and not all of them have been argued to be reflected in actual lexical distinctions. In my talk, I will first provide definitions that are sharp enough to allow us to administer linguistic tests, and I will then investigate for a couple of languages (English, German, Japanese) which of these categories relate to actual linguistic differences (specific linguistic items or constructions) and, most importantly, what patterns can be detected in how they relate to other morpho-syntactic or semantic properties of the linguistic expressions in question.

The third keynote speaker was Paul McNamara from the University of New Hampshire. His talk was titled "Toward a More Fine-Grained Framework for Some Fundamental Moral Notions", and is described by the following abstract:

Deontic logic, despite its merits, emerged in sin, sins that also infested ethical theory. One was the conflation of indifference with optionality, a conflation which tacitly, but inexorably, rules out action beyond the call of duty. Another pervasive conflation, in both deontic logic and ethical theory, has been that of must with ought. Perhaps with the exception of good, ought has been the most discussed moral term in 20th Century ethical theory, based on a mistaken but pervasive bipartisan presupposition that ought has the right continuity with traditional concerns with permissibility and impermissibility, and thus expresses obligatoriness. Arguably, yet another conflation is that of action beyond the call with supererogation, and permissible sub-optimality with suberogation. Other important notions, like the least you can do, praiseworthiness, and blameworthiness have been either ignored or at best underexplored. We will survey some of these issues, and explore some simple semantic structures that provide enough expressive power and conceptual discrimination to model all these notions without conflation, and to generate plausible logical relationships between them. We begin with a simple framework for *indifference*, and then show that the semantical and logical framework for *optionality* (and *contingency*) are just a special case. We then add a framework for *must*, and the *can* and *can't* of permissibility and impermissibility. Next, we order elements in the structures and add a framework for *ought*, and one for *the least one can do*. A simple semantic structure repeatedly emerges naturally from intuitive and independent reflections on either a) indifference vs. optionality, b) must vs. ought, c) the least one can do, d) action beyond the call, or e) permissible sub-optimality, so that interlocking considerations generate a cumulative case argument for the aptness of the structures and for the logical interrelationships hypothesized. We sketch generalizations and expansions of the framework (e.g. to include supererogation and suberogation, or conditional versions of the notions). Whatever refinements are needed, the simple overarching approach appears to be on track, and promises a much better synchronization of deontic systems with the rich array of concepts from commonsense morality of special interest to moral philosophers.

The fourth keynote speaker was Krister Segerberg from Uppsala University. His talk was titled "Information, Belief, Metaphor", and is described by the following short abstract:

A primitive theory of information is formulated and is then related to a simple doxastic logic.

The fifth keynote speaker was Bryan Skyrms from the University of California, Irvine. His talk was titled "Emergence of Meaningful Signals", and is described by the following abstract:

David Lewis proposed signaling games as a model that demonstrates the viability of signals with conventional meaning. The "meaning" of a signal is a function of the equilibrium of the game at which play resides. This raises the question of how equilibrium is selected, and indeed whether equilibrium is ever reached. We discuss this in the context of adaptive dynamics, both in the original context of Lewis signaling, and in more general contexts which relax assumptions of common interest and binary signaling.

We are grateful to everyone whose hard work made this conference possible. Most of all, we are grateful to our invited speakers, and to all the authors of the presented papers. Special thanks go to the members of the Program Committee and (when they liaised with one) the external referees for their service in reviewing papers and advising us on the program. They were all forced to work on a very tight timescale to make this volume a reality. We would also like to thank the members of the Local Organizing Committee, especially Erik Weber, Christian Straßer and Frederik Van De Putte, for taking care of all the countless details that a conference like this requires. We would also like to thank Leendert van der Torre and Jeff Horty, chair and vice chair of the ΔEON Steering Committee, respectively, for their advice and continuing goodwill. Finally we are indebted to Springer, and Alfred Hofmann, Anna Kramer and Elke Werner in particular, for their support in getting these proceedings published.

We record a deep sense of loss at the passing of Ingmar Pörn in February. He was a leading figure in the development of deontic logic. We dedicate this volume to his memory. An obituary by Andrew J.I. Jones opens this volume.

July 2014

Fabrizio Cariani
Davide Grossi
Joke Meheus
Xavier Parent

Organization

Program Chairs

Fabrizio Cariani	Northwestern University, USA
Davide Grossi	University of Liverpool, UK
Joke Meheus	Ghent University, Belgium
Xavier Parent	University of Luxembourg, Luxembourg

Program Committee

Maria Aloni	University of Amsterdam, The Netherlands
Guillaume Aucher	University of Rennes 1/Inria, France
Mathieu Beirlaen	Universidad Nacional Autónoma de México, México
Guido Boella	University of Torino, Italy
Jan Broersen	Utrecht University, The Netherlands
Mark A. Brown	Syracuse University, USA
Fabrizio Cariani	Northwestern University, USA
José Carmo	University of Madeira, Portugal
Roberto Ciuni	Ruhr-Universität Bochum and Humboldt Foundation, Germany
Robert Demolombe	IRIT, France
Lou Goble	Willamette University, USA
Guido Governatori	NICTA, Australia
Norbert Gratzl	Ludwig-Maximilians University, Germany
Davide Grossi	University of Liverpool, UK
Jörg Hansen	University of Leipzig, Germany
Andreas Herzig	IRIT-CNRS, France
Jeff Horty	University of Maryland at College Park, USA
Andrew Jones	Imperial College London, King's College London, UK and Norwegian Academy, Norway
Stefan Kaufmann	University of Connecticut, USA
Gert-Jan Lokhorst	Delft University of Technology, The Netherlands
Emiliano Lorini	IRIT, France
Joke Meheus	Ghent University, Belgium
Ron van der Meyden	University of New South Wales, Australia
John-Jules Meyer	Utrecht University, The Netherlands
Eric Pacuit	University of Maryland at College Park, USA
Xavier Parent	University of Luxembourg, Luxembourg

Henry Prakken	University of Utrecht and University of Groningen, The Netherlands
Antonino Rotolo	CIRSFID, University of Bologna, Italy
Oliver Roy	University of Bayreuth, Germany
Giovanni Sartor	EUI/CIRSFID, Italy
Ken Satoh	National Institute of Informatics and Sokendai, Japan
Marek Sergot	Imperial College London, UK
William Starr	Cornell University, USA
Audun Stolpe	Norwegian Defence Research Establishment (FFI), Norway
Christian Straßer	Ghent University, Belgium
Allard Tamminga	University of Groningen, The Netherlands
Paolo Turrini	Imperial College London, UK
Leendert van der Torre	University of Luxembourg, Luxembourg
Malte Willer	University of Chicago, USA
Tomoyuki Yamada	Hokkaido University, Japan

Additional Reviewers

Albert Anglberger
Tina Balke
Olga Pacheco
Floris Roelofsen

Katsuhiko Sano
Filipe Santos
Satoshi Tojo

Chairs of the Local Organizing Committee

Christian Straßer	Ghent University, Belgium
Erik Weber	Ghent University, Belgium

Local Organizing Committee

Inge De Bal
Tjerk Gauderis
Merel Lefevere
Julie Mennes
Christian Straßer
Frederik Van De Putte
Erik Weber
Dietlinde Wouters

Sponsoring Institutions

Research Foundation – Flanders (FWO), Belgium
Faculty of Arts and Philosophy, Ghent University, Belgium

Table of Contents

Ingmar Pörn – *In Memoriam*

Andrew J.I. Jones

King's College London and Imperial College London
andrewji.jones@kcl.ac.uk

Ingmar Pörn died at his home in Finland in February 2014, at the age of 78. During his academic career, Pörn carried out research in a number of areas of Philosophy, including significant contributions to the logics of action, interaction and norms, on which this note will focus.

Pörn's principal English-language publications on DEON-related topics belong to the first twenty years of his research output; they are listed in the references below. It is in particular the two books, *The Logic of Power*, 1970 [1], and *Action Theory and Social Science – Some Formal Models*, 1977 [5], that provide the clearest guide to his approach to the theory of action and norms. The later book – in my opinion his finest work in this domain – not only refines and develops the modal-logical analyses of [1], but also lays the foundations of a formal theory of practical reasoning, and applies a number of *other* formal models, in addition to those based on modal logic, to the characterization of a range of fundamental sociological and social-psychological concepts.

Following the approach described in [3], the opening chapter of [5] modifies the logic of action of [1], the key change being the addition of a notion of *counteraction conditionality*. Accordingly, sentences of the form 'agent *a* brings it about that *p*' are defined in terms of the following conjunction: '*p* is necessary for something that *a* does and but for *a*'s action it might not be the case that *p*'.[1] The 'brings it about that' modality thus defined is classical, but not normal, in the sense of [15]; the negative condition secures the result that the bringing about of logical truths is beyond the scope of anyone's agency. Furthermore, the positive condition conjoined with the *negation of* the negative condition affords a means of defining 'it is unavoidable for *a* that *p*', the dual of which, Pörn suggested, expresses a notion of agent ability.[2]

Chapter 2 of [5] introduces a novel approach to the logic of belief, based on a 3-valued propositional logic. One interesting property of this logic is that, where *p* logically implies *q*, *a*'s believing *q* is a logical consequence of *a*'s believing *p* only if

---

[1] As Dag Elgesem has observed [14, p.75], this characterization of 'brings it about that' in terms of a pair of positive and negative conditions resembles the later approach adopted by Belnap *et al.* in their analysis of STIT. Chapter 2 of [14] makes some interesting comparisons of Pörn's action logic and STIT.

[2] In [9] expressions of the form 'Ought *p*' are also defined in terms of a conjunction of positive and negative conditions, requiring the truth of *p* in all ideal versions of a given world and the falsity of *p* in at least one sub-ideal version of that world. Corresponding to the action-theoretic notion of 'unavoidability', [9] went on to define the deontic necessity of *p* in terms of the truth of *p* in all ideal and all sub-ideal versions of a given world. In these respects, the approach of [9] parallels that of [5, ch.1].

F. Cariani et al. (Eds.): DEON 2014, LNAI 8554, pp. 1–4, 2014.
© Springer International Publishing Switzerland 2014

the descriptive signs occurring in q are among those that occur in p. So, for instance, 'a believes that b brings it about that p' logically implies 'a believes that p' – the T schema holds of course for the action operator – whereas 'a believes that p' does not logically imply 'a believes (p or q)'. Furthermore, in this same chapter, Pörn followed [16] in employing both the modality 'Shall', to represent *directive* norms, and an *evaluative* modality designated by 'Ought'. 'Shall' and the action operator are used to define norms and normative positions[3], and with the addition of the belief modality provide a means of formally characterizing such notions as 'a intends to bring it about that q if p', 'a decides to bring it about that q if p' and 'b expects a to bring it about that q if p'. And in terms of 'Ought' and belief, Pörn then articulates the formal structure of wants, valuations and attitudes, suggesting that there are "....far-reaching affinities between the theoretical treatment of norms and that of attitudes" [5, p.41]. The chapter also contains a number of fascinating conjectures concerning the ways in which the four modalities may be brought together to provide the foundations of a formal theory of practical reasoning, and is a wonderful illustration of the richly expressive power afforded by a language based on just a small set of component modal building-blocks. Yet, to my knowledge, the resulting theory has received very little detailed attention from the DEON community.

While the modal-logical language continues to play a key role in the remaining four chapters of [5], it is supplemented by a number of *other* types of formal model. Chapter 3 investigates the structure of activities, proceedings and organizations, drawing on techniques from the theory of grammars and automata, including so-called developmental grammars first introduced by Lindenmayer ([17] and [18]) to model the cellular interactions of developing organisms. Automata theory figures again in chapter 4, in which the focus turns to the characterization of control, influence and agent-interaction.[4] Then, in chapter 5, the central concept of cybernetics – the information-feedback control loop – is applied to the study of time-varying (dynamic) aspects of interactions in the context of social systems. Three later sections of that chapter go on to discuss the phenomenon of interdependent decision from a games-theoretic perspective, with particular focus on the Prisoner's Dilemma. Chapter 6 concludes the book with a discussion and analysis of the nature of action-explanations.

The above summary serves, I hope, to highlight both the broad area of topics covered in [5] and the range of formal techniques Pörn brought to their systematic treatment. It is worth recalling that it is now nearly forty years since the final draft chapters of that book were written. At that time, work in theoretical AI was still in a

---

[3] The 'position-generation' method figured quite prominently in Pörn's work. In [11] he applied it to the analysis of the cognitive and volitional components of emotions, using logics for belief, knowledge and desire.

[4] Emiliano Lorini and Giovanni Sartor kindly allowed me to see a draft of their recent paper on STIT-logic and the concept of influence [19]. They argue that the characterization of influence in STIT-logic faces serious obstacles; but the solution they propose, I suspect, applies only to a rather narrowly defined class of influence relations, whereas Pörn's 'brings it about that' logic offers the prospect of developing a rather comprehensive taxonomy of different types of influence. A comparative investigation of these issues will be the topic of some planned collaboration.

relatively early stage of development and, more particularly, the emergence of the research field now known as 'Autonomous Agents and Multiagent Systems' (AAMAS) was still two decades away. Viewed from that perspective, it is fair to say that [5] was way ahead of its time. However, with only a few exceptions, the book has attracted little attention from the AAMAS community, despite its very obvious relevance to that community's concern with the study of norm-governed systems and agent-interaction, and despite the fact that AAMAS research has used some of the same formal tools as those applied by Pörn. There may be many reasons for this neglect; but among them, I suggest, is the widespread tendency in AAMAS work towards designing computational models – which are alleged to be models of aspects of social phenomena and processes – in the *absence of* clear, underlying conceptual models.[5]

....

In the period 1969-1977 it was my very great privilege to work closely with Ingmar Pörn, first as his research student and then as a colleague, in the Department of Philosophy at the University of Birmingham, UK. After we had both left Birmingham for Scandinavia, and Ingmar had succeeded Erik Stenius in the Swedish-speaking Chair of Philosophy at the University of Helsinki, we continued our collaboration; and later on, in the 1990's, I was very happy to be able to invite Ingmar to Norway, for a couple of periods as Visiting Professor in the Department of Philosophy at the University of Oslo. By that time it was clear that Ingmar's primary research focus had moved away from the logics of action and norms, and towards issues in the Philosophy of Health and Medicine – in addition to his continued, long-standing interest in the writings of Søren Kierkegaard.

Ingmar will be fondly remembered by all those fortunate enough to have experienced his charm, his humour and his incisiveness in philosophical discussion, including those DEON 'veterans' who met him at the Amsterdam, Oslo, Sesimbra and Bologna workshops. It is to be hoped that his work will eventually receive the attention and recognition it so richly deserves.

London, April 2014

References

1. Pörn, I.: The Logic of Power. Basil Blackwell, Oxford (1970)
2. Pörn, I.: Elements of Social Analysis. Philosophical Studies published by the Department of Philosophy at the University of Uppsala, Sweden, No. 10 (1971)
3. Pörn, I.: Some Basic Concepts of Action. In: Stenlund, S. (ed.) Logical Theory and Semantic Analysis. Essays Dedicated to Stig Kanger on his Fiftieth Birthday. Synthese Library, vol. 63, pp. 93–101. D. Reidel Publishing Company, Dordrecht (1974)
4. Pörn, I.: A Remark on Determinism and Agency. In: Kausalitet, pp. 69–70. Proceedings edited by the Department of Philosophy. University of Oslo, Norway (1976)

---

[5] This theme is developed in some detail in [20].

5. Pörn, I.: Action Theory and Social Science – Some Formal Models. Synthese Library, vol. 120. D. Reidel Publishing Company, Dordrecht (1977)
6. Pörn, I.: Davidson and Tuomela on the Characterization Problem for Actions. Reports from the Department of Philosophy, pp. 59–64. University of Helsinki, Finland (May 1980)
7. Pörn, I.: A Note on Lennart Åqvist's Epistemic Obligation Paradox. In: Philosophical Studies published by the Department of Philosophy at the University of Uppsala, Sweden, vol. 34, pp. 76–79 (1982)
8. Pörn, I.: Deontic Detachment. Bulletin of the Section of Logic 13, 60–63 (1984)
9. Jones, A.J.I., Pörn, I.: Ideality, Sub-Ideality, and Deontic Logic. Synthese 65, 275–290 (1985)
10. Jones, A.J.I., Pörn, I.: 'Ought' and 'Must'. Synthese 66, 89–93 (1986)
11. Pörn, I.: On the Nature of Emotions. In: Needham, P., Odelstad, J. (eds.) Changing Positions – Essays Dedicated to Lars Lindahl on the Occasion of his Fiftieth Birthday. Philosophical Studies published by the Philosophical Society and the Department of Philosophy, University of Uppsala, Sweden, vol. 38, pp. 205–214 (1986)
12. Jones, A.J.I., Pörn, I.: A Rejoinder to Hansson. Synthese 80, 429–432 (1989)
13. Pörn, I.: On the Nature of a Social Order. In: Fenstad, J.E., Frolov, T., Hilpinen, R. (eds.) Logic, Methodology and Philosophy of Science VIII, pp. 553–567. Elsevier Science, New York (1989)
14. Elgesem, D.: Action Theory and Modal Logic, Department of Philosophy, University of Oslo, Norway (1993)
15. Chellas, B.F.: Modal Logic – An Introduction. Cambridge University Press, Cambridge (1980)
16. Kanger, S.: Law and Logic. Theoria 38, 105–132 (1972)
17. Lindenmayer, A.: Mathematical Models for Cellular Interactions in Development. Journal of Theoretical Biology 18, I: 280-299, II: 300-315 (1968)
18. Lindenmayer, A.: Developmental Systems without Cellular Interactions, their Languages and Grammars. Journal of Theoretical Biology 30, 455–484 (1971)
19. Lorini, E., Sartor, G.: A STIT Logic Analysis of Social Influence. Paper to appear in the Proceedings of AAMAS 2014, Paris, France (May 2014)
20. Jones, A.J.I., Artikis, A., Pitt, J.V.: The Design of Intelligent Socio-technical Systems. Artificial Intelligence Review 39(1), 5–20 (2013)

Deontic Diversity

Sven Ove Hansson

Division of Philosophy, Royal Institute of Technology
100 44 Stockholm, Sweden
soh@kth.se
home.abe.kth.se/~soh

Abstract. It is commonly assumed that deontic logic concerns "the" logic of normative concepts. However, a close look at actual usage shows that the structural patterns of deontic notions differ between different usages. Some of these differences are difficult to discern in natural language, but may be easier to keep apart with the more precise tools of a formal language. We should use the resources of deontic logic to discover and distinguish between different meanings of the deontic terms in natural language. Some of the ingrained disagreements on postulates in deontic logic may be resolvable if we recognize that the different viewpoints correspond to different meanings of the normative terms of ordinary language.

Keywords: deontic logic, Seinsollen, degrees of ought, moral rules, situation-specific norms, prima facie norms, action guidance, derived obligations, subjective ought.

1 Introduction

Deontic logic has usually been devoted to attempts to determine a single correct logic for the common deontic terms. It is the purpose of this contribution to show that a pluralistic approach may be more appropriate. A careful analysis of common usage will show that the normative terms of natural language are used in a variety of meanings, and that these meanings are in part characterized by differences in logical properties. The focus in this presentation will be on prescriptive terms such as "ought" and "morally required". Corresponding analyses of prohibitive and permissive terms such as "forbidden" respectively "allowed" will be left for another occasion. The same applies to conditional (dyadic) normative sentences, i.e. normative if-sentences.

In an important sense, philosophy like all other linguistic activities are bound by the limits of what can be expressed in our natural languages. But these limits are not fixed; to the contrary they are constantly being pushed back as we develop new concepts and distinctions together with new linguistic means to express them. Using the tools of philosophy we can develop new and improved ways to categorize and distinguish between different usages even of common everyday terms, including those used to express normative requirements, such as "ought", "obligatory", "must", "duty", "should", "have to", and "morally

F. Cariani et al. (Eds.): DEON 2014, LNAI 8554, pp. 5–18, 2014.

required". In what follows, we will investigate differences between these terms, but also (and perhaps even more importantly) between different usages of one and the same term.

2   Normative and Non-normative Usages

Some of the terms used to express norms are also used to express non-normative notions. This applies in particular to "must" and "ought". Beginning with "must", the most important distinction in its usage can be illustrated with the following examples:

"You must help her."
"You must be wrong."

It is only in the first case that "must" expresses a norm. In the second case it expresses a necessity. These two usages are reflected in different logics. Let $\overline{M}$ represent "must". The following property:

$$\overline{M}p \vdash p \tag{1}$$

(If p must be the case, then it follows logically that p is the case) holds for "must" in the necessitative but not in the normative sense. This is a rather crucial difference, and the two senses of "must" are standardly treated as two distinct concepts with different logics (represented by $\Box$ respectively O).

The normative and non-normative meanings of "ought" can be exemplified as follows:

"You ought to help your destitute brother."
"There ought to be no injustice in the world."

The first sentence is normative in the sense of recommending (or commanding) someone what to do. The second is a value statement rather than a normative one. We should distinguish between the *normative ought* (ought-to-do, Tunsollen) and the *ideal ought* (ought-to-be, Seinsollen) that expresses an evaluation of some possible state of the world. As Richard Robinson observed, statements of the latter type "are not prescriptive at all, either prudentially or morally, but express valuations. Such is 'Everybody ought to be happy'. This is not a prescription or command to anybody to act or to refrain." [1, p. 195] This non-normative usage is specific for "ought", and does not apply to prescriptive predicates in general.

From the viewpoint of logical formalization, it is a major difference between the normative and the ideal ought that the arguments of the normative operator (i.e. p in Op) always represent a voluntary action, whereas those of the ideal "ought" are not subject to that restriction. The ideal ought is usually applied to non-actions, but it can also be applied to actions:

"You ought to sing in tune."
"The world ought to be such that you sing in tune."

The "ought" of the former statement is normative. The sentence implies that the person in question is able to sing in tune, and enjoins her to do so. The "ought" of the latter statement indicates that nature has not endowed her with that ability. These examples should be sufficient to show that the normative and the non-normative "ought" differ in meaning. They should be kept apart, just like the the normative and the non-normative "must". In what follows I will assume that the O of deontic logic represents a normative concept.

3   The Strength of Norms

Moral requirements differ in stringency (strength). This is fairly obvious on an intuitive level. Every child knows the difference in stringency between the two instructions "do not speak with food in your mouth" and "do not erase the hard disk on mom's computer". There are various ways to explicate the difference: the more stringent requirement is more important, its violation is worse, it is less easily overridden etc. [2, p. 151] These explications of strength do not necessarily coincide. For instance, the most important of two moral requirements may not be the one that is least easily overridden. However, for many (and arguably most) purposes the distinction between different types of strength may not be worth the considerable complication of the formal structure that its introduction will lead to.

Often, one and the same normative predicate is used to express moral requirements with different strengths, for instance:

"You must buy flowers for her."
"You must throw out the life-buoy at her."

However, there is a difference in the stringency usually associated with the prescriptive predicates of natural language. "Must"' is more stringent than "ought", and "ought" is more stringent than "should". [3] [4, pp. 131-133] The following definitions are useful for comparing prescriptive predicates in terms of their strengths:

Definition:
Let O_1 and O_2 be two predicates that both take sentences as arguments. Then:
 O_1 *includes* O_2 if and only if it holds for all sentences p that if O_2p, then O_1p.
 O_1 *properly includes* O_2 if and only if O_1 includes O_2 but O_2 does not include O_1.
Furthermore, if O_1 and O_2 are prescriptive predicates[1] then:
 O_1 is *at least as strong as* O_2 if and only if O_2 includes O_1.
 O_1 is *stronger than* O_2 if and only if O_2 properly includes O_1.

---

[1] We exclude atypical prescriptive predicates such as "it is recommended but not required that".

The relation "at least as strong as" is not necessarily complete, i.e. there can be prescriptive predicates O_1 and O_2 such that for some p and q we have $O_1 p$, $\neg O_1 q$, $\neg O_2 p$, and $O_2 q$.

Most deontic systems contain only one prescriptive predicate, and therefore they cannot express differences in strength. However, some authors have introduced two degrees of obligatoriness, usually said to represent "must" and "ought". [5,6] Proposals have also been put forward that allow for a whole series of O operators, representing different degrees of strength. [4,7,8] In my view, the strength of a norm is such an important characteristic that we need to keep track of it, even though natural language only expresses it in irregular and imprecise ways. We can also expect differences in logical properties between deontic predicates of different strengths; in particular the agglomeration postulate

$$Op \ \& \ Oq \to O(p\&q) \tag{2}$$

is more plausible for stronger predicates, since it will less often lead to inconsistencies.

4   The Scope of Norms

A particularly important distinction concerns the scope of normative statements, i.e. what situations they refer to. It is essential to distinguish between those norms that refer to situations in general and those that refer to a particular situation. Unfortunately, this distinction is not obvious, since the English language (like many others) employs the same linguistic forms for both purposes:

(4.1) "You have to leave this room immediately."
(4.2) "You have to pay back money that you borrow."
(4.3) "You must stop beating Puss."
(4.4) "You must be kind to all animals."

Examples (4.1) and (4.3) report *situation-specific* norms, i.e. (overall or only prima facie) norms that obtain in a particular situation, in these examples in the present state of the world. No conclusion can be drawn from these statements about what holds in other situations. In contrast, (4.2) and (4.4) refer to what is obligatory in general, i.e. in all situations. They can be called *transsituational norms*.[2] I have previously used the term "veritable norms" for situation-specific norms. It would perhaps be natural to use the term "actual" about obligations referring to the present situation, but such usage should be avoided since that term has been used by Ross and others for overall obligations. [12, p. 20]

---

[2] This distinction was made in [9]. Similarly, Carlos Alchourrón distinguished between "a norm for a single possible circumstance (which may be the actual circumstance)" and a norm for "all possible circumstances" [10], and David Makinson distinguished between norms "in all circumstances" and norms "in present circumstances" [11].

Situation-specific norms most commonly refer to the present situation, but they may also refer to a past or future situation:[3]

(4.5) "Two years ago my son had to be home at nine o'clock in the evenings."

(4.6) "Participants in next year's finals will be required to bring an ID."

(4.2) and (4.4) both exemplify the most common form of transsituational norms, namely *normative rules*. They state that a certain norm holds in all states of the world in which it is applicable.

The distinction between situation-specific and transsituational norms is easily confused with that between overall norms (non-overridden norms) and prima facie norms (norms that may or may not be overridden). However, situation-specific norms can be either overall or prima facie:

(4.7) "All things considered, you are morally required to pay for the damage that your dog caused yesterday."

(4.8) "Basically, you have to be there in person, but due to the circumstances you are excused."

Confusion between the two distinctions situation-specific/transsituational and prima facie/overall has sometimes led to difficulties with the latter. For instance, Searle claimed that almost no obligation can be overall since "any obligation is subject to being overridden by special considerations in particular circumstances" [13]. What this shows, however, is only that rules about obligations cannot refer to overall obligations. It says nothing against the existence of situation-specific overall obligations.

As already mentioned, the distinction between situation-specific and transsituational norms is often overlooked since we do not make it in a standardized way in natural language. English (like many other languages) employs the same normative terms for both purposes. Markers such as "always" and "now" are sometimes used:

(4.9) "You must always be kind to animals."

(4.10) "You must pay the entrance fee now in order to enter the museum."

In these and most other cases the markers are dispensable. It is usually from the context and not from the words that we can understand if a norm is situation-specific or transsituational. Deontic logic follows natural language in using the same expressions for situation-specific and transsituational norms. Hence Op is used both to express that something (p) is obligatory in the present situation and that it is obligatory in general. As I hope now to have made clear, this is a troublesome conflation. However, I am not aware of any single logical principle

---

[3] Conditional situation-specific norms may also refer to hypothetical situations. "If she had been seriously ill, you would have been required to pay her hospital bills." Such hypothetical situations may be placed in the past, present, or future.

that holds for all situation-specific but no transsituational prescriptions, or the other way around. These are two rather large groups of normative utterances, and it may be easier to find characteristic logical principles for some of their subgroups than for these two large groups themselves.

5   Overridden, Overall, and Action-Guiding Norms

Prima facie and overall duties. The distinction between a prima facie duty and an overall duty was introduced by W. David Ross to account for the existence of conflicting duties. When I have conflicting duties "what I have to do is to study the situation as fully as I can until I form the considered opinion (it is never more) that in the circumstances one of them is more incumbent than any other; then I am bound to think that to do this prima facie duty is my duty sans phrase in the situation." [12, p. 19]

According to one version of the distinction, a person who has a prima facie obligation does not really have an obligation, but only seems to have one. This view may have some support in Ross's own text, but it is not a tenable position. It would mystify the existence of moral conflicts and make unexplainable the existence of residual obligations [14,15].

A multitude of phrases have been used for non-overridden duties: "duty sans phrase", "overall duty", "duty proper", "actual duty", and "absolute duty". The phrase "overall" is the most common one.

Moral reasoning does not always lead directly from prima facie to overall norms. There may be intermediate duties, derived from the immediate prima facie duties but finally overridden by other considerations. Intermediate duties have been treated as prima facie duties for instance by Azizah Al-Hibri who defined prima facie obligations as those that an agent formulates "immediately or mediately, on the basis of one or more aspects of the situation under consideration" [16, p. 84].

There is a certain ambiguity in the philosophical usage of the phrase "prima facie". Sometimes it is used to denote overridden norms, in contrast to non-overridden ones. On other occasions it is used − often by the same authors − to denote the totality of overridden and non-overridden norms, so that a prima facie duty may also be an overall duty. I propose that we follow the last-mentioned practice, and use the word "overridden" for the first, more restricted sense of "prima facie".

The distinction between prima facie and overall obligations is relevant for a wide range of consistency principles and also for the principle of agglomeration. Let us begin with the consistency principles.

Consistency and compatibility. Two obligations Oa and Ob are *inconsistent* (in logical conflict) if and only if $a\&b$ is inconsistent, and they are *(practically) incompatible* if and only if $a\&b$ is impossible to realize in practice. For instance:

(a) "You should be in London when this contract expires."
(b) "You should not be in London when this contract expires."

(c) "You should be in New Delhi when this contract expires."
(d) "You should not steal."
(e) "You should not let your child starve."

(a) and (b) are logically inconsistent. (a and (c) are logically consistent but practically incompatible. (d) and (e) are logically consistent but under some circumstances nevertheless practically incompatible. It is common in deontic logic to treat all conflicts between norms as logical. This may seem to be an innocuous practice for instance when we consider the conflict between (a) and (c). But when we consider conflicts such as that between (d) and (e) – and probably, most conflicts in real life are more of that nature – it should be clear that this representation is deficient and that a separate representation of practical possibility may be needed.

These concepts can also be defined for larger sets of obligations:

Definition
A set $\{Oa_1, Oa_2, \ldots\}$ of obligations is *inconsistent* if and only if $\{a_1, a_2, \ldots\}$ is inconsistent. It is *incompatible* if and only if the realization of all elements of $\{a_1, a_2, \ldots\}$ is practically impossible.

A distinction should be drawn between on the one hand inconsistencies (incompatibilities) among the obligations of one and the same agent and on the other hand inconsistencies (incompatibilities) that involve actions by more than one agent. The former can be called *monoagent* inconsistencies (incompatibilities) and the latter *multiagent* inconsistencies (incompatibilities).

Case i: I am under an obligation to keep the window open. I am also under an obligation to keep it closed.

Case ii: I am under an obligation to keep the window open. You are under an obligation to keep it closed

As this example shows, both monoagent and multiagent inconsistencies (incompatibilities) are problematic, but the latter less so. Let us now consider what consistency properties are plausible for prima facie respectively overall obligations.

Prima facie obligations. It lies at the very core of the prima facie concept that both an action and its non-performance can be prima facie obligatory. Therefore, we cannot demand the joint consistency (or joint practical possibility) of prima facie obligations. Principles such as $\neg(Op \ \& \ O\neg p)$ cannot be expected to hold.

However, another consistency principle for prima facie obligations is plausible. It was indicated by Richard Brandt who said: "[T]here is no overall duty to do the impossible but there can be a prima facie obligation to do not what is inherently impossible but what cannot be done consistently with meeting all other prima facie obligations." [17, p. 368] Thus each prima facie duty should be inherently logically consistent and, presumably also singly practicable:

$$\text{If } Op \text{ then } p \text{ is logically consistent. (\textit{self-consistency})} \tag{3}$$

$$\text{If } Op \text{ then } p \text{ is practically possible. (\textit{self-practicability})} \tag{4}$$

Two types of overall obligations. The logic of obligatoriness all things considered will have to depend on whether or not we exclude the existence of moral dilemmas. In a logic that allows for moral dilemmas it should be possible to have obligations Op and Oq such that p and q are incompatible. The price we have to pay for allowing this is of course that such a system of norms cannot be action-guiding in a full sense. Telling someone to do what cannot be done is not much of guidance. We can distinguish between *action-guiding* and *discordant* systems of overall obligations; the former but not the latter always yield consistent advice.

The question whether or not moral dilemmas (conflicts among overall moral obligations) are allowed has mostly been treated as a choice between moral theories. But alternatively, it can be seen as two different notions of obligations that may have a place within one and the same moral viewpoint:

> MORALIST: You have a large debt that is due today. You should pay it.
> SPENDTHRIFT: It is impossible for me to do that. I do not have the money.
> MORALIST: I know that.
> SPENDTHRIFT: Yes, and I already know what my obligations are. Please, as a moralist, tell me instead what I should do.
> MORALIST: I have already told you. You should pay your debt. [18]

Here, the "should" of "You should pay your debt" is unsuitable for action-guidance since it requires something that Spendthrift cannot do. However, it would seem frivolous to describe this obligation as overridden (merely prima facie), in the same sense as your obligation to meet a friend in time for a concert is overridden if you have to help saving a life after an accident. As was noted by Michael Stocker, "it would be at best a bad joke for me to suggest that if I have squandered my money, then I no longer ought to repay my debts." [19] It makes sense to tell people like Spendthrift: "You really ought to do X, and there is no excuse for not doing it, but due to your own wrongdoings you cannot do what you ought to do." Such an admonition seems intelligible enough, but it would be self-contradictory if we ruled out the possibility of conflicting, non-overridden obligations.[4] (I leave open the interesting moral issue whether a person can have conflicting, non-overridden obligations through no fault of her own.)

But on the other hand, the shift in focus that Spendthrift asks for but Moralist refuses is sensible enough. We can talk about what a person "really" should do when she cannot satisfy all her non-overridden obligations. As deontic logicians we can do this by distinguishing between an operator O_d of discordant overall obligations, representing all her non-overridden obligations and an operator O_s of action-guiding moral obligations, representing what she ought to do given that she cannot satisfy all her non-overridden obligations [18].

The difference between prima facie obligations and discordant overall obligations seems difficult to capture in logical terms. In particular, they appear to satisfy (only) the same consistency requirements.

---

[4] At this point, the "ought implies can" principle is dismissed through the backdoor.

Action-guiding obligations. Obviously, the principles of self-consistency and self-practicability are valid for action-guiding obligations as well. However, for them these are not a strong enough consistency principles, since we also want the combined performance of overall obligations to be possible.

It is not immediately clear, however, how that requirement should be formulated. As the following example shows, there are two major alternatives:

> I have promised my child to buy a certain unique object at an auction, and you have made a similar promise to your child to buy the same object. We assume that both these promises give rise to obligations of fulfilment. We cannot both satisfy our obligations, but does the fact that you made this promise relieve me of my obligation to fulfil mine?

According to the principle of *universal compatibility* all obligations are compatible. According to the weaker principle of *agent-specific compatibility*, the overall obligations of each agent should be compatible, but those of different agents need not be so. Expressing compatibility with a modal possibility operator $\diamond$ and using $\&$ to denote the conjunction of all elements of a (finite) set of sentences, we can express this as follows[5]:

$$\diamond \& \{p \mid Op\} \ (universal \ compatibility) \tag{5}$$

If $A \subseteq \{p \mid Op\}$ and all elements of A represent actions by the same agent,
$$\text{then } \diamond \, \& A. \ (agent\text{-}specific \ compatibility) \tag{6}$$

Universal compatibility may be plausible according to classical utilitarianism, but not according to moral theories that allow for agent-specific commitments. Examples where universal compatibility appears to be a questionable ideal are not difficult to find. It may be claimed that the commanders-in-chief of two armies at war are both under an obligation to bring about the victory of their respective side. In a less belligerent context, a coach may consider himself consistent when telling each of several athletes that they ought to win a contest. The study of deontic compatibility has interesting connections with game theory.

Agglomeration. The principle of agglomeration says that moral requirements can be combined by conjunction:

$$Op \ \& \ Oq \rightarrow O(p\&q) \tag{7}$$

Agglomeration is accepted in Standard Deontic Logic and most other systems of deontic logic. It has a strong intuitive appeal. "If one ought to do each of two things, it seems quite natural to think that one also ought to do both of them." [20, p. 274]

---

[5] We have of course a choice between different criteria of possibility. They correspond to different operators $\diamond$, each of which can be combined with different O operators.

However, for prima facie obligations, the principle of agglomeration is not at all plausible. The combination of two prima facie duties can have have consequences or other properties that none of the duties has by itself:

> Both Alice's brother and her sister have a disease that in a few cases may lead to kidney failure and to the need for a transplantation. She has solemnly promised each of them that one of her kidneys will be available for transplantation if that should be medically called for. Let p denote that she donates one of her kidneys to her sister and q that she does so to her brother. Within a week, both her brother and her sister turn out to need a transplantation. Then Alice has the two prima facie duties Op and Oq, but she cannot reasonably be said to have a prima facie duty $O(p\&q)$.

In the presence of logically conflicting duties, agglomeration can even lead to violations of the principle of self-consistency. This is a principle that we have good reasons to uphold. It is one thing to have one moral requirement to be at the pub and another requirement not to be at the pub. This is a conundrum we can understand, and we can deliberate on various (partial and imperfect) solutions. It is something quite different to have a moral requirement to both be and not be at the pub. Such an obligation (if we take it seriously) is impossible to do anything about. We cannot even imagine an action that would in any way take us closer to complying with it.[6] If such obligations arise in a deontic system, they are anomalies of the formal system that should preferably be eliminated. We can conclude that the principle of agglomeration does not hold for prima facie obligations. For the same reason it does not hold for discordant (non-action-guiding) overall obligations (if such exist).[7]

On the other hand, there are strong reasons to accept agglomeration for action-guiding overall obligations by one and the same individual. If I ought to perform the two actions p and q, then $p\&q$ is a correct description of things I am under obligation to do. The only qualms that we may have about this concerns the oddity of combining completely unrelated obligations in this way. If you have an obligation to pay your taxes and also an obligation to clean your brother's car that you borrowed, it may seem somewhat odd to say that you have an obligation to pay your taxes and clean your brother's car. However, this is arguably a minor oddity of a type that has to be accepted in formal representations.

---

[6] In combination with the SDL postulate of necessitation this in its turn gives rise to *universal obligatoriness*, i.e. Oq for all q, which is of course an even more absurd conclusion. Having both a moral requirement to be at the pub and a moral requirement not to be there does not imply a moral requirement to put the pub on fire.

[7] A much weaker version may hold in both cases: If (i) $Op \& Oq$, (ii) $p\&q$ is consistent, and (iii) $p\&q\&r_1\& \ldots \&r_n$ is consistent whenever $Or_1\& \ldots \&Or_n$ and $r_1\& \ldots \&r_n$ is consistent, then $O(p\&q)$.

6 Basic and Derived Obligations

There are some things we have to do, not because they are duties in an immediate sense but because they follow from our duties:

> "If I am obliged to pay for some goods and carry them away, but fail for some reason to pay for them, I can hardly carry the goods away, claiming that I am keeping at least one of my obligations!" [21, p. 487]

> I work as a janitor at a bank. Usually I have no access to the money handled by the bank, but one day my boss orders me to fetch a box containing €10,000,000 and carry it to another bank office a couple of blocks away. Unknown to him, I suffer from weakness of will. I know that once I have the money in my hands it is in fact quite improbable that I will be able to resist the temptation to elope with it. Therefore, if I pick up the money (p) it is most unlikely that I will also hand it over to the other bank (q). Since it is part of my job to run errands for the bank, I certainly have the obligation represented by $O(p\&q)$. But do I have the obligation represented by Op? [7]

At stake here is the following principle:

$$\text{If } p \vdash q \text{ then } Op \vdash Oq \text{ } (necessitation) \tag{8}$$

There is an interesting parallel with epistemic logic, belief revision, and related areas, where the corresponding principle is assumed to hold for the belief predicate. Thus, since I believe that Copenhagen is the capital of Denmark, I also believe that either Copenhagen is the capital of Denmark or gold has the atomic number 2. The latter is of course not a belief that I seriously entertain. It is a "merely derived" belief, contrary to the "basic" beliefs that have an independent standing. The inclusion of all merely derived beliefs in formal belief representations is highly simplifying, but it also gives rise to some notoriously counter-intuitive results such as the recovery postulate in belief revision. [22,23]

Similarly, necessitation gives rise to most of the more important deontic paradoxes, including Ross's paradox ("If you ought to mail the letter, then you ought to either mail or burn it.") [24, p. 62] This postulate has often been criticized, essentially with the argument that "the fact that we can't help but bring about the necessary consequences of our action does not mean we have an *obligation* to bring them about." [25, p. 188] Several authors including myself have proposed deontic logics in which the paradoxes are avoided by exclusion of the necessitation principle. [7]

But I would now like to propose another approach to this problem: Just as we can distinguish between basic and derived beliefs, we can distinguish between basic and derived norms. However, for norms a somewhat broader definition may be appropriate. The derived norms should include not only that which follows logically from the basic (or explicit) norms but also that which follows by practical necessity. In informal normative discourse this notion is often expressed with the phrases "have to" and "must":

"I am required to pay back my loan to Robert by the end of this month. Therefore I have to take a loan in the bank."

"It is now your duty to prevent fires in the building. Therefore you must thoroughly learn these safety routines and regulations."

It seems sensible to distinguish between an operator O_b for basic moral requirements and an operator O_d that includes derived moral requirements. One way to construct them is to take O_b as primitive, and combine it with a necessity operator $\Box$ to derive O_d. Obviously, O_d but not O_b should satisfy necessitation.

7   Objective and Subjective Norms

Words denoting moral prescriptions can be construed as either objective or subjective with respect to matters of fact. In the objective sense, what I ought to do depends on what is actually true, whereas in the subjective sense it depends on what I believe to be true.

1. *The gas stove*: You turn on your stove in order to cook a meal that you are for some reason obligated to make. When you do this, an unexpected and extremely implausible malfunction results in an explosion in your neighbour's apartment, leading to his death. [26]

2. *The beneficial school shooting*: A desperate person shoots randomly at a large number of university students in a park. Since he is an exceptionally bad shooter he only manages to hit one of them. This turns out to be a student who would otherwise have committed a terror attack the next day, killing thousands of people. [27]

In case 1 it would be strange to claim that you ought to turn on the stove, although this seems to follow from a subjective interpretation of "ought". Referring to this example, Judith Thomson argued that it is doubtful whether there is at all a subjective sense of "ought". [26] But on the other hand, in case 2 it would be equally strange to claim that the school shooter ought to shoot randomly at students in the park, since he had no reason to believe that the major consequence of this would be the prevention of a horrendous terror attack.

Ordinary usage seems to merge the two interpretations, which gives rise to difficulties when reference is made to situations in which only one of them is plausible. In a more developed analysis, they should be treated as two distinguishable concepts of moral obligatoriness. According to the subjective concept, an action cannot be obligatory for me (relative to a specified norm or normative system) unless I have enough factual information to know that this is so. This is the sense of oughtness that is most relevant in discussions of responsibility and blameworthiness. The objective concept describes what I ought to (should) do, in a sense that is independent of my knowledge. When giving people advice, it is usually the most relevant of the two concepts. Although this is a classical distinction in moral philosophy [28, p. 177] [29, pp. 146-191], it has not attracted much

Table 1. Proposed logical characteristics of some categories of normative notions

Distinction	Logical characteristics
Different strengths	$O_1p \to O_2p$ (O_1 is at least as strong as O_2)
Prima facie or discordant overall vs. action-guiding overall	$\Diamond \& \{p \mid Op\}$ (action-guiding) $Op \ \& \ Oq \to O(p\&q)$ (action-guiding)
Basic vs. derived	If $p \vdash q$ then $Op \vdash Oq$ (derived)
Objective vs. subjective	$Op \to KOp \vee OKOp$ (subjective)

attention in deontic logic. Letting K denote knowledge (or a related epistemic concept), the subjective but not the objective ought satisfies

$$Op \to KOp \qquad (9)$$

or at least

$$Op \to KOp \vee OKOp \qquad (10)$$

An obvious topic for logical investigation is the logical derivability of subjective obligatoriness from objective obligatoriness and an epistemic concept such as knowledge.

8 Conclusion

Our conclusions concerning specific categories of normative notions are summarized in Table 1. More generally, I propose that

- instead of searching for "the" logic of normative concepts we should use logical tools to investigate the deontic diversity of natural language,
- normative notions are intermingled with notions of necessity, possibility and knowledge in complex ways that can be elucidated in formal languages that also contain representations of modal and epistemic concepts, and
- some of the controversies over specific axioms in deontic logic (such as necessitation and agglomeration) can be resolved by showing that they hold for some prescriptive notions but not for others.

References

1. Robinson, R.: Ought and ought not. Philosophy 46, 193–202 (1971)
2. Chisholm, R.: The ethics of requirement. American Philosophical Quarterly 1, 147–153 (1964)
3. Guendling, J.: Modal verbs and the grading of obligations. Modern Schoolman 51, 117–138 (1974)
4. Hansson, S.O.: The Structure of Values and Norms. Cambridge University Press, Cambridge (2001)
5. Jones, A., Pörn, I.: 'Ought' and 'Must'. Synthese 66, 89–93 (1986)

6. McNamara, P.: Must I do what I ought (or will the least I can do do?). In: Brown, M., Carmo, J. (eds.) Deontic Logic, Agency and Normative systems, DEON 1996: Third International Workshop on Deontic Logic in Computer Science, Sesimbra, Portugal, January 11-13, pp. 154–173. Springer, Berlin (1996)
7. Hansson, S.O.: Alternative semantic for deontic logic. In: Gabbay, D., Horty, J., Parent, X., van der Meyden, R., van der Torre, L. (eds.) Handbook of Deontic Logic and Normative Systems, vol. 1, pp. 445–497. College Publications, London (2013)
8. Dellunde, P., Godo, L.: Introducing grades in deontic logics. In: van der Meyden, R., van der Torre, L. (eds.) DEON 2008. LNCS (LNAI), vol. 5076, pp. 248–262. Springer, Heidelberg (2008)
9. Hansson, S.O.: Deontic logic without misleading alethic analogies, Parts I-II. Logique et Analyse 31, 337–370 (1988)
10. Alchourrón, C.: Philosophical foundations of deontic logic and the logic of defeasible conditionals. In: Meyer, J.J., Wieringa, R. (eds.) Deontic Logic in Computer Science, pp. 43–84. John Wiley & Son, Chichester (1993)
11. Makinson, D.: On a fundamental problem of deontic logic. In: McNamara, P., Prakken, H. (eds.) Norms, Logics and Information Systems. New Studies on Deontic Logic and Computer Science, pp. 29–53. IOS Press, Amsterdam (1999)
12. Ross, W.: The Right and the Good. Clarendon Press, Oxford (1930)
13. Searle, J.: Prima facie obligations. In: van Straaten, Z. (ed.) Philosophical Subjects: Essays presented to P.F. Strawson, pp. 238–259. Clarendon Press, Oxford (1980)
14. Pietroski, P.: Prima-facie obligations: ceteris paribus laws in moral theory. Ethics 103, 489–515 (1993)
15. Brummer, J.: The structure of residual obligations. Journal of Social Philosophy 27, 164–180 (1996)
16. Al-Hibri, A.: Conditionality and Ross's deontic distinction. Southwestern Journal of Philosophy 11, 79–87 (1980)
17. Brandt, R.: Ethical Theory: the problems of normative and critical ethics. Prentice-Hall, Englewood Cliffs (1959)
18. Hansson, S.O.: But what should I do? Philosophia 27, 433–440 (1999)
19. Stocker, M.: Moral conflicts: What they are and what they show. Pacific Philosophical Quarterly 68, 104–123 (1987)
20. McConnell, T.: Moral dilemmas and consistency in ethics. Canadian Journal of Philosophy 8, 269–287 (1978)
21. Purtill, R.: Paradox-free deontic logics. Notre Dame Journal of Formal Logic 16, 483–490 (1975)
22. Hansson, S.O.: Taking belief bases seriously. In: Prawitz, D., Westerståhl, D. (eds.) Logic and Philosophy of Science in Uppsala, pp. 13–28. Kluwer, Dordrecht (1994)
23. Hansson, S.O.: Belief contraction without recovery. Studia Logica 50, 251–260 (1991)
24. Ross, A.: Imperatives and logic. Theoria 7, 53–71 (1941)
25. Sayre-McCord, G.: Deontic logic and the priority of moral theory. Noûs 20, 179–197 (1986)
26. Thomson, J.: Rights, Restitution, and Risk: Essays in Moral Theory. Harvard University Press, Cambridge (1986)
27. Hansson, S.O.: Objective or subjective 'ought'. Utilitas 22, 33–35 (2010)
28. Price, R.: A Review of the Principal Questions in Morals, 3rd edn. (1787)
29. Ross, D.: Foundations of Ethics. Clarendon Press, Oxford (1939)

Open Reading without Free Choice

Albert J.J. Anglberger[1], Huimin Dong[2], and Olivier Roy[2]

[1] Munich Center for Mathematical Philosophy, LMU Munich, Germany
[2] Philosophy and Economics, University of Bayreuth, Germany

Abstract. The open reading of permission (OR) states that an action α is permitted iff every execution of α is normatively OK. Free Choice Permission (FCP) is the notorious principle turning permission of disjunction into conjunction of permissions $P(\varphi \vee \psi) \rightarrow P\varphi \wedge P\psi$. We start by giving a first-order logic version of OR that defines permission of action types in terms of the legality of action tokens. We prove that implies FCP. Given that FCP has been heavily criticized, this seems like bad news for OR. We disagree. We observe that this implication relies on a debatable principle involving disjunctive actions. We proceed to present alternative views of disjunctive actions which violate this principle, and which so block the undesired implication. So one can have the open reading without free choice and, as we argue towards the end of the paper, there are philosophical reasons why one should.

This paper is about two related principles pertaining to permissions. On the one hand we have *Open Reading* of permissions.[1]

---

[1] To our knowledge the term was coined by [6]. In what follows, we give a brief history about the development of the terms "Strong Permissions" and "Open Specification [9] (or Open Interpretation [6])". Strong Permission is first mentioned in [27], and an action is permitted in this sense if "the authority has considered its normative status and decided to permit it". But his later work [28] defines strong permissions satisfying a property that $P(A \vee B) = PA \wedge PB$, and names it as Free Choice Permission. Our FCP in this paper is the one direction of his. The open interpretation of an action expression is first mentioned in [9]. Roughly speaking, an action expression in an open sense is that "if an action expression is used it means informally that the action denoted by that action expression occurs, possibly in combination with other actions" [8]. Under such an "open" specification for actions, a strong permission here adopts the definition that an action is permitted if every way to perform this action never leads to a violation state. Formally, $P\alpha := [\alpha]\neg Violation$. Thus, the idea of strong permissions can return to [28]'s sense, namely the FCP property. Thus, strong permissions is equal to saying FCP in [28], and it is defined under an open specification of actions and it implies FCP in [9,8]. Open interpretation in [6] expands the idea of open specification into the openness with respect to the concrete description of the effects of actions. Our basic idea of Open Reading is rooted in [28, pp. 34-35], but not in its Strong Permission sense; also, the openness of OR is different from the one in [9,6], because we reject the Additivity while they accept it. Our OR focuses on the interaction between action tokens and action types, and then interprets permissions according to their relations. These are never explicitly expressed in the literature. More details will be presented in section 1.

F. Cariani et al. (Eds.): DEON 2014, LNAI 8554, pp. 19–32, 2014.

(OR). An action type α is permitted iff every token of α is normatively OK.

On the other hand, we have *Free Choice Permission*. This principle states that a permission to perform a disjunctive action is a permission to choose freely between the disjuncts.

(FCP). $P(A \vee B) \rightarrow P(A) \wedge P(B)$

FCP is "probably the most discussed issue in the logic of permission", and most of this discussion is heavily critical [12, p.207] We do not try to defend this principle here. In Section 1 we show, however, that it follows for OR. Is one then forced to reject OR? No. What one should do is change the underlying theory of action. This is the main point of this paper. Indeed, a quick look at the derivation of FCP from OR reveals that the culprit is what we call the principle of additivity:

(Add). If t is an action token of type φ then it is also of type $(\varphi \vee \psi)$.

The bulk of this paper is devoted to explaining why and when Add should be abandoned. In Section 2 we present three general reasons why Add fails. In Section 3 we show two concrete examples of the failure of Add. So it is possible to keep OR while rejecting FCP. But is this also a plausible view? Yes. We argue for that in Section 4.

1   A First-Order Derivation of FCP from OR

Deontic logic has primarily dealt with normative notions as applied to *generic actions* or *action types*, rather than *individual acts* or *act tokens*. The relation between generic actions and individual acts has not attracted much attention in deontic logic, although there are notions of normative concepts that can be interpreted as relating individual acts to generic actions. As we will see later, OR can be interpreted as being one of them. Here is its original formulation.

(Original OR). An action α is permitted iff every execution of α results in a state that is normatively OK.

Although mainly used in dynamic deontic logic (DDL), the idea behind the open reading of permission goes back (at least) to G. H. von Wright [28, p. 34-35]. In DDL, early formalizations of Original OR can already be found in J.-J. Ch. Meyer's seminal paper [20]. He associates permission with the dynamic logic box operator [ ] and its dual $\langle \ \rangle$, and defines a concept of *free choice permission* P_F as:

(PDL P_F). $P_F \alpha =_{df} P\alpha \wedge [\alpha]OK$, where OK expresses, that a state is normatively OK, and the "usual" concept of permission $P\alpha$ is defined as $\langle \alpha \rangle OK$ (Meyer used the negation of a violation constant instead of OK. For our purpose, though, this difference is irrelevant.)

J. Broersen [6] implements a version of Original OR without the additional requirement $\langle \alpha \rangle OK$, which makes it even closer to its natural language version.

(PDL OR). $P\alpha =_{df} [\alpha]OK$, where OK expresses, that a state is normatively OK (Broersen also used the negation of a violation constant instead of OK.)

As both authors observe (cf. [20, p. 121] and [6, p. 166]), this very association makes the so-defined permission predicate one that obeys free-choice:[2]

(PDL FCP). $P(\alpha \cup \beta) \rightarrow P\alpha \wedge P\beta$

In dynamic deontic logics α, β, ... usually are singular terms representing action types and permission therefore a defined predicate [20], [6]. Interestingly, in dynamic logic the theorem responsible for the derivability of PDL FCP from PDL OR is a special form of additivity, expressed in the formula:

(Add PDL). $< \alpha > \varphi \rightarrow < \alpha \cup \beta > \varphi$

As we will see later, this is not a coincidence. We will argue in a slightly more abstract setting, that every deontic logic that contains the Open Reading and an additivity principle (or rule) of a special form (expressible by a certain first-order proposition), leads to free-choice permission.

In our framework, we require generic actions to be describable by a proposition (a common assumption found in many deontic logics), and interpret "every execution of α" as "every individual act instantiating α". This gives a natural reinterpretation of Original OR, one that links generic actions to individual acts:

(OR). An action type φ is permitted iff every individual act instantiating φ is normatively OK.

Contrary to DDL, this results in a logic where permission once again can be treated as a sentential operator, and OK becomes a property applicable to individual acts rather than a propositional constant being true at certain states. Furthermore, OR *explicitly reduces* the permission of a generic action to the claim of certain individual acts being normatively OK or legal.[3] We believe this to be a very natural approach connecting singular and generic actions.

In a first step we now formalize OR. We extend a first-order language with individual constants $c_1, c_2, \ldots$ and two additional types of formulas:

(Inst) Formulas $Inst(t, \varphi)$, where t is a singular term and φ is an arbitrary formula not containing occurrences of $Inst$. $Inst(t, \varphi)$ is read as 't instantiates φ'.

---

[2] Although only stated implicitly, R. Trypuz and P. Kulicki make a similar observation in their algebraic account of actions [26].

[3] So this approach also reduces permissions and obligations to another property of act tokens, namely being "normatively OK" or legal. We leave it has a primitive notion here. One could explicate this notion in terms of being liable, or not, to blame or sanctions, as done in DDL.

(OK) Formulas $OK(t)$, where t is a singular term. $OK(t)$ is read as 't is normatively OK'.

OR can now be easily expressed using these notions:

(OR*) $P\varphi =_{df} \forall x(Inst(x, \varphi) \to OK(x))$

Note that we do not make any restrictions on $Inst$ and OK, we merely assume that individual actions can be the range of quantification, and that generic actions can be represented by formulas. Recall that OK is a predicate applicable to individual actions (action tokens) and not to states as in DDL.

What exactly the resulting deontic logic will look like depends on the theory of individual acts, the instantiating operator, and the OK predicate (as applied to individual acts). How well this approach performs certainly is a matter for more detailed investigations concerning the logic of $Inst$ and OK. At this stage, we leave this for further research. In what follows, we just want to discuss one principle for $Inst$, and take a look at some of its consequences. The three frameworks presented in next section might serve as a blueprint for developing a logic of $Inst$.

At this point it seems natural to strengthen this theory by postulating additional rules for more complex generic actions. At least at first sight, for disjunctive generic actions $\varphi \vee \psi$ we have two obvious and equally reasonable candidates.

For every singular term t:

$$(\vee Int1) \; \frac{Inst(t, \varphi)}{Inst(t, \varphi \vee \psi)} \qquad (\vee Int2) \; \frac{Inst(t, \varphi)}{Inst(t, \psi \vee \varphi)}$$

Although the resulting logic of action is still very weak, it is strong enough to prove FCP. Suppose $P(\varphi \vee \psi)$, which by OR* means $\forall x(Inst(x, \varphi \vee \psi) \to OK(x))$. Using $\forall x(Inst(x, \varphi) \to Inst(x, \varphi \vee \psi))$ (follows from $\vee Int1$) and transitivity of material implication we arrive at $\forall x(Inst(x, \varphi) \to OK(x))$, which (according to OR*) is equivalent to $P\varphi$. Analogously for $P\psi$ via $\vee Int2$.[4]

So if we want to avoid FCP we have to drop either OR* or both $\vee Int1$ and $\vee Int2$.[5] Dropping only one of $\vee Int1$ and $\vee Int2$ is not an option in our view. We think the following is a plausible principle for disjunctive actions.

$(\vee Sym) \; \forall x(Inst(x, \varphi \vee \psi) \leftrightarrow Inst(x, \psi \vee \varphi))$

Under that principle $\vee Int1$ and $\vee Int2$ turn out to be deductively equivalent. Another argument for dropping both $\vee Int1$ and $\vee Int2$ is that otherwise one still gets either $P(\varphi \vee \psi) \to P\varphi$ or $P(\varphi \vee \psi) \to P\psi$. For someone dissatisfied with FCP this is equally undesirable.

---

[4] This derivation seems to have been already observed, at least implicitly, for instance in Makinson's account of disjunctive permissions as "checklist conditionals" [16]. The contribution here is to make it explicit.

[5] Of course, one could also give up certain principles of first-order logic. We do not consider this option here though.

So if we want to avoid FCP we are left with a choice between dropping OR* or both $\lor Int1$ and $\lor Int2$, or both. Given what we have just said, we will lump $\lor Int1$ and $\lor Int2$ together and talk generally of a choice between OR and additivity Add, as presented in the introduction.[6]

If we drop Add, we cannot derive FCP from OR* anymore. This can be shown by a simple first order model, in which one utilizes the fact that $Inst(a, \varphi)$ is independent from $Inst(a, \varphi \lor \psi)$.[7]

The next two sections can be seen as a conceptual exploration into ways to keep OR while resisting FCP. We argue that there are plausible notions of disjunctive generic actions which allow that, i.e. for which Add does not hold. Of course, one could agree with us on Add but also insist on rejecting OR. So no problem with free choice, that person would say. In Section 4 we address this potential objector. In our view OR does express a plausible notion of permission, and so there are good reasons to look for ways to accept it without being committed to FCP.

2   Rejecting Additivity: General Reasons

In this section we answer two questions:

1. Why reject Add?
2. And if we do so, which algebraic properties do action types retain?

We do so by explaining how normality, resource sensitivity and relevance, once applied to action types, warrant a rejection of additivity. Of course dropping additivity is a well-known solution to avoid classical deontic paradoxes. The novelty of our approach is to provide additional reasons, beyond avoiding paradoxes, to reject additivity.

Normality, resource sensitivity and relevance are behind very well-known logical calculi. So rejecting Add does not force us into a logical no-man's land. The algebraic properties of these systems are known, and can be used to develop a full-fledged calculus of action types that does not include Add.

2.1   Normality

Statements about action types can be seen as referring to the normal instances of that type. Normality statements are made against a set of conventional assumptions, shared views or beliefs. This idea is by now widely accepted in AI [18], non-monotonic logic [17] and linguistics (see below). Applied to action types this blocks Add. A normal instance of "walking back home" might not be a normal instance of "walking back home or robbing a bank." Why? One could argue that normal instances of disjunctive action types are those where both disjuncts are or were live options. In that view then there might just not be any normal

---

[6] The reason for calling it additivity is the connection with linear logic. See Section 2.2.

[7] We explore this in more detail in section 5.

instances of "walking back home or robbing a bank". I might not be the kind of person who is normally in a state of mind where I can rob a bank. Less dramatically, the normal cases of walking home might be completely disjoint from those where robbing a bank is a live alternative. All the latter might be abnormal instances of the former. So if generic action types are taken to refer to normal tokens, additivity fails.

There is an extensive literature in formal semantics showing how to handle this phenomena when it comes to action sentences. [3], for instance, treats action types as generic. Generics do not express unrestricted quantifications over all instances. Constraints such as interests, our beliefs or shared knowledge constrain this quantification [3,29,25].

2.2   Resource Sensitivity

Resource sensitivity is the general idea that resources, be they linguistic or informational, cannot be reused or discarded at will. Just like normality, it is by now a fairly accepted idea that language and information are resource sensitive [4,13]. Linear logic [7] has been developed precisely to deal with this phenomena. There are two kinds of disjunctions in linear logic: additive and multiplicative. These are expressed by the following rules, in natural deduction systems:

$$\textbf{(par)} \ \frac{\Gamma, A, B}{\Gamma, A \,\mathbf{\mathcal{B}}\, B}$$

$$\textbf{(plus}_\textbf{L}\textbf{)} \ \frac{\Gamma, A}{\Gamma, A \oplus B} \qquad\qquad \textbf{(plus}_\textbf{R}\textbf{)} \ \frac{\Gamma, B}{\Gamma, A \oplus B}$$

The rule **par** is for the multiplicative disjunctive connective $\mathbf{\mathcal{B}}$, and the rule **plus** for the additive one $\oplus$. [8] Notice that the information contexts of premises in the multiplicative disjunction need all information of both alternatives. This is not the case for the additive one [10,11]. This reflects the fact of resource sensitivity. The same information inputs must be used in both alternatives for the additive disjunction. In the multiplicative case each input is used exactly once in producing the output [1].

This general idea applies to action as well. Additive disjunction expresses the choice of one action token between two possible token alternatives, where multiplicative disjunction expresses a dependency between two alternative tokens [10,11]. For instance, in a Chinese restaurant, an action token consisting of

---

[8] We also have a Sequent Calculus version of the rules **par** for the multiplicative disjunctive connective $\mathbf{\mathcal{B}}$, and the rules **plus** for the additive one $\oplus$ as follows:

$$\textbf{(par}_\textbf{L}\textbf{)} \ \frac{\Gamma_1, A \vdash \Delta_1 \qquad \Gamma_2, B \vdash \Delta_2}{\Gamma_1, \Gamma_2, A \,\mathbf{\mathcal{B}}\, B \vdash \Delta_1, \Delta_2} \qquad\qquad \textbf{(par}_\textbf{R}\textbf{)} \ \frac{\Gamma \vdash A, B, \Delta}{\Gamma \vdash A \,\mathbf{\mathcal{B}}\, B, \Delta}$$

$$\textbf{(plus}_\textbf{L}\textbf{)} \ \frac{\Gamma, A \vdash \Delta \qquad \Gamma, B \vdash \Delta}{\Gamma, A \oplus B \vdash \Delta} \qquad \textbf{(plus}_\textbf{R}, i = 1, 2) \ \frac{\Gamma \vdash B_i, \Delta}{\Gamma \vdash B_1 \oplus B_2, \Delta}$$

All of these still reflect the fact of resource sensitivity in our topic.

an offer of baked oysters is not a token of an offer of a choice between baked oysters or pizza. The cook knows nothing about Italian pizza, or the restaurant lacks the ingredients for Italian dishes. So this is a case of a multiplicative, non-additive disjunction.

2.3 Relevance

Additivity can also be rejected by invoking relevance. This idea has been explored in [24], [23]. According to these authors the root of (some) classical paradoxes is the rules governing (classical) disjunction introduction. In deontic logic the paradigmatic example is Ross's paradox. Instead of using non-normal modalities, the authors in [24], [23] suggest that inference steps using rules like Add are problematic because they introduce a certain kind of *irrelevance*:

> "[...] one should distinguish between *validity* in the sense of mathematical logic and *appropriateness* with respect to *applied arguments*. The paradoxes rest on certain *irrelevant* deductions, which are, although mathematically valid, nonsensical and often enough harmful in applied arguments." [23, p. 399]

Generally speaking, an inference rule "From φ infer ψ" is prone to introducing irrelevance iff ψ contains a subformula which can be replaced by an arbitrary formula *salva validitate*. In the case of classical deontic logic, blocking inference rules prone to introducing irrelevance excludes Add and as a consequence also Ross's Paradox. If this criteria is applied to the logic of our $Inst$ operator (Section 1), $\lor Int1$ and $\lor Int2$ have to be given up.

Why should we accept this relevance criterion? Here are three arguments from [23]. First this solves a whole variety of paradoxes from very different areas, e.g. deontic logic, philosophy of science, epistemology. In particular, if we are interested in applying logic, this criteria seems to do a pretty good job. Second, there even seems to be empirical evidence that subjects reason in line with this criteria. Schurz reports conducting a small experiment with students untrained in logic (see [23, p. 429–430]). As it turns out, these students regard (logically) valid inferences *with* irrelevant parts occurring in their conclusion as intuitively invalid. On the other hand, valid arguments *without* irrelevant parts appearing in their conclusion are considered to be intuitively valid. This gives us at least partial evidence for believing that everyday reasoning is closer to logical reasoning including this criteria than to logical reasoning without it. Finally, we conjecture that this also holds for reasoning about actions, and that Add can be rejected on these grounds. The following is a revealing instance:

$$\frac{Inst(a, p)}{Inst(a, p \lor \bot)} \ ,$$

where p stands for some generic and possible action (e.g. smoking), and a for an instance of p (e.g. John's smoking of his first cigarette after lunch). Taking into account Schurz's empirical results, we think it is very unlikely that logically untrained subjects would regard a derivation from

"John's smoking of his first cigarette after lunch is an instance of smoking"
to
"John's smoking of his first cigarette after lunch is an instance of smoking *or* walking and not-walking"

to be intuitively valid. Whether or not this is so remains to be seen. If this turns out to be true, there is a notion of disjunctive actions (e.g. the one presumably used in everyday reasoning) that does not validate Add. But Schurz's results already show that relevance does play a role in reasoning in general, and that Add could be rejected on that ground.

3   Rejecting Additivity: Two Concrete Cases

In this section we exemplify our claim that additivity should be rejected. We observe that the agency operator "deliberative stit" is not additive. Additivity also fails when one sees disjunctive actions as non-trivial mixings. So having the Open Reading without Free Choice Permission does not commit to an outlandish theory of action. The principle fails in well-known cases.

There is another reason why it is important to show concrete cases of non-additive disjunctive action types. It is not clear what constitutes a disjunctive action type in the first place. PDL gives a simple answer to that. These are non-deterministic choices. An action of type $\alpha \cup \beta$ is a non-deterministic choice between an action of type α and an action of type β. If one can become rich by "going to college or robbing a bank", then one can do so by either "going to college" or "robbing a bank." And also the other way around.

So in PDL the vague notion of disjunctive action types is made precise by reducing it to a disjunction of types. This reduction is reflected by the following PDL validity:

(ND-Choice PDL). $(<\alpha > \varphi \vee < \beta > \varphi) \leftrightarrow <\alpha \cup \beta > \varphi$

If all disjunctive action types were analyzable this way, additivity would hold. The examples below show that this is not the case. There are natural readings of disjunctive action types for which additivity, and thus the interpretation as non-deterministic choice, fail.

3.1   Deliberative Stit

According to the so-called deliberative *stit*, or *dstit*, agent i sees to it that φ in history h and moment m whenever "φ is guaranteed by a present choice of i", while i could have done otherwise [5, p.37]. Choices at a given moment m are represented by a partition of the set of histories compatible with m. Figure 1 shows a small example. Four histories are compatible with m_1. Agent i has two possible choices or options: c_1 and c_2. In c_1 she forces that the future will lie in histories h_1 or h_2. In c_2 she rules out these two histories and forces h_3 or h_4. Write $Choice_i^m(h)$ the cell of i's choice partition at m that contains h. A formula

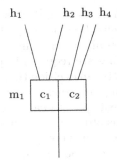

Fig. 1. Branching time with choices

$[i \ dstit : \varphi]$ is true at moment-history pair m, h whenever φ is true at all pairs m, h' with $h' \in Choice_i^m(h)$, and there is a pair m, h'' in which φ is false.

Stit theory is about consequences of actions, not directly action types. A formula of the form $[i \ dstit : \varphi]$ doesn't mean that i performs a φ action. Rather, it says there is a concrete choice available to i through which she can achieve φ.

But this is not to say that no action types can be described using the $dstit$ operator. First, atomic propositions in can be interpreted as describing state where an action of a given type is executed. Say p is "Bob is smoking." Then [Bob $dstitp$] is the statement that Bob sees to it that the current history is one where he is smoking. So $stit$ formulas can describe action types in this indirect way. Second, some action types can be described by the consequences they bring about. Polluting and breaking are good examples, just like the more positive actions of healing and rescuing. In this case $stit$ formulas directly describe a specific action type.

So there are at least two different ways to read off action types out of $stit$ formulas. All that remains to observe is that additivity doesn't hold for them. Let p only be true everywhere except at m_1, h_3 in Figure 1, and q be true only at m_1, h_4. Then $[i \ dstit : p]$ is true at m_1, h_1, but not $[i \ dstit : p \vee q]$. There is no moment-history pair that falsifies both p and q. So $dstit$ is not a normal modality. This fact is well-known, c.f. [5, p.40], and should not come as a surprise since the operator can be defined as a combination of two normal modalities [15]. The important point for the present argument is that we have a well-known theory of action in which additivity fails.

3.2   Non-trivial Mixing

Given t and t' two actions available to an agent, a mixed action μ is the action of doing t with probability p and t' with probability $1 - p$. Generally, a mixed action over a (measurable) set T is a probability distribution over that set. We call an action *non-trivially* mixed on a set T when it assigns non-zero probability to all the elements of T, i.e. when it has full support over T. Non-trivial mixings require two things. First, each of its components must be a real alternative to the agent. Second, the agent must be able to randomize between them.

Mixed actions are fundamental to game and decision theory. In game theory (non-trivial) mixed strategies are required in proving the existence of Nash equilibria in some games (c.f. [21]). In decision theory dominance reasoning sometimes involves non-trivial mixing. And such actions have recently been put to use to answer purported counter-examples to causal decision theory (see e.g. [14]). One can interpret mixed actions objectively or, in games, as the beliefs of others regarding one's own action. We focus here on the objective interpretation. Here the idea is that playing a mixed action involves real randomizing, eventually using some appropriate device. A typical example is random security searches at airports.

Non-trivial mixing of actions be seen as one kind of disjunctive action, and this is sufficient to invalidate additivity. Take a token t that is an instance of smoking. This token is not necessarily an instance of a non-trivial mixing between smoking and, say, jogging. There might be no token of jogging available to the agent in the first place, and the agent might not be able to randomize. She might just lack the required device to do so. So it is not true that for any token of type φ this token is also an instance of type $\varphi \vee \psi$. In some cases these disjunctive actions will be non-trivial mixings, for which additivity fails.

4   Why Keep the Open Reading

Up to now the paper could been seen as an investigation of the logical space between OR and FCP. We found plenty. This is an important observation in itself. But one could still argue that this investigation is philosophically moot because OR is dubious in the first place. We argue in this section that this is not the case. Not only can one keep OR, one *should* keep it. To be more precise, our claim is the following:

OR reduces permissions of action types to the legality or normative status of action tokens. Under that reading, to say that an action type φ is permitted in a given situation is to say that executing a φ action is sufficient for legality. The companion idea is that obligations give necessary conditions for legality: φ is obligatory iff no non-φ action is legal. This is the Andersonian-Kangerian Reduction of obligation [19]. In our view, combination of the two ideas has intuitive appeal (c.f. [22,2]). Obligations give necessary and sufficient conditions for legality.[9] As an historical aside, it is worth mentioning that von Wright seems to have endorsed a view of permissions close to OR:

> "It seems to me that the most natural way of understanding the phrase is this: "It is permitted that p", no circumstances being specified, means that it is permitted that p, no matter what the circumstances are, i.e. in *all* circumstances." [28, pp. 34-35]

---

[9] Note that here we lose the usual duality between O and P.

There are two obvious objection to OR. The first one goes as follows. Suppose smoking is permitted outside the building, and that I could smoke outside while robbing my fellow smokers. Surely smoking and robbing my fellow smokers is not permitted, because robbing is forbidden. But any smoking and robbing token is a smoking token. So OR cannot be correct.

There are three ways to reply to this first objection. First is to say that all the objection shows is that, with OR, permissions require circumscription. If smoking while robbing is not permitted then it is not smoking that was permitted in the first place. It is smoking and not robbing. Unless a skeptic can show that it is not possible to fully circumscribe permitted action types, the open reading remains in force.

The second reply is that the open reading can be seen as a *prima facie* or default reading. Faced with abnormal or exceptional cases like smoking and robbing one should revise the system of permissions so as to exclude these cases. As before, in the revised system the open reading regains its status as default reading.

Finally, one can avoid the objection by restricting the universal quantification to the set of executions that are not otherwise illegal or not normatively OK. Smoking while robbing is not a legal instance of smoking, because it is an instance of robbing, which is otherwise illegal. The objection points to the danger of unrestricted quantification. Once appropriately restricted the open reading remains. Smoking is permitted if and only if all instances of smoking are legal, which are not deemed illegal through being an instance of a different action type. In other words, in this amended sense, permission provides sufficient conditions for legality, everything else being legal. This reply has the advantage over the first one in that it is less prone to the skeptical rebuttal. The "everything else" clause can remain implicit or be seen as expressing the *prima facie* character of permissions. It doesn't require explicit circumscription. But it is still the open reading.[10]

So OR has intuitive appeal, especially in combination with the Andersonian-Kangerian Reduction for obligation, and the obvious objection less grip than one would think. Is there a positive argument for OR? Yes. One could argue for it in terms of applicability or "decidability"[11] for the assessment of concrete situations. Is token t legal or illegal? E.g. was *that* a legal driving maneuver? Under the classical Andersonian-Kangerian reduction for permission the judge only has a negative test for assessing whether an action is legal or not. Under the open reading the judge has a pair of tests: is t an instance of a forbidden type? If no then it is not illegal. But is it legal then? Some legal systems have

---

[10] There is another objection looming. It uses conditional permissions. If I have no income, I am permitted not to pay taxes. Otherwise not paying taxes is illegal. So it false that all instances of not paying taxes are OK or legal. And this is neither because some of these instances are otherwise illegal, nor because the permitted action type is not circumscribed enough. Could the open reading be adapted to capture such conditional permissions? We think so, but leave it for future work. We thank an anonymous referee of DEON for raising this point.

[11] In a non-technical sense of "decidable".

closure conditions stating that everything not explicitly forbidden is permitted. But not all systems have that. In that case OR provides the second step needed for assessing legality, namely checking whether the token at hand is an instance of a permitted type. So OR gives additional tools for deciding the legality of specific actions, and this is a reason to accept it.

5   Conclusion

This paper makes three main points:

1 Any deontic system containing the open reading and additivity for action types must validate Free Choice Permission.
2 There are good philosophical reasons to reject additivity, and the property fails in well-known concrete theories of agency. So one can have the open reading without free choice.
3 There are good philosophical reasons to accept OR. So one should accept the open reading, arguably also without free choice.

Of course, there are several issues that deserve further investigation. One point concerns the logic of agency one would get by incorporating each of the philosophical reasons to reject Add that we presented in Section 2. We observed that for each of these we have known logical systems, and that there are different ways of constructing these: one can start with a non-boolean disjunction, for which Add already fails on the level of PC formulas (e.g. relevance), and through that reject Add also on the levels of *Inst* formulas. Here a disjunction has the same meaning in all contexts. The second option is to start with a boolean disjunction, but interpret *Inst* as a non-normal modality in such a way that Add fails (e.g. stit). In this setting, *Inst* contributes significantly to the meaning of disjunctive generic actions, and what we mean by a disjunction depends on whether it occurs inside or outside an *Inst* operator. Prima facie, we regard both options to be worth further research. But how these systems would behave as explicit logic of agency is an interesting open question.

Finally, in answering an obvious objection we mentioned in section Section 4 a possible qualification of OR, namely using an "everything else being legal" clause. The logic of permissions defined this way remains to be investigated. It would be particularly interesting to study the forms of FCP that it would licence in the presence of Add. We leave these questions open for now, content with the observation that one can (and should!) have the open reading without free choice.

Acknowledgments. We would like to thank the three referees of DEON 2014, the participants of the first meeting of the Bavarian Deontic Group, and in particular Nathan Woods, for helpful comments and suggestions.

References

1. Abramsky, S.: Computational interpretations of linear logic. Theoretical Computer Science 111(1), 3–57 (1993)
2. Anglberger, A., Gratzl, N., Roy, O.: The logic of obligations as weakest permissions (2013) (manuscript)
3. Asher, N., Pelletier, F.: Generics and defaults. In: van Bentham, J., ter Meulen, A. (eds.) Handbook of Logic and Language. Elsevier (1997)
4. Asudeh, A.: Linear logic, linguistic resource sensitivity and resumption, eSSLLI (2006)
5. Belnap, N., Perloff, M., Xu, M.: Facing the future: Agents and choice in our indeterminist world (2001)
6. Broersen, J.: Action negation and alternative reductions for dynamic deontic logics. Journal of Applied Logic 2, 153–168 (2004)
7. Di Cosmo, R., Miller, D.: Linear logic. In: Zalta, E.N. (ed.) The Stanford Encyclopedia of Philosophy, Fall 2010 edn. (2010)
8. Dignum, F., Meyer, J.J., Wieringa, R.: Free choice and contextually permitted actions. Studia Logica 57(1), 193–220 (1996)
9. Dignum, F., Meyer, J.J.: Negations of transactions and their use in the specification of dynamic and deontic integrity constraints. In: Semantics for Concurrency. Workshops in Computing, pp. 61–80. Springer, London (1990)
10. Girard, J.Y.: Linear logic. Theoretical Computer Science 50(1), 1–102 (1987)
11. Girard, J.Y.: Linear logic: Its syntax and semantics. In: Girard, J.Y., Lafont, Y., Regnier, L. (eds.) Advances in Linear Logic, vol. 222. Cambridge University Press (1995)
12. Hansson, S.: The varieties of permissions. In: Gabbay, D., Horty, J., Parent, X., van der Meyden, R., van der Torre, L. (eds.) Handbook of Deontic Logic and Normative Systems, vol. 1, College Publication (2013)
13. Van Benthem, J.: Language in Action: categories, lambdas and dynamic logic. MIT Press (1995)
14. Joyce, J.M.: Regret and instability in causal decision theory. Synthese 187(1), 123–145 (2012)
15. Kracht, M., Wolter, F.: Normal monomodal logics can simulate all others. Journal Symbolic Logic 64(1), 99–138 (1999)
16. Makinson, D.: Stenius' approach to disjunctive permission. Theoria 50, 138–147 (1984)
17. Makinson, D.: Bridges from classical to nonmonotonic logic. College Publications (2005)
18. McCarthy, J.: Epistemological problems of artificial intelligence. In: IJCAI, vol. 77, pp. 1038–1044 (1977)
19. McNamara, P.: Deontic logic. In: Zalta, E.N. (ed.) The Stanford Encyclopedia of Philosophy, Fall 2010 edn. (2010)
20. Meyer, J.J.C.: A different approach to Deontic Logic: Deontic Logic Viewed as a Variant of Dynamic Logic. Notre Dame Journal of Formal Logic 29, 109–136 (1988)
21. Osborne, M.J., Rubinstein, A.: A course in game theory. MIT Press (1994)
22. Roy, O., Anglberger, A.J.J., Gratzl, N.: The logic of obligation as weakest permission. In: Ågotnes, T., Broersen, J., Elgesem, D. (eds.) DEON 2012. LNCS, vol. 7393, pp. 139–150. Springer, Heidelberg (2012)
23. Schurz, G.: Relevant Deduction: From Solving Paradoxes Towards a General Theory. Erkenntnis 35, 391–437 (1991)

24. Schurz, G., Weingartner, P.: Paradoxes solved by simple relevance criteria. Logique et Analyse 113, 3–40 (1986)
25. Simons, M.: Dividing things up: The semantics of or and the modal/or interaction. Natural Language Semantics 13(3), 271–316 (2005)
26. Trypuz, R., Kulicki, P.: On deontic action logics based on boolean algebra. Journal of Logic and Computation (2013) (forthcoming)
27. von Wright, G.H.: Norm and Action - A Logical Enquiry. Routledge (1963)
28. von Wright, G.H.: An Essay in Deontic Logic and the General Theory of Action. North-Holland Publishing Company (1968)
29. Zimmermann, T.: Free choice disjunction and epistemic possibility. Natural Language Semantics 8(4), 255–290 (2000)

'Must', 'Ought' and the Structure of Standards

Gunnar Björnsson[1] and Robert Shanklin[2]

[1] Umeå University, Umeå, Sweden
gunnar.bjornsson@umu.se
[2] Santa Clara University, Santa Clara, CA, USA
rshanklin@scu.edu

Abstract. This paper concerns the semantic difference between strong and weak necessity modals. First we identify a number of explananda: their well-known intuitive difference in strength between 'must' and 'ought' as well as differences in connections to probabilistic considerations and acts of requiring and recommending. Here we argue that important extant analyses of the semantic differences, though tailored to account for some of these aspects, fail to account for all. We proceed to suggest that the difference between 'ought' and 'must' lies in how they relate to scalar and binary standards. Briefly put, *must(φ)* says that among the relevant alternatives, φ is selected by the relevant binary standard, whereas *ought(φ)* says that among the relevant alternatives, φ is selected by the relevant scale. Given independently plausible assumptions about how standards are provided by context, this explains the relevant differences discussed.

Keywords: necessity modals, ought, must, Kratzer, von Fintel, Iatridou.

1 Introduction

Many philosophers take 'ought' to be the canonical term for asserting and discussing moral obligations and requirements. Indeed, the first entry for 'ought' in many well-known dictionaries identifies it as a word for duty or moral obligation. However, this proves problematic, as revealed by a (now) classic example:

(1) Employees must wash hands. Non-employees really ought to wash their hands, too. [9]

It is not 'ought,' but rather 'must' that indicates an obligation or duty, or what we are required to do. In (1), while it is clear that employees are required to wash their hands, 'ought' seems to indicate a weaker claim on the non-employee—something more like a recommendation or exhortation.

Since both 'must' and 'ought' (and closely related expressions like 'have to' and 'should') play central roles in moral judgments and moral reasoning, it is important to ascertain what 'ought' does indicate, if not duties or obligations, as well as to understand the difference(s) between 'must' and 'ought.'[1] That is the purpose of this paper.

---

[1] The fact that the distinction occurs across a variety of languages furthermore suggests that it tracks some stable and important cognitive distinction [10].

F. Cariani et al. (Eds.): DEON 2014, LNAI 8554, pp. 33–48, 2014.
© Springer International Publishing Switzerland 2014

We suggest that the difference should be understood in terms of *differently structured standards*. 'Ought' relates to a scalar standard—a ranking of alternatives as better or worse. 'Must', by contrast, relates to a requiring standard—one that rules out all alternatives not satisfying a certain condition. This distinction, we argue, can explain the variety of differences between the two locutions. At the same time, it is both parsimonious, relying on independently motivated assumptions and existing resources within semantics, as well as conservative, being compatible with standard general approaches to the analysis of modals, and compatible with substantive views in normative ethics about what, in particular, ought to be done or ought to be the case. (For earlier discussions of the distinction, see [2,3,4,5], [9,10,11,12], [14], [17,18], [20].) In what follows, we spell out some relevant explananda and indicate why we find some extant accounts of these wanting (section 2), present our proposal and how it accounts for the explananda (section 3), and address some complications (section 4).

2   Explananda

We start with some of the data that should be captured by accounts of 'ought' (and its close relative 'should') and 'must' (and its relative 'have to'):[2]

Different Flavors. Both 'ought' and 'must' famously come in different "flavors", relating to different kinds of modalities:

MORAL: "One ought to help one's friends." / "One must not murder."
PRUDENTIAL: "You ought to lock the garage; car thieves are active in the area!" / "We've been hit twice by violent burglars; we must protect ourselves."
BOULETIC: "Oh man, she ought to be here—she'd love this!" / "You simply must see the Rembrandt exhibit while it's in town."
TELEOLOGICAL: "To get to Harlem, you ought to take the A-Train." / "Actually, the subway broke down, the cabs are on-strike, and the heliport is closed for renovation; to get to Harlem, you have to walk."
EPISTEMIC: "He ought to be home within 10 minutes; he left an hour ago." / "He must have arrived; I see he checked-in on Facebook."

The variety of modal flavors is of course familiar. Different modal claims clearly relate to different sorts of considerations. On a generic analysis of *must(φ)* in terms of quantifications over possibilities, it means *in all relevant possibilities, φ*. Within such an analysis, the different flavors correspond to different ways of selecting the relevant possibilities, ways that are in turn determined by context. In the case of epistemic 'must', for example, the relevant possibilities might be those compatible with the evidence; in the case of moral 'must', the possibilities might be those that are morally best among the possibilities an agent can bring about at a time.

---

[2]   In this paper, we assume that what goes for 'ought' goes for 'should,' and that what goes for 'must' goes for 'have to,' though in reality matters are more complicated, with regard to both connotations and syntax. We set aside such complications in pursuit of an understanding of broader differences between the two categories of expressions, working on the assumption that they encode two importantly different kinds of thought.

What is important here is that the distinction between 'ought' and 'must' is felt across these flavors: exchanging one locution of the other in the examples above makes a striking difference across the board. In trying to account for the difference, our default assumption should be that the difference stems systematically from a difference in meaning between the two locutions.

Intuitive Difference in Strength. As we have already noted, 'ought' seems weaker, in some sense, than 'must'. One way of bringing out intuitive differences in strength is to substitute one for the other in a given sentence, such as:

(2)   When you are in town, you *must/have to/should/ought to* see the new Rembrandt exhibit.

If one thinks that the Rembrandt exhibition is great and wants to recommend seeing it on this ground, the intuitive strength of one's recommendation depends on whether we use 'must/have to' or 'ought/should': using the former would seem to express a *stronger* recommendation.

A difference in strength is also suggested by the fact that it often seems reasonable to say that someone should or ought to do something while denying that she has to, but not the other way around:

(3)   You ought to attend class every day, but you don't have to.

(4)   # You have to attend class every day, but I'm not saying that you should.

(5)   She ought to help her neighbor, but she doesn't have to.

(6)   # She must help her neighbor ...but it's not as if she ought to.

As (1), (3) and (5) illustrate, it might be perfectly natural to say that someone ought to do something while denying that he must, but as witnessed by (4) and (6), the reverse is problematic.

The most straightforward way of understanding differences in strength is in terms of *logical* strength: $must(\varphi)$ implies $ought(\varphi)$, but not the other way around. This needs an obvious qualification, however, as both 'ought' and 'must' come in different flavors. Depending on how fine a distinction we make between these flavors, the two locutions might have different flavors within examples like (3) through (6). Consider again (1) ("Employees must wash hands. Non-employees really ought to wash their hands, too"). Here, 'must' might be understood as legal or policy-based, whereas 'ought' is more naturally understood as moral. Moreover, if there is a shift, then clearly we can have cases where both *have to/must(φ)* and *should/ought(~φ)* are felicitous (even assuming that *should not(φ)* implies *not should(φ)*):

(7)   I've now read the regulations: you must hand in the documents by the end of today. But you really shouldn't. We can save lives if we hold on to them until tomorrow.

The datum here, then, is that *when must(φ) and ought(φ) have the same flavor*, the former seems stronger. Exactly how this is spelled out obviously depends both on how sameness of flavor is to be understood and on the semantics of the two locutions.

Perhaps the best-known attempt to represent the differences in strength between 'ought' and 'must' comes from Kai von Fintel and Sabine Iatridou [9,10]. Following

Angelica Kratzer [7,8], they take *must(φ)* to mean (simplifying somewhat) *φ holds in all the highest ranking accessible possible worlds.* The conversational background provides a "modal base" determining the set of worlds accessible from a world *w*, and an "ordering source" determining the ranking of worlds. In the case of moral 'must', the ordering source is a set of propositions describing a morally ideal situation; in the case of a legal 'must', a legally ideal situation. Their suggestion is that *ought(φ)* is similar, but that it takes a second ordering source which orders the accessible worlds favored by the primary ordering source. So if we say that

(8)   Liz ought to ψ; in fact she must.

and if 'must' and 'ought to' take the same primary ordering source, we are saying that

(9)   Liz ψs in all accessible worlds favored by O that are also favored by O'; in fact, she ψs in all accessible worlds favored by O.

Since the second ordering source (O') restricts the worlds that φ are said to hold in, the ought-claim in the first conjunct is weaker than the must-claim in the second.

Obviously, the proposal straightforwardly captures a difference in strength between the two locutions (cf. [12]).[3] But there seem to be (a) cases where two ordering sources are at play but 'must' still seems appropriate, and (b) cases where 'ought' seems clearly appropriate even though it is unclear what primary ordering source might be in play.

For an example of (a), suppose that we are considering whether to schedule a seminar on Monday, Tuesday, or Wednesday. Learning that the speaker can't make it

---

[3]   von Fintel and Iatridou [9] suggest that 'anankastic' oughts—sentences of the form 'if you want X, you ought to Y' or 'to X, you ought to Y'—are best understood to involve two ordering sources. That would let the explicit goal (X) operate on the first, thus ensuring that it isn't trumped by other goals such that *to X you ought to Y* comes out as true even when Ying would in no way promote X. They furthermore think that this is best explained under the assumption that 'ought' takes two ordering sources generally. But we do not see why clauses like "if you want to X" or "in order to X" cannot equally well work to *introduce* a privileged ordering source.

In a more recent paper [10], von Fintel and Iatridou also note that in many languages, weak necessity modals are expressed using a combination of strong necessity modals and temporally unmotivated past tense morphology characteristic of counterfactuals. They take this to suggest that weak necessity modals operate with two ordering sources, but the connection they propose between the past morphology and an extra ordering source seems largely ad hoc. In the case of counterfactuals, the past tense morphology does not introduce an extra ordering source restricting the relevant possibilities, but instead relaxes constraints on the possible to include what might be epistemically impossible. We should expect it to do something similar here and speculate that in the case of necessity modals, it indicates a widening of the considerations grounding the relevant selection of accessible alternatives: whereas 'must' encodes a binary condition decisively favoring some alternatives over others, 'ought' encodes a scale given which such a condition would be one among many possible conditions determining an alternatives position on that scale (see Section 3). For an example of such weakening by past morphology at work in the case of modals, consider the two close synonyms of 'ought' in Swedish: 'bör' and its morphologically past 'borde'. Both are weaker than the equivalent of 'must' ('måste'), but 'borde' is weaker than 'bör', indicating more uncertainty or less decisive reasons.

on Monday, we cross that day off our list, and remembering that we had dearly promised to leave Tuesday open for a departmental meeting, we cross Tuesday off our list too, leaving Wednesday. Even though we have two salient operative sources for our selection of the remaining alternative, it would be natural for us to conclude that:

(10)   We have to schedule the seminar for Wednesday.

At least, it seems that this would be natural if we took our previous promise to be clearly decisive.

For an example of (b), suppose that we hear of some natural disaster. Thinking about the urgent needs, we might naturally say:

(11)   We ought to contribute to disaster relief.

It is not clear, however, what the primary ordering source would be in this case, or that we need to identify one to know very well what was meant by (11). At the very least, it is not clear that the considerations triggering our utterance involve focusing on anything other than the fact that it is *better* if we contribute to disaster relief.

For these reasons, the number of salient ordering sources cannot itself be what distinguishes 'must' from 'ought': instead, there is something about the sources or our attitude towards them that favors one of the locutions over the other. And if we had an account of that difference in ordering sources, it might well be that we could explain the difference in strength without taking 'ought' to operate with a secondary ordering source.

Requirements vs. Recommendations. As illustrated by several of our examples above, 'must' is naturally used to express requirements, whereas 'ought' is naturally used to express something like recommendations. This is something that an analysis of these terms should let us explain. However, a few words are needed about the strength of the connections between 'ought' and recommendations, and between 'must' and requirements.

At one extreme, one might think of these connections as mere connotations, attached to the locutions by historical accident rather than grounded in the semantics. This seems implausible, as the connections seem to hold cross-linguistically. At the other extreme, one might take them as part of the meaning of the terms. For example, Mike Ridge analyses 'must' as relating to standards that require certain actions or states of mind and 'ought' as relating to standards that recommend. Recommending and requiring standards are in turn understood in terms of the distinction between kinds of speech acts: in recommending something, we are typically disposed to tolerate that someone ignores our recommendations; in requiring something, by contrast, we are disposed to insist on compliance, and impose sanctions for non-compliance ([14], ch. 1, § 3).

As will be clear, we think that the connection between 'must' and 'ought' and these speech acts is no mere coincidence, but we doubt that it is part of the meaning of the terms. The first problem is that the connection seems insufficiently tight to ground a difference in meaning. Speakers seem to use 'must' in a variety of contexts where they are not disposed to impose sanctions or insist on compliance in relation to the standards invoked. For example it is unclear whether we should expect any more insisting or sanctions for non-compliance from someone uttering (2) with 'ought' or 'should' rather than 'must'.

A different and perhaps more serious problem arises in the case of epistemic modals and uses of 'must' that express nomological or logical necessity. The problem is that what logically, nomologically or epistemically ought or must be the case often need not be an action:

(12)   It must/ought to be snowing in Stockholm by now.

(13)   When temperature increases, either volume or pressure must increase too.

Obviously, we do not normally recommend or require that it be snowing in Stockholm or that volume or pressure increase. Ridge suggests that statements expressing epistemic modals (such as (12)) say that what the relevant standards require or recommend is that we believe a certain proposition (e.g. that it is snowing in Stockholm). That might seem plausible (though perhaps less so for nomological or logical modals, as he acknowledges). But it introduces a compositionality problem: whereas (3) through (2) represent the actions recommended or required by the relevant standards, (12) or (13) do not in any clear way represent any *believing*, only a content that can be believed. Somehow, in the case of practical *must(φ)* or *ought(φ)*, the relevant standards would concern *φ*, whereas in the epistemic (or nomological or logical) case, it would apply to *our believing φ*. A more uniform account would be preferable.

For these reasons we do not think that it is part of the meaning of 'must' and 'ought' that they express or otherwise semantically relate to different levels of intolerance of non-compliance. But it is clear enough that 'must' is particularly well suited to express requirements in the sanction-implicating sense, and this is something that calls out for an explanation.

Probability, Conditionality, and Collective Commitments. Looking at epistemic uses of 'ought' and 'must,' one might think that the difference between the terms has to do with certainty or uncertainty. If one says that it *must* be snowing in Stockholm right now, one might seem to imply that we can be certain that it is snowing, or, more carefully, certain enough to consider the matter closed and not up for debate. If one says that it *ought* to be snowing, the implication is instead that it is probable or believable that it is snowing, or that a default assumption that it is snowing is in order, and one seems to leave the matter open for discussion. An analysis of the difference between 'ought' and 'must' should help us understand this difference.

As with the distinction between recommendations and requirements, one might think that the tendency of 'must' to express states of certainty or of considering a matter closed is more or less tightly connected to the meaning of the term, and similarly for the tendency of 'ought' to express probabilistic judgments or default assumptions. Suppose, for example, that we say that someone *must* or *ought* morally to lend a helping hand, thereby expressing that moral considerations require or support that action. Here one might further think that the ought-judgment, as opposed to the must-judgment, *semantically leaves open* that there might be stronger moral reasons not to lend a helping hand, or that some other action might also achieve whatever ends or satisfy whatever ideals we are concerned with, though with lower probability, or perhaps that it leaves open that other parties of the conversation do not share the priorities on which one based one's judgment.[4]

We do not see that the phenomenon generalizes in that way, however. Many think that there are cases where, all told and without remaining uncertainty, moral reasons favor but do not require a certain act. Morally speaking, it is what the agent should or ought to do, but not what he must do; it is morally recommended, but not mandatory. The difference here seems to be between the kinds of reasons involved, not their certainty or unqualified nature. Or take a prudential example, where we are faced with the choice between two routes to work, Highway 9 and Route 17, offering different driving conditions and different scenery. Given the current traffic, the weather and our mood, we judge that Route 17 is somewhat better all things considered. Though the difference is relatively small (while Route 17 is a little longer, it is prettier and a little less bumpy), we agree about the relative weights of these considerations and the facts involved, and so consider the matter settled. Now compare:

(14)   We ought to/should take Route 17.

(15)   ? We must/have to take Route 17.

(14) strikes us as perfectly felicitous, whereas (15) seems out of place. (Must we take Route 17? No, that is putting it too strongly. We may take Highway 9, though it wouldn't be as good.) But suppose that $must(\varphi)$ unconditionally represented φ as selected by considerations of the relevant flavor and presupposes that we are collectively committed to the priorities involved. Suppose also that $ought(\varphi)$ semantically leaves open the possibility that considerations supporting φ are outweighed, or represent φ as having merely probabilistic support, or takes priorities that the parties

---

4   Stephen Finlay has defended the suggestion that $must(\varphi)$ means that φ holds in all the relevant possibilities (where the relevant possibilities in the case of practical or bouletic modals are those in which some relevant end is realized), whereas $ought(\varphi)$ means roughly that φ is more likely than other relevant possibilities (see e.g. [3,4] and [5], §3.2). Aynat Rubinstein [15,16] distinguishes two kinds of priorities on which (non-epistemic) modal claims are based: those that support 'must' (i.e. provide a primary ordering source, in the von Fintel & Iatridou framework) are ones to which conversational participants are presupposed to be committed; those that support 'ought' (provide secondary ordering sources) lack that presupposition. Whereas Finlay takes 'ought' to leave room for uncertainty about the achievement of the relevant end, what Rubinstein takes 'ought' to leave unsettled (in the conversational context) are the preferences involved, i.e. more like the ends of Finlay's account. Similarly, and simplifying quite a bit, Alex Silk [17] suggests that $ought(\varphi)$ is distinguished from $must(\varphi)$ in that $ought(\varphi)$ represents φ as holding in all relevant possibilities *conditional* on the applicability of the ordering source, whereas $must(\varphi)$ represents φ holding in all relevant possibilities unconditionally. Much earlier, Jones and Pörn [12] proposed that $must(\varphi)$ indicates that $ought(\varphi)$ holds unconditionally or inescapably, under relevant ideal *and* non-ideal conditions. Unfortunately for our purposes, they say little about how relevant non-ideal conditions are selected, or how this might apply to epistemic 'ought', and the plausibility of the suggestion crucially depends on getting that selection just right. (We thank an anonymous referee for pointing us to Jones and Pörn's proposal.)

of the conversation are not committed to. Then contrary to what we find, (15) should have been perfectly fine, and (14) too weak.[5]

For these reasons, we think that while 'ought' is often better suited for contexts of uncertainty and that this calls for an explanation, it is not part of the semantics of 'ought' that it leaves open that some alternative is better, all facts considered.

3   SOS: 'Ought', 'Must', and the Structure of Standards

In the previous section, we listed phenomena that an account of the difference between 'ought' and 'must' should account for, and indicated problems for extant analyses of the difference to do so. Though we cannot pretend to have shown that these problems cannot be dealt with, we do hope to show how our own proposal can account for the phenomena in comparatively straightforward ways, and that it is worthy of further consideration.

The basic intuition behind our proposal is this. In thinking that something *must* or *has to* be case, we are thinking that, among relevant alternatives, it uniquely satisfies some salient condition. In the case of teleological 'must', it is the only alternative compatible with achieving the relevant goal; in the case of epistemic 'must', it is the only alternative compatible with the evidence; in the case of a moral 'must', it is the only that satisfies some moral requirement, and so forth. In thinking that something *ought to* or *should* be the case, by contrast, we have in mind considerations seen as providing overall sufficient support for selecting that alternative. In the former case, we have in mind a requiring condition or standard, or requirement; in the latter a scalar standard of some sort, providing considerations based on which we can see alternatives as more or less supported.

Obviously not all scales and requirements ground oughts and musts. We do not think that something ought or must epistemically be the case because it is most *un*likely or because it is the only alternative *in*compatible with the evidence. Similarly, we do not conclude that we ought to do something on the ground that this is most unlikely to give us what we want, or most likely to give us what we do not want. Generally, the scales and requirements on which we ground ought and must-judgments are ones that we take to be relevant in deliberation about what proposition to realize in action, have a positive attitude towards, or believe, depending on whether we are engaged in practical, evaluative or epistemic deliberation. Differently put, the standards that ground our judgments are standards for practical, evaluative or epistemic *endorsement* of propositions. This is not to say that we make ought and must-judgments only when we are in the business of forming beliefs, intentions, or attitudes, for standards can be applied from other points of view than the first-person

---

5   Finlay ([5] Ch. 6 §6) suggests a pragmatic explanation of the difference. We currently think that our account is more straightforward, and avoids other problems with Finlay's account, in particular problems with accounting for how alternatives are compared not only with respect to likelihood of achieving an end, but with respect to how likely they are to provide amounts of various valued quantities. Finlay ([5]: Ch. 7) has an extensive discussion of this problem, but we are not yet convinced that he can fully handle he problem.

present-tense deliberation. We can apply them in deliberating on behalf of someone else who has to make a decision (as potential advisors), or from the point of view of an unspecified agent in a hypothetical situation, perhaps with different beliefs or access to different information than we have. Nor is it to say that the standards in question must be standards that we ourselves endorse in full detail. We can make practical ought or must-judgments in relation to goals that we do not ourselves assign any practical authority, thinking what the movie villain ought to do to avoid the police, and we can reason theoretically from premises that we do not in fact accept. Still, the interpretation and use of 'ought' and 'must' seems to operate under the expectation that standards encoded by 'ought' and 'must' are possible standards for practical, evaluative or epistemic endorsement.

Here, then, is the basic idea of our proposal. First, both *ought(φ)* and *must(φ)* select some alternative for (practical, preferential, theoretical) endorsement at the exclusion of others. If we think of alternatives as propositions, the simplest case is one where the alternatives consist of a proposition, φ, and its negation, $\sim\varphi$. In this case, *must(φ)* and *ought(φ)* imply $\sim$*must($\sim\varphi$)* and $\sim$*ought($\sim\varphi$)*, respectively (ignoring dialetheism). Second, *ought(φ)* and *must(φ)* differ as to the grounds, or standard, of selection. 'Ought' semantically encodes a scalar standard—a 'scale'—which selects an alternative based on its position on that scale. 'Must' encodes a binary standard or a condition—a 'requirement'—which selects an alternative fulfilling that condition.[6] Call this the 'Structure of Standards', or 'SOS' account of the difference between 'ought' and 'must':

OUGHT(φ): Among the relevant alternatives, φ is selected by the relevant scale.
MUST(φ): Among the relevant alternatives, φ is selected by the relevant requirement.

We do not here endorse a specific way of understanding alternatives, but put in the most familiar terms of quantification over possible worlds, we can think of the relevant alternatives as sets of possible worlds, and the selection of an alternative φ as a

---

[6] Compare Sloman's early [18] suggestion that practical *ought(φ)* means that φ is, or is a necessary condition for, the best of the possibilities in some contextually determined class Z, whereas *must(φ)* means that φ is the only alternative.

The proposal in this paper also shares obvious similarities with Daniel Lassiter's recent highly interesting proposal that modals in general relate to scales, that *ought(φ)* means (roughly) that φ exceeds some contextually salient threshold on some contextually salient scale (e.g. of probability or expected value) to a significantly higher degree among salient alternatives, whereas *must(φ)* means (roughly) that it is the only relevant alternative passing a very high threshold ([11] Ch. 6). (We thank an anonymous reviewer for pointing us to Lassiter's dissertation.) Much of Lassiter's discussion strikes us as illuminating and plausible, but our ambition here is somewhat different than Lassiter's. Our primary goal is to say something general (and necessarily schematic) about the different contributions of 'ought' and 'must' that might explain differences in behavior between the two locutions across the various flavors, whereas Lassiter aims to provide detailed truth conditions for epistemic and deontic modals, respectively. Though we lack space to show this here, we think that given plausible assignments of scalar and binary standards, the general account outlined here can accommodate crucial aspects of Lassiter's explanations.

restriction of the union of these to those in which φ holds. Prior to the selection encoded by a particular *must(φ)* or *ought(φ)* judgment, relevant alternatives will typically have been restricted in various way. In the case of practical oughts, for example, alternatives will have been restricted to those that the relevant agent is capable of bringing about at some specific time; in the theoretical counterparts, global skeptical alternatives might have been ruled out. Some such 'preselections' would correspond roughly to Kratzerian modal bases. Others will be based on something akin to Kratzerian ordering sources: perhaps we have already restricted our attention to alternatives in which we achieve some goal, and select among those using, say, a requirement that a promise be held, or a scale of degrees of convenience. A crucial difference between this proposal and that of von Fintel and Iatridou is that what determines whether 'ought' or 'must' is the appropriate locution is the structure of the salient selection (or ordering) source—is it a requirement or a scale?—not the number of ordering sources at play.[7]

For illustration, consider again:

(1)   Employees must wash hands. Non-employees really ought to wash their hands, too.

When interpreting the 'must'-sentence in (1), one identifies as best one can the requirement or the rough kind of requirement made most salient by the occurrence of 'must' in the context: in this case that it is a practical requirement, and perhaps more precisely one backed up by company policy, or perhaps legislation. The sentence is then understood as expressing that the requirement in question selects the alternative that employees wash their hands over the alternative that they do not. When interpreting the 'ought'-sentence, one instead identifies the scale or kinds of scales that are made most salient by the occurrence of 'ought' in the context: in this case perhaps a scale of social or moral desirability, a scale on which hygiene might affect the ranking of alternatives. The sentence is then understood as expressing that considerations on this scale select the alternative in which non-employees wash their hands.

In the rest of this section, we explain how this proposal accounts for the data; in the section that follows we discuss some complications.

Different Flavors. On the SOS proposal, both oughts and musts come in different flavors because encoded standards of both the requiring and scalar kind come in different flavors: moral, prudential, bouletic, teleological, and epistemic. At least intuitively, most of us take morality to require us to act or not to act in certain ways, and to favor actions and states of affairs as morally better than others. Prudence might similarly require some actions and favor others as better than the alternatives. This makes for moral and prudential musts and oughts. Similarly, the achievement of certain ends or the satisfaction of certain desires might require certain actions or states of affairs,

---

[7]   SOS assumes no particular account of how, exactly, alternatives are selected or ranked given salient considerations of a certain type (moral, prudential, epistemic, etc.). We thus take it to be compatible with a variety of existing and possible suggestion (for relevant recent work the selection of alternatives, see e.g. [1], [6], [19]). More generally, we think that the proposal could be worked out in a variety of frameworks for modeling the content of modals, not only the broadly Kratzerian approach used here for illustration because of its familiarity.

and some actions might be more rational or desirable means to certain ends than others, making for teleological and bouletic musts and oughts. Finally, epistemic alternatives might be selected by the requirement that they be compatible with evidence, or by their degree of likelihood or believability thus making for the distinction between epistemic musts and oughts. SOS thus promises to account for the distinction between musts and oughts in the variety of areas where it is encountered, based on an intuitively available distinction between relevant kinds of grounds.

Intuitive Difference in Strength. SOS does not itself tell us that 'must' is stronger than 'ought', as 'must' and 'ought' encode differently structured standards. Still, when focusing on cases where $must(\varphi)$ and $ought(\varphi)$ have the same flavor, it gives us reasons to expect cases of $ought(\varphi)$ but not $must(\varphi)$ as well as reasons not to expect cases of $must(\varphi)$ but not $ought(\varphi)$, or cases of $must(\varphi)$ but $ought(\sim\varphi)$.

The key here is to understand what it is to take $must(\varphi)$ and $ought(\varphi)$ to have the same flavor. A natural first proposal is that it is to see the relevant requirement and the relevant scale as *simultaneously relevant to the selection from the same set of relevant alternatives*. They might be relevant to the same choice between propositions to believe on the basis of evidence (as in the epistemic case), or the same choice between propositions to wish true based on whether they satisfy relevant desires (as in the case of bouletic modals). Or they might be relevant to the same choice between actions on the basis of features relevant to morality, to prudence, or to the achievement of specified goals (as in the case of moral, prudential or teleological modalities).

To see the requirement and the scale as providing a coherent set of considerations is in effect to see them as grounding a *scale* in the sense we are operating with here: a set of possible considerations based on which we select alternatives that have sufficient support. But notice that if we see the requirement as a consideration that determines an alternative's position on a scale and continue to see it *as a requirement*—as a condition which selects some alternatives in favor of others—then we will see the scale as ranking the alternatives that satisfy the requirement higher than the alternatives that do not. From this it follows that if something uniquely satisfies the relevant requirement it must also be the highest-ranking alternative relative to this scale. Consequently, no other alternative *ought to be* relative to that scale, and insofar as we take it to be sufficiently supported by the considerations on the scale, it will also be seen as what ought to be, relative to that scale. But conversely, nothing prevents one among several alternatives that satisfy the requirement to be uniquely selected by further, non-requiring considerations relevant to the scale, giving us a case where something ought to be the case, though it doesn't have to be. Given SOS, this is the explanation of why $must(\varphi)$ seems logically stronger than $ought(\varphi)$: it is an effect of what it is for a condition to be a *requirement* and what it is for a scale and a requirement to be seen as *of the same flavor*.

Another way in which SOS predicts intuitive differences in strength of recommendations emerges in contexts where (e.g.) one recommends seeing the Rembrandt exhibition using 'must/have to' or 'should/ought to'. Using the former seems to express a stronger recommendation than the latter: the question is why. According to SOS, if one uses 'have to', one is treating the relevant considerations as grounding a requirement, i.e. as constituting a condition that *itself* rules out other alternatives. By contrast, if one uses 'ought to', one is treating the relevant considerations as ranking the alternative in question (seeing the exhibit) best, so that perhaps the winning alternative came out by a slim margin. For the same considerations to ground a 'have to' rather than an 'ought to' is to treat it them, in a straightforward way, as decisive.

Requirements vs. Recommendations. If we understand the speech act of requiring as involving a disposition to insist on compliance and impose sanctions for non-compliance, it is natural that it will be tied to requirements of the sort encoded by 'must' on the SOS account, i.e. to whether some binary condition is satisfied. Issues of vagueness to the side, insistence and sanctions are most naturally or even necessarily tied to binary conditions, considerations that are either violated or satisfied: without such a condition it is unclear what is insisted upon, or to what the sanctions are tied. To recommend something, by contrast, is to express that it is appropriate for some relevant purpose. In some cases, it might be that the recommended alternative is appropriate in virtue of being the only alternative satisfying some salient requirement: if our question concerns what to do when in town, we might think that missing out on the Rembrandt exhibition disqualifies any alternative, i.e. treat seeing the exhibition as a requirement, and so express our recommendation of this action using 'must'. In other cases, however, we recommend one alternative over others because it ranks higher on some relevant non-binary scale, and in these cases the recommendation will be expressed using 'ought'. So SOS correctly predicts that acts of requiring are tied to 'must' and 'have to' rather than to 'ought' and 'should', whereas recommendations can be expressed using either sort of expression, depending on the ground for the recommendation.

Probability, Conditionality, and Collective Commitments. We do not take the phenomena considered thus far to necessarily be beyond the ken of alternative accounts of the difference between 'ought' and 'must.' Contextualist accounts can make room for a variety of flavors, and accounts that take 'ought' to involve some element of probability, conditionality or lack of agreement about priorities might be able to handle differences in strength and relations to recommendations and requirements. However, we think that SOS is particularly well suited to account for phenomena motivating the latter sorts of accounts while leaving room for cases involving neither uncertainty nor hedging.

Given SOS, it is clear why 'ought' is preferred to 'must' when the modal judgment is grounded in considerations that might be outweighed or undermined by further considerations, including probabilistic considerations. The reason is exactly that in such cases the modal judgment is grounded not in some requirement, but in considerations that raise the score of the alternative in question on the relevant scale. In the case of epistemic modals, we will judge that something must be the case when its not being the case violates the requirement of compatibility with the evidence. But when the possibility in question is merely highly likely, other possibilities meet the requirement of compatibility and all we can say is that it ought to be the case. In the case of practical modals, we will judge that something must be done when it is the only alternative that satisfies the relevant requirements, but when one alternative is selected because it strikes a better balance of risks and opportunities, we will judge that it ought to be done. Similarly, in the case of bouletic modals, when we take something to be the only satisfactory alternative, we think that it must be the case, but when we just take something to be more satisfactory than the alternatives such that further considerations might change that balance, or because it strikes a better balance of risks and opportunities, we think that it ought to be the case.

While explaining why 'ought' is preferred to 'must' under circumstances of uncertainty, SOS allows that 'ought' might be preferable even in cases without uncertainty. Recall the case where we are considering what route to take, and that we agree, without any significant remaining uncertainty, that Route 17 is on the balance a little better than Highway 9. We might now naturally conclude that we *ought* to take Route 17 though it would be unnatural to conclude that we *must*. Taking Route 17 is selected by a salient scalar standard weighing various considerations, but there is no salient requirement that rules out taking Highway 9.

4   Non-requiring Thresholds and Scale-Based Requirements

It is not our business in this paper to propose a fully-fledged analysis of any one particular flavor of 'ought' or 'must'. But epistemic uses of 'ought' might raise a question about the SOS proposal. At first blush, the proposal applies nicely to epistemic 'ought' and 'must': intuitively, we think that something must (epistemically) be the case when we think that it is the only alternative satisfying the requirement of compatibility with the evidence, and we think that something ought (epistemically) to be the case if it is sufficiently well supported by the evidence, i.e. scores high enough on some scale of evidential support. The problem is that having a sufficiently high score on an evidential scale itself seems to be a requirement: a requirement for *rational believability*, say. If it is, SOS might seem to predict that 'must' would be felicitous whenever 'ought' is, collapsing the distinction.

Notice that it doesn't help here to say that it is a requirement that refers to a threshold *on a scale*, for many requirements that ground musts do too: guests must leave a bar after a certain time (time provides a scale), and drivers must keep a certain distance to other vehicles (distance is another scale). Nor do we think that it helps to say that the thresholds that ground ought-judgments as opposed to must-judgments are essentially comparative, selecting the alternative that scores *highest* on the relevant scale. It is of course true that many ought-judgments do seem to select the highest-scoring alternative: it is often the case that we ought to do something because it is the best alternative. Unfortunately, epistemic ought doesn't seem to be grounded in comparisons in the required way, instead relying on thresholds (perhaps of a vague and context dependent nature): in cases where alternative A is 45% likely and B 55% likely, we are generally not warranted in saying that B ought to be the case, though one alternative is clearly more likely.[8]

Even if epistemic ought could be understood as selecting the most likely alternative, another problem remains: comparisons on scales can ground requirements and

---

[8]   On Finlay's account, epistemic *ought(φ)* (and indeed all oughts) means (roughly) *φ is most likely*. Elsewhere we raise problems for this view and Finlay's attempts to explain away certain counterexamples. Since our concern here is to argue that SOS is tenable even if a highest likelihood account of epistemic ought is incorrect, we do not repeat the arguments here: should they be mistaken our view here has one less problem to deal with.

must-judgments. Suppose that we judge that some action is the best alternative available to us. Given SOS we will also naturally judge this as what we ought to do. But one might think that there is a rational requirement to do what is optimal. If one does, then it should make sense to say that we not only ought to do it, but that we have to, rationally speaking. This, we think, sounds just right: because it is best, we ought to do it, and if we are rationally required to do what is best, we have to do it, rationally speaking. But whether we think that there is a rational optimality requirement or not, the very selection of one option over others because it is optimal—the selection that we have said is operative in practical ought judgments—itself employs an optimality requirement: suboptimal alternatives are rejected. On the SOS proposal, one might think, this would mean that we should be willing to apply 'must' whenever we apply 'ought', which we clearly are not.

The solution to both these problems, about epistemic ought and about requirements of optimality, lies in the fact that not all requirements are the most salient requirements in a given context. The relevant distinction between requiring and scalar standards concerns the *salient* structure of the considerations grounding the selection of some relevant alternative. When we ask what requirement a given use of 'must' will convey, what matters is thus the relative salience of different requirements, which is affected by how easily we can think of the requirements and how informative or relevant the idea is that a certain alternative satisfies that requirement.

First apply this to the question of why the threshold that grounds epistemic oughts doesn't ground epistemic musts. To answer this question, we should ask what requirement is most salient in an epistemic reasoning. Here, compare the requirement that an alternative is compatible with whatever information is taken for granted (i.e. treated as evidential ground) with the requirement that it reaches above some threshold of evidential support required for believability. Both requirements are important, but the second is much less *clearly* binary in that it allows for more borderline cases, and thus less striking as a requirement. Because of this, when we ask in an epistemic setting whether something must be the case, the SOS proposal suggests that 'must' will pick out the former requirement rather than the latter.[9]

Next consider the question about why the optimality requirement apparently operative in practical ought-judgments does not ground practical must-judgments. Again, the question is what the most salient requirements are when we are making the judgment, now in contexts of practical deliberation. On the one hand we have requirements on action backed up by preferences, emotional reaction, moral conscience and law, along with a variety of formal and informal sanctions. On the other hand, we have a general requirement to pick the best alternative, a requirement that is implicitly operative whenever we make a practical ought-judgment. Here, we suggest, the former sorts of requirements should be much more salient. For example, when we deny that we *must* or *have to* take Route 17 though we think that we *ought to*, the

---

[9] The condition that alternatives be compatible with the evidential ground can be understood as requiring *logical* compatibility. However, it might more plausibly be understood as requirement that they not be rendered insignificantly likely by the evidential ground. If so, our proposal would have as a consequence something close to what Lassiter ([11], pp. 89–92) takes to be required to account for the connection between epistemic 'must' and claims about likelihood.

requirement to do what is best just does not spring to mind, and it is unlikely that it will except in philosophical contexts.[10] For these reasons, the existence of optimality requirements does not undermine the SOS proposal.

5   Conclusion

What we have offered here does not comprise full analyses of 'ought' and 'must'. We have not proposed a formal semantics for either locution, and have left open whether a full analysis should be purely truth-conditional or involve expressivist elements. Furthermore, we have only briefly discussed some of the pragmatics involved in the production and interpretation of the relevant modal claims, and have said nothing about how to understand disagreement about what ought to be done or ought to be the case among interlocutors who have different standards in mind or access to different evidence.

For these reasons, our proposal is best seen as a kind of analysis of the modal semantics and pragmatics of the two locutions and their relatives—one that we think best explains their different behaviors. Contrary to a common assumption, it is not 'ought' but rather 'must' and 'have to' that are typically used to talk about obligations. 'Ought' is used to express something weaker, such as recommendations or exhortations. This difference, we argued, is naturally and plausibly understood in terms of different kinds of standards: 'ought' and 'should' encode salient scalar standards for selecting alternatives, whereas 'must' and 'have to' encode binary standards.

The type of analysis we propose not only offers an explanation of this difference in strength across the various "flavors" of 'ought' and 'must.' It also sheds light on what relations ought-judgments and must-judgments bear and do not bear to uncertainty and acts of recommending and requiring, without imposing implausibly strong constraints on either the role of probability or on the illocutionary acts performed using these locutions.

Acknowledgments. We thank Barry Schein, Alex Silk, Justin Snedegar, Mark Schroeder, Steve Finlay, and participants at seminars at Lund University and University of Gothenburg for comments on early versions of this paper. Part of Björnsson's work on this paper was funded by the Swedish Research Council.

References

1. Cariani, F.: 'Ought' and Resolution Semantics. Noûs 47, 534–558 (2013)
2. Finlay, S.: Oughts and Ends. Philosophical Studies 143, 315–340 (2009)
3. Finlay, S.: What Ought Probably Means, and Why You Can't Detach It. Synthese 177, 67–89 (2010)
4. Finlay, S.: Confusion of Tongues: A Theory of Normative Language (forthcoming Oxford U.P, 2014)

---

[10] Compare: when thinking about why a house burnt down, we are unlikely to focus on the fact that the air contained oxygen, even though our thinking about the matter would change drastically if we no longer assumed that it did.

5. Hacquard, V.: Aspects of Modality. Ph.D. Thesis. MIT (2006)
6. Katz, G., Portner, P., Rubinstein, A.: Ordering Combination for Modal Comparison. In: Proceedings of SALT, vol. 22, pp. 488–507 (2012)
7. Kratzer, A.: What 'must' and 'can' must and can mean. Linguistics and Philosophy 1, 337–355 (1977)
8. Kratzer, A.: The notional category of modality. In: Eikmeyer, H.J., Rieser, H. (eds.) Words, worlds, and contexts: New approaches in word semantics, pp. 38–74. de Gruyter, Berlin (1981)
9. von Fintel, K., Iatridou, S.: How to say ought in foreign: The composition of weak necessity modals. In: Guéron, J., Lecarme, J. (eds.) Time and Modality, pp. 115–141. Springer (2008)
10. von Fintel, K., Iatridou, S.: What to Do If You Want to Go to Harlem: Anankastic Conditionals and Related Matters,
http://mit.edu/fintel/www/harlem-rutgers.pdf (retrieved)
11. Lassiter, D.: Measurement and Modality: The Scalar Basis of Modal Semantics. Ph.D. Thesis, NYU (2011)
12. Jones, A.J.I., Pörn, I.: 'Ought' and 'Must'. Synthese 66, 89–93 (1986)
13. Portner, P.: Imperatives and modals. Natural Language Semantics 15, 351–383 (2007)
14. Ridge, M.: Impassioned Belief. Oxford U.P., Oxford (2014)
15. Rubinstein, A.: Figuring out What We Ought to Do: The Challenge of Delineating Priorities. U. of Pennsylvania Working Papers in Linguistics 19, 169–178 (2013)
16. Rubinstein, A.: Roots of Modality. U. of Massachusetts Amherst dissertation (2012)
17. Silk, A.: Modality, weights, and inconsistent premise sets. SALT 22, 43–64 (2012)
18. Sloman, A.: 'Ought' and 'Better'. Mind 79, 385–394 (1970)
19. Snedegar, J.: Reason claims and constrastivism about reasons. Philosophical Studies 133, 231–242 (2013)
20. Wertheimer, R.: The Significance of Sense. Cornell University Press, London (1976)

A Preference-Based Semantics
for CTD Reasoning

Erica Calardo[1], Guido Governatori[2,3], and Antonino Rotolo[1]

[1] CIRSFID, University of Bologna, Italy
[2] NICTA, Software Systems Research Group, Australia
[3] Queensland University of Technology, Brisbane, Australia

Abstract. In [8] the authors developed a logical system based on the definition of a new non-classical connective $\otimes$ capturing the notion of reparative obligation. The system proved to be appropriate for handling well-known contrary-to-duty paradoxes but no model-theoretic semantics was presented. In this paper we fill the gap and define a suitable possible-world semantics for the system for which we can prove soundness and completeness. The semantics is a preference-based non-normal one extending and generalizing semantics for classical modal logics.

1 Introduction

One of the main research themes in deontic logic is about reasoning with contrary-to-duty (CTD) obligations [3]. In this perspective, it is widely acknowledged that the crisis of Standard Deontic Logic is historically and technically related to the formulation of some notorious paradoxes centering around the regulation of the violation of obligations.

The deontic logic literature on CTD reasoning is immense. However, two fundamental mainstreams have emerged as particularly interesting.

A first line of inquiry is mainly semantic-based. Moving from well-known studies on dyadic obligations, CTD reasoning is interpreted in settings with ideality or preference orderings on possible worlds or states [14]. The value of this approach is that the semantic structures involved are quite flexible: depending on the properties of the preference or ideality relation, different deontic logics can be obtained. This semantic approach has been fruitfully renewed in the '90 for example by [18,21], and most recently by works such as [13,20], which have confirmed the vitality of this line of inquiry.

The second mainstream is mostly proof-theoretic. Examples, among others, are various systems springing from Input/Output Logic [16,17] and the Gentzen system proposed in [8]. Both perspectives refer to the slogan "no logic of norms without attention to the normative systems in which they occur" [15], which draws inspiration from the pioneering works by [19] and [1]. This line of investigation is based on the intuition that any obligation can be explained in terms of a consequence relation of what is explicitly stated as obligatory in a normative system. While Input/Output approach mainly works by imposing some constraints

F. Cariani et al. (Eds.): DEON 2014, LNAI 8554, pp. 49–64, 2014.

on the manipulation of conditional norms, [8] is based on the introduction of the new non-classical binary operator $\otimes$: the reading of an expression like $a \otimes b$ is that a is primarily obligatory, but is this obligation is violated, the secondary obligation is b. The intuition behind this construction is that CTD obligations are a special kind of exception. For instance, the expression

$$Invoice \rightarrow PayBy7days \otimes Pay5\%Interest \otimes Pay10\%Interest$$

can be intuitively viewed as a compact representation of the following (where $\Rightarrow$ stands for any defeasible conditional)

$$Invoice \Rightarrow \mathsf{OBL}PayBy7days$$
$$\mathsf{OBL}PayBy7days, \neg PayBy7days \Rightarrow \mathsf{OBL}Pay5\%Interest$$
$$\mathsf{OBL}Pay5\%Interest, \neg Pay5\%Interest \Rightarrow \mathsf{OBL}Pay10\%Interest$$

The logic for $\otimes$ proved to be flexible for several applied domains, such as in business process modeling [12], normative multi-agent systems [5], temporal deontic reasoning [11], and reasoning about different types of defeasible permission [10].

Nevertheless, no semantic model-theoretic analysis of the operator $\otimes$ has been so far provided. In this paper we fill the gap and define a suitable possible-world semantics for this operator. Such semantics is a preference-based non-normal one extending and generalizing neighbourhood frames for classical modal logics. In this perspective, our contribution may also offer useful insights for establishing connections between the two mentioned mainstreams on CTD reasoning.

The layout the paper is as follows. Section 2 presents the basic logical system for $\otimes$ by recalling some intuitions from [8] as well as by integrating the original logic with some new schemata. Section 3 defines a multi-preference neighbourhood semantics suitable for the system. Section 4 illustrates logic and semantics with a real-life scenario. Sections 5 and 6 provide, respectively, some characterization and completeness results. Further developments for future work are outlined in Section 7. Some conclusions end the paper.

2   The Logic of $\otimes$

Let us briefly summarize, adjust, and extend in this section the logic for the CTD operator $\otimes$ presented in [8].

The language consists of a countable set of atomic formulas. Well-formed-formulas are then defined using the usual Boolean connectives and the binary connective $\otimes$, which is intended to formalize CTD statements. The language of [8] is integrated here by adding the deontic operator O and P denoting, respectively, standard unary obligation and permission.[1]

---

[1] The original Gentzen system presented in [8] was based on a binary consequence relation $\vdash_\mathsf{O}$: an expression $\Gamma \vdash_\mathsf{O} a$ meant that, whenever the set of well-formed formulas Γ occurs, a is obligatory. We will not assign in the remainder a direct deontic meaning to $\vdash$, thus explicitly introducing the operator O. The permission operator P was not considered in the logic of [8].

The intended interpretation of an expression like $a \otimes b$ is that b is a deontic reparation of a or, more explicitly, that a is obligatory, but if this obligation is violated, then b becomes obligatory. Hence the operator $\otimes$ captures the combination of primary and CTD obligations into unique provisions.

The language is formally defined as follows:

Definition 1 (Language). *Let Prop $= \{a, b, \dots\}$ be a countable set of atomic propositions. Let $X \in \{O, P\}$ be the set of unary deontic operators, such that $P =_{def} \neg O\neg$, and $\otimes$ be the CTD binary operator.*

- *All atomic propositions are well formed formulas (wffs);*
- *If a and b are wffs, all Boolean expressions made using a and b are wffs;*
- *If a is a wff and $\otimes$ does not occur in a, then Xa is a wff;*
- *If $a_1, \dots a_n$ are Boolean expressions and $\otimes$ does not occur in $a_1, \dots, a_n$, then $a_1 \otimes \cdots \otimes a_n$ is a wff;*
- *Nothing else is a wff.*

The basic logical system for $\otimes$ consists of the following axiom schemata and inference rules.

$$\bigotimes_{i=1}^{n} a_i \equiv \left(\bigotimes_{i=1}^{k-1} a_i \right) \otimes \left(\bigotimes_{i=k+1}^{n} a_i \right) \qquad \text{(where } a_j \equiv a_k,\ j < k) \quad (\otimes\text{-contraction})$$

$$a \otimes (b \otimes c) \equiv (a \otimes b) \otimes c \qquad\qquad (\otimes\text{-associativity})$$

$$(a \otimes \neg a) \equiv \top \qquad\qquad (\otimes\text{-}\top 1)$$
$$(a \otimes \top) \equiv \top \qquad\qquad (\otimes\text{-}\top 2)$$
$$(\top \otimes a) \equiv \top \qquad\qquad (\otimes\text{-}\top 3)$$

$$(a \otimes \bot) \equiv Oa \qquad\qquad (\otimes\text{-}\bot 1)$$
$$(\bot \otimes a) \equiv Oa \qquad\qquad (\otimes\text{-}\bot 2)$$
$$\bigotimes_{i=0}^{n} a_i \otimes \bigotimes_{j=0}^{m} b_j \equiv \left(\bigotimes_{i=0}^{n} a_i \right) \otimes \bot \otimes \left(\bigotimes_{j=0}^{m} b_j \right) \qquad (\otimes\text{-}\bot 3)$$

A few comments are in order.

The first equivalence ($\otimes$-contraction) corresponds to duplication and contraction: for example, $a \otimes b \otimes a$ is equivalent to $a \otimes b$. Intuitively, if I'm obliged not to cause any damage, but if I cause any, then I have the obligation to compensate, and, if don't compensate, then I'm obliged not to cause any damage; this just means that my primary obligation is not to cause any damage and my secondary obligation is to compensate.

The meaning of ($\otimes$-associativity) is self-evident. Let us only remark that this implies that nested $\otimes$-formulas are meaningless in our language. This is in fact reflected in our previous Definition 1 when we have formally excluded from the set of wffs $\otimes$-expressions such as $\otimes a \otimes \neg(b \otimes c)$ and only accepted expressions like $\neg(a \otimes b \otimes c)$ or $(a \otimes b) \wedge (c \otimes d)$.

Schema ($\otimes$-$\top$1) says that, if my primary obligation is a and my secondary one is $\neg a$, although there is an order of preference, whatever I'm doing will be deontically acceptable (either ideal or sub-ideal). Again, if I'm obliged not to cause any damage, but, if I don't do that I'm obliged to cause damages, then I have a trivial normative provision. Hence $a \otimes \neg a$ is equivalent to $\top$. Analogously, ($\otimes$-$\top$2) and ($\otimes$-$\top$3) hold, as one of the two obligation is always satisfied.

Schemata ($\otimes$-$\perp$1), ($\otimes$-$\perp$2), and ($\otimes$-$\perp$3) can be justified as follows. First of all, bear in mind that an expression like Oa can be intuitively viewed as an $\otimes$-formula of length 1.[2] Indeed, if my primary obligation is a and my secondary one is $\perp$, since the latter cannot be satisfied in any possible world, then the expression $a \otimes \perp$ is equivalent to having a simple obligation Oa. Similar considerations apply to ($\otimes$-$\perp$2) and ($\otimes$-$\perp$3).

Introduction and elimination rules for $\otimes$ are as follows:

$$\frac{a \otimes \left(\bigotimes_{i=1}^{n} b\right) \otimes c \qquad \neg b_1 \wedge \cdots \wedge \neg b_n \rightarrow \bigotimes_{i=1}^{m} d_i}{a \otimes \left(\bigotimes_{i=1}^{n} b\right) \otimes \left(\bigotimes_{i=1}^{m} d_i\right)} \qquad (\otimes\text{-I})$$

$$\frac{\bigotimes_{i=1}^{n} a_i \otimes b \otimes \bigotimes_{i=1}^{m} c_i \qquad \bigotimes_{i=1}^{n} a_i \otimes \neg b}{\bigotimes_{i=1}^{n} a_i \otimes \bigotimes_{i=1}^{m} c_i} \qquad (\otimes\text{-E})$$

where $\bigotimes_{i=1}^{n} a_i \otimes b \otimes \bigotimes_{i=1}^{m} c_i \not\equiv \top$ and $\bigotimes_{i=1}^{n} a_i \otimes \neg b \not\equiv \top$.

Let us illustrate the introduction rule ($\otimes$-I) by considering the well known 'dog, sign and fence' scenario [3]. The scenario contains the following four statements

1. There ought to be no dog;
2. If there is no dog, there ought to be no warning sign;
3. If there is a dog, there ought to be a warning sign;
4. If there is a dog and no warning sign, there ought to be a high fence.

The scenario is formalised as follows:

1. $O\neg dog$
2. $\neg dog \rightarrow O\neg sign$
3. $dog \rightarrow Osign$
4. $dog \wedge \neg sign \rightarrow Ofence$

Clearly $\otimes$-I is applicable for 1. and 3. from which we derive

$$\neg dog \otimes sign.$$

At this point we can use the newly derived formula and 4. as the premise of $\otimes$-I to conclude

$$\neg dog \otimes sign.$$

As we have just seen the inference rule ($\otimes$-I) generates chains of CTDs in order to deal iteratively with violations of compensatory obligations.

---

[2] We will see that it is not technically obvious how to capture this intuition.

The rule ($\otimes$-E) operates in the opposite direction by removing in $\otimes$-formulas those propositions that are negated in other true $\otimes$-formulas. Here is a concrete example:

$$\frac{Pay\_Taxes \otimes Pay\_Interest \otimes Foreclosure \qquad Pay\_Taxes \otimes \neg Pay\_Interest}{Pay\_Taxes \otimes Foreclosure} \tag{1}$$

As extensively explained in [8], this shows that we have to distinguish between genuine normative conflicts from apparent ones. By normative conflict we mean any situation ruled by opposite norms and which results in an impossible state of affairs; or, in other words, a situation in which the normative content of all relevant norms cannot be fulfilled, ending inevitably in a violation that cannot be compensated by any other CTD.

3   Multi-preference Semantics

Let us introduce the semantic structures that we use to interpret $\otimes$-formulas. In fact, they are just an extension of neighbourhood frames for classical modal logics.

Definition 2. *A* multi-preference frame *is a tuple* $\mathcal{F} = \langle W, \mathcal{C} \rangle$ *where:*

- *W is a countable non empty set of worlds;*
- *$\mathcal{C}$ is a neighbourhood function with the following signature*

$$\mathcal{C} \colon W \mapsto 2^{((2^W)^n)}$$

such that for each $w \in W$, for any ordered n-tuple $\langle X_1, \ldots, X_n \rangle$ of subsets of W in $\mathcal{C}_w$ the following holds:
- *if $i \neq j$, then $X_i \neq X_j$*
- *Or $\bigcup_{1 \leq i \leq n} X_i = W$.*

In general, a multi-preference frame is nothing but a structure where the standard neighbourhood function is replaced by a function that establishes an order between elements (i.e., sets of worlds) of each neighbourhood associated to every world. Figure 1 offers a pictorial representation of the intuition. The two conditions on sequences of sets of worlds are that there are no repetitions or that, if this is not the case, the union set of all sets in the sequence is W.

Given a formula $\bigotimes_{i=1}^{n} a_i$ we stipulate

$$\bigotimes_{i=1}^{n} a_i = \begin{cases} Oa_1 & n = 1 \\ a_1 \otimes \cdots \otimes a_n & n > 1 \end{cases}$$

The following definitions introduce the notion of redundancy and the operations of *zipping* and *s-zipping*, i.e., operations that, respectively, remove repetitions or redundancies occurring in $\otimes$-chains and in sequences of sets of worlds. Intuitively, these operations are necessary because, despite the fact the our language allows for building expressions like $a \otimes b \otimes a$ or $a \otimes \bot$, these last must be semantically evaluated using the sequences of sets of worlds $\langle \|a\|_V, \|b\|_V \rangle$ and $\langle \|a\|_V \rangle$ (see axiom schemata ($\otimes$-contraction) and ($\otimes$-$\bot$1)-($\otimes$-$\bot$3)).

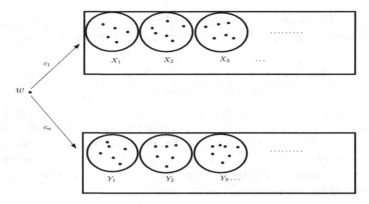

Fig. 1. Multi-preference basic structure: $X_1, X_2, X_3, \cdots \subseteq W$ and $Y_1, Y_2, Y_3, \cdots \subseteq W$

Definition 3. *A formula A is* redundant *iff $A = \bigotimes_{i=1}^{n} a_i$, $n > 1$ and*

- $\exists a_j, a_k,\ 1 \leq j, k \leq n$ *such that $a_j \equiv a_k$;*
- $\exists a_j,\ 1 \leq j \leq n$ *such that $a_j \equiv \bot$.*

Definition 4. *Let $A = \bigotimes_{i=1}^{n} a_i$ be any redundant formula. We say that the non-redundant B is* zipped *from A iff B is obtained from A by applying recursively the operations below:*

1. *If $n = 2$, i.e., $A = a_1 \otimes a_2$, and $a_1 \equiv a_2$, then B, the zipped from, is Oa_1;*
2. *Otherwise, if $n > 2$, then for $1 \leq k \leq n$, if (i) $a_j \equiv a_k$, for $j < k$, or (ii) $a_k \equiv \bot$, delete $\otimes a_k$ from the sequence.*

Let $X = \langle X_1, \ldots, X_n \rangle$ such that $X_i \in 2^W$ ($1 \leq i \leq n$). We analogously say that Y is s-zipped *from X iff Y is obtained from X by applying the operations below:*

1. *If $n = 2$ and $X_1 = X_2$, then its s-zipped from Y is $\langle X_1 \rangle$;*
2. *Otherwise, if $n > 2$, then for $1 \leq k \leq n$, if $X_j = X_k$, for $j < k$, or $X_k = \emptyset$, delete X_k from the sequence.*

Definition 5 (Models with sequences). *A model $\mathcal{M}$ is a couple $\langle \mathcal{F}, V \rangle$ where $\mathcal{F}$ is a frame and V is a valuation such that:*

- *for any non-redundant $\bigotimes_{i=1}^{n} a_i$, $\models_w^V \bigotimes_{i=1}^{n} a_i$ iff there is a $c_j \in \mathcal{C}_w$ such that $c_j = \langle \|a_1\|_V, \ldots, \|a_n\|_V \rangle$;*
- *for any redundant $\bigotimes_{i=1}^{n} a_i$, $\models_w^V \bigotimes_{i=1}^{n} a_i$ iff*
  - *$\bigotimes_{f=1}^{k} a_f$ is zipped from $\bigotimes_{i=1}^{n} a_i$, and*
  - *$\models_w^V \bigotimes_{f=1}^{k} a_f$.*
- *$\models_w^V Oa$ iff there $c_l \in \mathcal{C}_w$ such that:*
  - *$c_l = \langle \|a_1\|_V, \ldots, \|a_n\|_V \rangle$;*
  - *for some $k \leq n$, $X_k = \|a\|_V$;*
  - *for $1 \leq j < k$, $w \notin X_j$.*

Figure 2 pictorially illustrates the types of models used for evaluating $\otimes$-formulas. In fact, we use only finite sequences of worlds closed under s-zipping. A formula $\bigotimes_{i=1}^{n} a_i$ is true iff the corresponding appropriate finite sequence of sets of worlds (without redundancies) is in $\mathcal{C}_w$. Notice that the evaluation clause for Oa works using sequences of length 1 or with longer sequences whenever a is the k's element of the $\otimes$-chain and the previous a_j are such that $w \notin \|a_j\|_V$, i.e., the previous obligations have been violated in w.

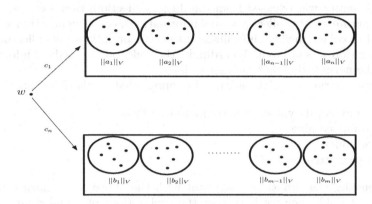

Fig. 2. Multi-preference models where finite sequences are used to evaluate the formulas $\bigotimes_{i=1}^{n} a_i, \ldots, \bigotimes_{i=1}^{m} b_i$

4   An Example

In [8] we showed that the $\otimes$-formalism is able to avoid most well-known CTD puzzles, such as Chisholm's and Forrester's paradoxes, Belzer's Reykjavik scenario and Makinson's Möbius strip example. These results can be trivially extended to the new variant of the logic presented here. Let us thus illustrate the the logic its semantics by considering a fresh deontic scenario, which seems to be problematic in most, if not all established formalisms dealing with CTD reasoning [9].

Suppose that a Privacy Act contains the following norms:

1. The collection of personal information is forbidden, unless acting on a court order authorising it.
2. The destruction of illegally collected personal information before accessing it is a defence against the illegal collection of the personal information.
3. The collection of medical information is forbidden, unless the entity collecting the medical information is permitted to collect personal information.

In addition, the Act specifies what personal information and medical information are, and they turn out to be disjoint.

Suppose an entity, subject to the Act, collects some personal information without being permitted to do so; at the same time they collect medical information. The entity recognises that they illegally collected personal information (i.e., they collected the information without being authorised to do so by a Court Order) and decides to remediate the illegal collection by destroying the information before being accessing it. Are they compliant with the above privacy act? Given that the personal information was destroyed the entity was excused from the violation of the first section (illegal collection of personal information). However, even if the entity was excused from the illegal collection, they were never entitled (i.e., permitted) to collect personal information[3], consequently they were not permitted to collect medical information; thus the prohibition of collecting medical information was in force. Accordingly, the collection of medical information violates the norm forbidding such an activity.

The logical structure of the act can be represented by the following norms:

1. a is forbidden, its violation is compensated by b.
2. a is permitted given c.
3. d is forbidden.
4. If a is permitted, so is d.

Let us consider the situations compliant with the above set of norms. Clearly, if c does not hold, then we have that the prohibitions of a and d are in force. Therefore, a situation where $\neg a$, $\neg c$, and d hold is fully compliant (irrespective whether b holds or not). If c holds, then the permission of a derogates the prohibition of a (situations with either a or $\neg a$ are compliant with he first two norms); in addition, the permission of a allows us to derogate the prohibition of d. Accordingly, situations with either d or $\neg d$ comply with the third norm. Let us go back to scenarios where c does not hold, and let us suppose that we have a. This means that the prohibition of a has been violated; nevertheless the set of norms allows us to recover from such violation by b. However, to have a violation we have to have either an obligation or a prohibition that has been violated: in this case the prohibition of a. Given that prohibition of a and permission of a are mutually incompatible, we must have, to maintain a consistent situation, that a is not permitted. But if a was not permitted, d is not permitted either; actually, according to the third norm, d is forbidden.

To sum up, a scenario where $\neg c$, a, b and $\neg d$ hold is still compliant (even if to a lesser degree given the compensated violation of the prohibition of a). In any case, no situation where both $\neg c$ and d hold is compliant.

The scenario can be reconstructed in our logic by meeting the desiderata[4]:

---

[3] If they were permitted to collect personal information, then the collection would have not been illegal, and they did not have to destroy it.

[4] The scenario uses strong permissions, i.e., the permissions derogating the obligations to the contrary. To accomplish this we have to specify that 2 overrides 1, and 4 overrides 3. This explains the Boolean structure of 1 and 3. The focus of this paper is not how to implement defeasibility, thus we just adopt the simplest procedure to handle this aspect.

1. $\neg c \rightarrow (\neg a \otimes b)$;
2. $c \rightarrow \mathsf{P}a$;
3. $\mathsf{O}\neg a \rightarrow \mathsf{O}\neg d$;
4. $\mathsf{P}a \rightarrow \mathsf{P}d$.

If $\neg c$, a, b and $\neg d$, then the deontic provision from 1 is sub-ideally satisfied by b; $\mathsf{P}a$ and $\mathsf{P}d$ cannot be obtained; $\mathsf{O}\neg d$ is obtained in 3 and satisfied. If $\neg c$ and d, then $\mathsf{O}\neg d$ is neither satisfied nor compensated. Let us also analyze the scenario and these cases semantically:

Let us consider a model $\mathcal{M}$ with a world w such that

- $w \notin \|c\|_V$, $w \notin \|d\|_V$, $w \in \|a\|_V$ and $w \in \|b\|_V$; and
- $\mathcal{C}_w = \{\langle \|\neg a\|_V, \|b\|_V \rangle, \langle \|\neg d\|_V \rangle\}$.

It is easy to verify that 1.–4. above are all true at w. Let us see the reasons why they are true. 1.) Since the sequence $\langle \|\neg a\|_V, \|b\|_V \rangle \in \mathcal{C}_w$, from which $\models^V_w \neg a \otimes b$ follows. 2.) trivially, since $w \notin \|c\|_V$. 3.) Similarly to 1. the sequence $\langle \|\neg d\|_V \rangle \in \mathcal{C}_w$; notice that we have $\models^V_w \mathsf{O}\neg s$ given that $\langle \|\neg a\|_V, \|b\|_V \rangle \in \mathcal{C}_w$, in addition $w \in \|a\|_V$, from which it follows $\models^V_w \mathsf{O}\neg d$. 4.) As we have already seen, $\models^V_w \mathsf{O}\neg a$, thus from the definition of P we get $\not\models^V_w \mathsf{P}a$; similarly for $\not\models^V_w \mathsf{P}d$.

In the scenario corresponding to w we are (weakly) compliant because $\mathsf{O}\neg a$ is violated (i.e., $w \in \|a\|_V$), but $\mathsf{O}b$ (which compensates $\mathsf{O}\neg a$) is complied with. Also $\mathsf{O}\neg d$ is complied with. Consider now a possible world y which is like w but $y \in \|d\|_V$. In y we are not compliant: we have a violation of $\mathsf{O}\neg d$ (which is not compensable).

5 Characterization Results

Let us consider the following inference rule:

$$\frac{\vdash \bigwedge_{i=1}^{n}(a_i \equiv b_i)}{\vdash (\bigotimes_{i=1}^{n} a_i) \equiv (\bigotimes_{i=1}^{n} b_i)} \qquad (\otimes\text{-RE})$$

It should be intuitively clear that $(\otimes\text{-RE})$ generalizes for $\otimes$-formulas the weakest inference rule for modal logics, i.e., the closure of $\Box$ (here O) under logical equivalence [4]:

$$\frac{\vdash a \equiv b}{\vdash \mathsf{O}a \equiv \mathsf{O}b} \qquad (\text{RE})$$

Lemma 1. *($\otimes$-RE) and (RE) hold in the class of all multi-preference frames, i.e., on the class of all multi-preference frames,*

- *if $\bigwedge_{i=1}^{n}(a_i \equiv b_i)$ is valid, then $(\bigotimes_{i=1}^{n} a_i) \equiv (\bigotimes_{i=1}^{n} b_i)$ is valid ($\otimes$-RE);*
- *if $a \equiv b$ is valid, then $\mathsf{O}a \equiv \mathsf{O}b$ is valid (RE).*

Proof (Sketch). The result for $(\otimes\text{-RE})$ trivially follows from the fact the valuation clause for any $\otimes$-formula $\bigotimes_{i=1}^{n} a_i$, at any world w and with any valuation

V, requires the existence of a sequence $c \in \mathcal{C}_w$ of truth sets $\langle \|a_1\|_V, \ldots, \|a_n\|_V \rangle$. (RE), too, trivially follows as a special case of ($\otimes$-RE): indeed, by stipulation $\bigotimes_{i=1}^n a_i = \mathsf{O}a_i$ when $n = 1$ and, semantically, $\mathsf{O}a$ is true in w iff there is a finite sequence $\langle X_1, \ldots, X_n \rangle$, where $n \geq 1$, such that $\|a\|_V = X_k$ and $w \notin X_1, \ldots, X_{k-1}$.

Also ($\otimes$-contraction), ($\otimes$-associativity), and ($\otimes$-$\perp$1)-($\otimes$-$\perp$3) hold in general:

Lemma 2. *($\otimes$-contraction), ($\otimes$-associativity), and ($\otimes$-$\perp$1)-($\otimes$-$\perp$3) are valid in the class of all multi-preference frames.*

Proof (Sketch). Consider ($\otimes$-contraction). First of all, remember that, by construction, all sequences $\langle X_1, \ldots, X_n \rangle$ for which we do not have that $\bigcup_{i=1}^n X_i \neq W$ are such that $X_k \neq X_j, \forall k, j \in \{1, \ldots, n\}$. Since sequences in frames are closed under the operation of s-zipping (see Definitions 4 and 5, i.e., the sequences used to evaluate $\otimes$-formulas do not contain repetitions), thus every redundant $\otimes$-formula is uniquely evaluated by one sequence without repetitions.

($\otimes$-associativity) holds for similar reasons: just consider how $\otimes$-formulas are recursively evaluated.

Finally, consider ($\otimes$-$\perp$1)-($\otimes$-$\perp$3). Indeed, all $\otimes$-formulas $\bigotimes_{i=1}^n a_i$ where at least one $a_k \equiv \perp$, $1 \leq k \leq n$, is redundant and so is evaluated by considering a sequence s-zipped from a sequence $\langle \|a_1\|_V, \ldots, \|a_k\|_V = \emptyset, \ldots, \|a_n\|_V \rangle$.

Finally, schemata ($\otimes$-$\top$1)-($\otimes$-$\top$3) hold, too:

Lemma 3. *($\otimes$-$\top$1)-($\otimes$-$\top$3) are valid in the class of all multi-preference frames.*

Proof (Sketch). The proof follows from considering sequences $\langle X_1, \ldots, X_n \rangle$ where $\bigcup_{i=1}^n X_i = W$. This makes trivially valid ($\otimes$-$\top$1); ($\otimes$-$\top$2) and ($\otimes$-$\top$3) are also valid because $\|a\|_V \cup W = W$ for any formula a.

Notice that the following axiom schema holds, too:

$$((\bigotimes_{i=1}^n b_i) \otimes c \otimes \bigotimes_{j=1}^n d_j)) \wedge (\bigwedge_{i=1}^n \neg b_i)) \to \mathsf{O}c \qquad \text{(O-Detachment)}$$

Lemma 4. *(O-Detachment) is valid in the class of all multi-preference frames.*

Proof (Sketch). The proof trivially follows from the definition of the operator O and the valuation clause for it.

The class of all multi-preference frames cannot validate introduction and elimination rules for $\otimes$, which require extra semantic conditions.

Conditions for characterizing ($\otimes$-I) are as follows:

Definition 6. *Let $\mathcal{F} = \langle W, \mathcal{C} \rangle$ be a frame. We say that $\mathcal{F}$ is $\otimes$-expanded iff for any $w \in W$ and $c_i, c_j \in \mathcal{C}_w$ such that $c_i = \langle X_1, \ldots, X_n \rangle$, $c_j = \langle Y_1, \ldots Y_m \rangle$, $\forall l : 1 \leq k < l < f \leq n$, $w \in W - X_l$, then there exists $c' \in \mathcal{C}_w$ such that c' is s-zipped from $\langle X_1, \ldots, X_k, Y_1, \ldots, Y_m \rangle$.*

Lemma 5. *($\otimes$-I) holds in the class of expanded multi-preference frames, i.e., on the class of expanded multi-preference frames, if*

$$a \otimes (\bigotimes_{i=1}^{n} b) \otimes c \tag{2}$$

$$\neg b_1, \ldots, \neg b_n \to \bigotimes_{i=1}^{m} d_i \tag{3}$$

are valid, then

$$a \otimes (\bigotimes_{i=1}^{n} b) \otimes (\bigotimes_{i=1}^{m} d_i) \tag{4}$$

is valid.

Proof (Sketch). By reductio, suppose that (2) and (3) are valid and that (4) is not. This means that there is a world w such that

1. $w \in \|\neg b_1\|_V \cap \ldots \cap \|\neg b_n\|_V$,
2. $\exists c_i \in \mathcal{C}_w$ such that c_i is s-zipped from $\langle \|a\|_V, \|b_1\|_V, \ldots, \|b_n\|_V, \|c\|_V \rangle$, and
3. $\exists c_j \in \mathcal{C}_w$ such that c_j is s-zipped from $\langle \|d_1\|_V, \ldots, \|d_m\|_V \rangle$, but
4. there is no $c_k \in \mathcal{C}_w$ such that c_k is s-zipped from $\langle \|a\|_V, \|b_1\|_V, \ldots, \|b_n\|_V, \|d_1\|_V, \ldots, \|d_m\|_V \rangle$.

From 2, it is clear that there is a subsequence of c_i' which is s-zipped from $\langle \|a\|_V, \|b_1\|_V, \ldots, \|b_n\|_V \rangle$. We concatenate c_i' and c_j creating the sequence $c_i'c_j$. If $c_i'c_j$ is s-zipped from itself we are done, since, in conjunction with 1, 2, 3 and the fact that the from is $\otimes$-expanded give us that $c_i'cj \in \mathcal{C}_w$ and $c_i'cj$ is s-zipped from $\langle \|a\|_V, \|b_1\|_V, \ldots, \|b_n\|_V, \|d_1\|_V, \ldots, \|d_m\|_V \rangle$. Contradiction. If $c_i'c_j$ is not s-zipped form itself, this means that there some elements of c_j which appear in c_i'. We create c_j' by remove such elements from c_j and we concatenate c_i' and c_j', obtaining the sequence $c_i'c_j'$. Again, by 1, 2, 3 and the fact that the frame is $\otimes$-expanded we have that $c_i'c_j' \in \mathcal{C}_w$ and it is mundane to verify that $c_i'c_j'$ is s-zipped from $\langle \|a\|_V, \|b_1\|_V, \ldots, \|b_n\|_V, \|d_1\|_V, \ldots, \|d_m\|_V \rangle$. Contradiction again.

Before giving the semantic conditions for ($\otimes$-E) we introduce some auxiliary concepts. Trivially any sequence induces a relation. Accordingly, for every world $w \in W$ and for every $c_i \in \mathcal{C}_w$ where $c_i = \langle X_1, \ldots, X_n \rangle$ we have a relation $\#_i$ such that $(X_j, X_{j+1}) \in \#_i$ for $1 \le j \le n$.

Definition 7. *Let $\mathcal{F} = \langle W, \mathcal{C} \rangle$ be a frame. We say that $\mathcal{F}$ is $\otimes$-contracted iff for any $w \in W$ and $c_i, c_j \in \mathcal{C}_w$ such that $c_i = \langle X_1, \ldots, X_n \rangle$, $\bigcup_{1 \le i \le n} X_i \ne W$, $c_j = \langle X_1, \ldots, X_{k-1}, W - X_k \rangle$, $(W - X_k) \cup \bigcup_{1 \le i \le k-1} X_i \ne W$ then*

$$\{c' \colon \#' = (c_i - \{(X_{k-1}, X_k), (X_k, X_{k+1})\}) \cup$$
$$\cup \{(X_{k+s}, X_k), (X_k, X_{k+s+1})\}_{1 \le s < n-k}\} \subset \mathcal{C}_w$$

Lemma 6. *($\otimes$-E) is characterized by the class of $\otimes$-contracted multi-preference frames.*

Proof (Sketch). Suppose by reduction that it does not hold. Thus there is a valuation V and a world w such that

(a) $\models_w^V \bigotimes_{i=1}^n a_i \otimes b \otimes \bigotimes_{i=1}^m c_i$;
(b) $\models_w^V \bigotimes_{i=1}^n a_i \otimes \neg b \otimes \bigotimes_{i=1}^f d_i$; but
(c) $\not\models_w^V \bigotimes_{i=1}^n a_i \otimes \bigotimes_{i=1}^m c_i$

(c) implies that there is no $c' \in \mathcal{C}_w$ which is s-zipped from $\langle \|a_1\|_V, \ldots, \|a_n\|_V,$ $\|c_1\|_V, \ldots, \|c_m\|_V \rangle$. From (a) we know that there is a $c_i \in \mathcal{C}_w$, such that $c_i = \langle X_1, \ldots, X_k, X, Z_1, \ldots, Z_s \rangle$ which is s-zipped from $\langle \|a_1\|_V, \ldots, \|a_n\|_V, \|b\|_V,$ $\|c_1\|_V, \ldots, \|c_m\|_V \rangle$; from (b) we know that there is a $c_j \in \mathcal{C}_w$ such that $c_j = \langle \|X_1\|_V, \ldots, \|X_k\|_V, W - \|X\|_V \rangle$ such that c_j is s-zipped from $\langle \|a_1\|_V, \ldots, \|a_n\|_V, \|\neg b\|\rangle$; since the $\bigotimes_{i=1}^n a_i \otimes \neg b \not\equiv \top$ we infer that b does not appear in $\bigotimes_{i=1}^n a_i$ and then $X = \|b\|_V$.

We consider two cases (i) b does not appear in $\bigotimes_{i=1}^m c_i$, and (ii) b appears in $\bigotimes_{i=1}^m c_i$. (i) is trivial since given that b does not appear, $X \neq Z_u$ ($1 \leq u \leq s$), and thus $c_i' = \langle X_1, \ldots, X_k, Z_1, \ldots, Z_s \rangle$ is in $\mathcal{C}_w$ given that the frame is $\otimes$-contracted and c_i' is s-zipped from $\langle \|a_1\|_V, \ldots, \|a_n\|_V, \|c_1\|_V, \ldots, \|c_m\|_V \rangle$. Contradiction.

6   The Minimal System $\mathsf{E}^\otimes$

Let us consider the minimal system $\mathsf{E}^\otimes$, which is given by adding to the classical propositional calculus the modal schemata $\otimes$-contraction, $\otimes$-associativity, $\otimes$-$\top$1, $\otimes$-$\top$2, $\otimes$-$\top$3, $\otimes$-$\bot$1, $\otimes$-$\bot$2, $\otimes$-$\bot$3, O-Detachment, and it is closed under the rules $\otimes$-**RE**, **RE**, and **MP**.

Definition 8 ($\mathsf{E}^\otimes$-Canonical Models). *A multi-preference model with sequences $\mathcal{M} := \langle W, \mathcal{C}, V \rangle$ is a canonical model for $\mathsf{E}^\otimes$ if and only if:*

1. $W := \{w \mid w \text{ is } \mathsf{E}^\otimes\text{-maximal}\}$
2. *for any propositional letter $p \in Prop$, $\|p\|_V := |p|_{\mathsf{E}^\otimes}$, where $|p|_{\mathsf{E}^\otimes} := \{w \in W \mid p \in w\}$*
3. *Let $\mathcal{C} := \bigcup_{w \in W} \mathcal{C}_w$, where for each $w \in W$, $\mathcal{C}_w := \{\langle \|A_1\|_V, \ldots, \|A_n\|_V \rangle \mid A_1 \otimes \cdots \otimes A_n \in w\} \cup \{\langle A \rangle_V \mid \mathrm{O}A \in w\}$, where each A_i is a meta-variable for a Boolean formula and $A_1 \otimes \cdots \otimes A_n$ is zipped.*

Notice that since each $\mathcal{C}_w$ contains only ordered sequences of truth sets obtained by *zipped* formulas, the following condition holds true: for any ordered n-tuple $\langle X_1, \ldots, X_n \rangle$ of subsets of W in $\mathcal{C}_w$ m if $i \neq j$, then $X_i \neq X_j$. Moreover, $\mathcal{C}$ contains only s-zipped sequences.

Lemma 7 (Truth Lemma). *$A \in w$ if and only if $\models_w A$.*

Proof (Sketch). Given the construction of the canonical model, this proof is quite straightforward and it can be given by induction on the length of a formula A. We consider only the modal cases.

Assume A has the form $a_1 \otimes \cdots \otimes a_n$ and that is redundant (clearly the case for non redundant formulas is easier and does not need to be considered here). Suppose $a_i \otimes \cdots \otimes a_n \in w$. Then, by Axiom $\otimes$-contraction and $\otimes$-$\perp 3$, we have that the formula $b_1 \otimes \cdots \otimes b_j$, the *zipped* form of A, is also in w. By definition of canonical model we have that there is a sequence $\langle \|b_1\|_V, \ldots, \|b_j\|_V \rangle \in \mathcal{C}_w$. Following from the semantic clauses given to evaluate $\otimes$-formulas, it holds that $\models_V^w a_1 \otimes \ldots \otimes a_n$.

Now suppose that $\models_V^w a_1 \otimes \cdots \otimes a_n$. By definition, there is a zipped formula $b_1 \otimes \cdots \otimes b_j$ such that $\models_V^w b_1 \otimes \cdots \otimes b_j$. Thus, $\mathcal{C}_w$ contains an ordered j-tuple $\langle \|b_1\|_V, \ldots, \|b_j\|_V \rangle$. By definition of $\mathcal{C}_w$ it follows that $b_1 \otimes \cdots \otimes b_j \in w$ and by the axioms $\otimes$-contraction and $\otimes$-$\perp 3$, all the *unzipped* forms of $b_1 \otimes \cdots \otimes b_j$ are also in w, including $a_1 \otimes \cdots \otimes a_n$.

If, on the other hand, A has the form Ob and $Ob \in w$, then $\|b\|_V \in \mathcal{C}_w$ and, by definition $\models_V^w Ob$. Conversely, if $\models_V^w Ob$, then there is an s-zipped sequence $\langle \|c_1\|_V, \ldots, \|c_n\|_V, \|b\|_V, \|d_1\|_V, \ldots, \|d_m\|_V \rangle \in \mathcal{C}_w$ and for $1 \leq i \leq n$, $w \notin \|c\|_i$. Thus, since any c_i is Boolean and w is maximal, $\neg c_1, \ldots, \neg c_n \in w$. Moreover $\bigotimes_{i=1}^n c_i \otimes b \otimes \bigotimes_{j=1}^m d_j \in w$. Hence by the O-Detachment axiom and **MP**, $Ob \in w$.

Corollary 1. *The logic* $\mathsf{E}^\otimes$ *is sound and complete with respect to the class of multi-preference frames with zipped sequences.*

7 Semantics with Multi-relational Frames

In this section we briefly outline a possible extension of the framework presented above. It consists in working with structures validating stronger modal inference rules and schemata, in particular

$$O\top \tag{N}$$

$$\frac{\vdash \bigwedge_{i=1}^n (a_i \to b_i)}{\vdash (\bigotimes_{i=1}^n a_i) \to (\bigotimes_{i=1}^n b_i)} \tag{$\otimes$-RM}$$

and

$$\frac{a \to b}{Oa \to Ob} \tag{O-RM}$$

This means moving to a class of logic called N-monotonic [2]. Hence, we can employ a simple generalization of Kripke semantics with a countable set of accessibility relations [7]. This basic semantic setting is here enhanced by adding preferences over sets of worlds.[5]

---

[5] See [6] for a similar construction, which is however used for different purposes.

7.1   Multi Relational Frames with Sequences

Let $\mathsf{NM}^\otimes$ be the system obtained by adding $\mathbf{N}$, ($\otimes$-RM), and (O-RM) to $\mathsf{E}^\otimes$.

Definition 9. *A frame is a tuple $\mathcal{F} = \langle W, \mathcal{R}, f, \mathcal{C} \rangle$ where:*

- *W is a possibly infinite set of worlds;*
- *$\mathcal{R}$ is a countable set of binary relations over W;*
- *Let f be a function $f : W \longrightarrow \mathcal{P}(\mathcal{P}(W))$ be a function assigning to each world $w \in W$ a set of finite ordered n-tuples of subsets of W, i.e., for each $w \in W$, $\mathcal{C}_w := \{\langle X_1, \ldots, X_i \rangle, \langle Y_1, \ldots Y_j \rangle, \ldots\}$.*
- *$\mathcal{C} := \bigcup_{w \in W} \mathcal{C}_w$.*

Notation and abbreviations. Let $R_i(w) := \{v \in W \mid wR_iv\}$, $\|A\|_V := \{w \in W \mid \models^V_w A\}$.

Definition 10 (Multi relational models with sequences)
A model $\mathcal{M}$ is a couple $\langle \mathcal{F}, V \rangle$ where V is a valuation such that:

- *$\models^V_w a_1 \otimes \cdots \otimes a_n$ where $a_1 \otimes \cdots \otimes a_n$ is zipped, iff there is a finite sequence $\langle X_1, \ldots, X_n \rangle \in \mathcal{C}_w$ such that for $1 \leq j \leq n$:*
  - *$X_j = \|a_j\|_V$*
  - *For each j, $1 \leq j \leq n$, there is a relation R_i such that $R_i(w) \subseteq X_j$*
- *$\models^V_w a_1 \otimes \cdots \otimes a_n$ where $a_1 \otimes \cdots \otimes a_n$ is redundant, iff $\models^V_w b_1 \otimes \cdots \otimes b_j$, where $b_1 \otimes \cdots \otimes b_j$ is its zipped form.*
- *$\models^V_w OA$ iff there is a finite sequence $\langle X_1 \ldots X_n \rangle \in \mathcal{C}_w$ such that for $1 \leq j \leq n$:*
  - *$X_j = \|a_j\|_V$ for some proposition a_j;*
  - *$R_m(w) \subseteq X_j$ for some $R_m \in \mathcal{R}$*
  - *for some $k \leq n$, $X_k = \|A\|_V$*
  - *for $1 \leq j < k$, $w \notin X_j$.*

7.2   Completeness Sketch

Definition 11 (NM$^\otimes$-Canonical Models). *Let $\mathcal{M} := \langle W, \mathcal{R}, \mathcal{C}, V \rangle$ be a multi-relational model with sequences. $\mathcal{M}$ is a canonical model for $\mathsf{NM}^\otimes$ if and only if:*

1. *$W := \{w \mid w$ is $\mathsf{NM}^\otimes$-maximal$\}$*
2. *for any propositional letter $p \in Prop$, $\|p\|_V := |p|_{\mathsf{NM}^\otimes}$, where $|p|_{\mathsf{NM}^\otimes} := \{w \in W \mid p \in w\}$*
3. *For each $w \in W$ and each natural number n, let $\mathcal{C}_w := \{\langle \|a_1\|_V, \ldots, \|a_n\|_V \rangle \mid a_1 \otimes \cdots \otimes a_n \in w$ and it is zipped $\} \cup \{\langle A \rangle_V \mid OA \in w\}$, where a_i is Boolean. Let $\mathcal{C} := \bigcup_{w \in W} \mathcal{C}_w$.*
4. *$\mathcal{R} := \{R^\star_{A_j} \mid A_j \in Fma(\mathcal{L})\} \cup \{R^i_{A_j} \mid A_j \in Fma(\mathcal{L})\}_{i \in \mathbb{N}}$*

   (a) For each $w \in W$, $R^\star_{A_j}(w) := \|A_j\|_V$ iff there is some $B_1 \otimes \cdots \otimes B_n \in w$ such that $B_i = A_j$;

(b) *For each $w \in W$, consider any subsequence $\langle \|A_1\|_V, \ldots, \|A_k\|_V \rangle$ such that $\langle \|A_1\|_V, \ldots, \|A_n\|_V \rangle \in \mathcal{C}_w$ and $k \leq n$ and $\mathsf{O}A_k \in w$, and $\neg A_i \in w$ for $1 \leq i \leq k-1$. Let $\mathcal{C}_w^1, \mathcal{C}_w^2, \ldots$ be an enumeration of such subsequences. Then for each $\mathcal{C}_w^i$ set the following: $R_{A_k}^i(w) := \|A_k\|$.*

Lemma 8 (Truth Lemma). *$A \in w$ if and only if $\models_w A$.*

Proof. Again, let us consider only the modal cases. Assume a redundant formula $a_1 \otimes \cdots \otimes a_n \in w$. Then its zipped form $b_1 \otimes \cdots \otimes b_j \in w$ by the axioms $\otimes$-contraction and $\otimes$-$\perp 3$ and modus ponens; also, by construction, the n-tuple $\langle \|b_1\|, \ldots, \|b_j\| \rangle \in \mathcal{C}_w$ and there are relations $R_{b_1}^\star, \ldots, R_{b_j}^\star$ such that for each i, $R_{b_i}^\star(w) = \|b_i\|$. Thus, $\models_V^w a_1 \otimes \cdots \otimes a_n$, its expanded form.

Analogously, suppose $\models_V^w a_1 \otimes \cdots \otimes a_n$. Then $\models_V^w b_1 \otimes \cdots \otimes b_j$, i.e., its zipped form. This means that there are relations $R_i, \ldots, R_j$ such that $R_i(w) \subseteq \|b_i\|$ for each i and $\langle \|b_1\|, \ldots, \|b_j\| \rangle \in \mathcal{C}_w$. By definition of $\mathcal{C}_w$, it holds that $b_1 \otimes \cdots \otimes b_j \in w$ and by the axioms $\otimes$-contraction and $\otimes$-$\perp 3$ and modus ponens there are also all its expanded forms, including $a_1 \otimes \cdots \otimes a_n$.

Suppose $\mathsf{O}A$ is in w. By definition, $\mathcal{C}_w$ contains the 1-tuple $\langle \|A\| \rangle$ and for some i, $R_A^i(w) = \|A\|$ and hence $\models_V^w \mathsf{O}A$.

Conversely, if $\models_V^w \mathsf{O}A$, then by definition there is some finite n-tuple $\langle \|b_1\| \ldots \|b_n\| \rangle \in \mathcal{C}_w$ such that for $1 \leq j \leq n$:

- $R_m(w) \subseteq \|b_j\|$ for some $R_m \in \mathcal{R}$
- for some $k \leq n$, $\|b_k\| = \|A\|_V$
- for $1 \leq j < k$, $w \notin \|b_j\|$.

Thus, the formula $b_1 \otimes \cdots \otimes b_k \otimes \cdots \otimes b_n \in w$ by definition of $\mathcal{C}_w$, $\neg b_1, \ldots, \neg b_i \in w$ and by the O-Detachment axiom $\mathsf{O}b_i \in w$. But $b_i \equiv A$, so by the **RM** rule and modus ponens, $\mathsf{O}A \in w$ too.

8 Conclusion

This paper offered a semantic study of the $\otimes$ operator originally introduced in [8]. We showed that a suitable logical system can be characterized in a class of structures extending neighbourhood frames with sequences of sets of worlds. In this perspective, our contribution may offer useful insights for establishing connections between the proof-theoretic and model theoretic approaches to CTD reasoning.

A number of open research issues are left for future work. Among others, we aim at going beyond basic completeness results with multi-preference structures for the classical case by considering introduction or elimination rules for $\otimes$, for which we only presented characterization results. Second, an extensive investigation should be done when we move to logics closed under logical implication (see Section 7). Finally, we expect to enrich the language and allow for nesting of $\otimes$-expressions, thus having formulas like $\neg(a \otimes b) \otimes c$; although we argued in [8] that the meaning of those formulas is not clear, they pose anyway interesting technical problems.

Acknowledgements. NICTA is funded by the Australian Government and the Australian Research Council through the ICT Centre of Excellence program. Antonino Rotolo was supported by the Unibo FARB 2012 project *Mortality Salience, Legal and Social Compliance, and Economic Behaviour: Theoretical Models and Experimental Methods.*

References

1. Alchourrón, C.E.: Philosophical foundations of deontic logic and the logic of defeasible conditionals. In: Meyer, J.-J., Wieringa, R.J. (eds.) Deontic Logic in Computer Science. Wiley, New York (1993)
2. Calardo, E.: Non-normal Modal Logics, Quantification, and Deontic Dilemmas, PhD thesis, University of Bologna (2013)
3. Carmo, J., Jones, A.J.I.: Deontic logic and contrary to duties. In: Gabbay, D.M., Guenther, F. (eds.) Handbook of Philosophical Logic, 2nd edn. Kluwer Academic Publishers, Dordrecht (2002)
4. Chellas, B.F.: Modal Logic, An Introduction. Cambridge University Press (1980)
5. Dastani, M., Governatori, G., Rotolo, A., van der Torre, L.W.N.: Programming cognitive agents in defeasible logic. In: Sutcliffe, G., Voronkov, A. (eds.) LPAR 2005. LNCS (LNAI), vol. 3835, pp. 621–636. Springer, Heidelberg (2005)
6. Goble, L.: Preference semantics for deontic logic — Part II: Multiplex models. Logique et Analyse 47, 113–134 (2004)
7. Goble, L.: Preference semantics for deontic logic part I–Simple models. Logique et Analyse 46, 383–418 (2003)
8. Governatori, G., Rotolo, A.: Logic of violations: A Gentzen system for reasoning with contrary-to-duty obligations. Australasian Journal of Logic 4, 193–215 (2006)
9. Governatori, G.: Thou shalt is not you will. Technical Report 8026, NICTA (2014)
10. Governatori, G., Olivieri, F., Rotolo, A., Scannapieco, S.: Computing strong and weak permissions in defeasible logic. J. Philosophical Logic 42(6), 799–829 (2013)
11. Governatori, G., Rotolo, A.: Justice delayed is justice denied: Logics for a temporal account of reparations and legal compliance. In: Leite, J., Torroni, P., Ågotnes, T., Boella, G., van der Torre, L. (eds.) CLIMA XII 2011. LNCS, vol. 6814, pp. 364–382. Springer, Heidelberg (2011)
12. Governatori, G., Sadiq, S.: The journey to business process compliance. In: Handbook of Research on Business Process Modeling, ch. 20, pp. 426–454. IGI Global (2009)
13. Hansen, J.: Conflicting imperatives and dyadic deontic logic. J. Applied Logic 3(3-4), 484–511 (2005)
14. Hansson, B.: An analysis of some deontic logics. Nous (3), 373–398 (1969)
15. Makinson, D.: On a fundamental problem of deontic logic. In: McNamara, P., Prakken, H. (eds.) Norms, Logics, and Information Systems. IOS (1999)
16. Makinson, D., van der Torre, L.: Input/output logics. J. Philosophical Logic 29(4), 383–408 (2000)
17. Makinson, D., van der Torre, L.: Constraints for input/output logics. J. Philosophical Logic 30(2), 155–185 (2001)
18. Prakken, H., Sergot, M.J.: Contrary-to-duty obligations. Studia Logica 57(1), 91–115 (1996)
19. Stenius, E.: Principles of a logic of normative systems. Acta Philosophica Fennica 16, 247–260 (1963)
20. van Benthem, J., Grossi, D., Liu, F.: Priority structures in deontic logic. Theoria (2013)
21. van der Torre, L.: Reasoning about obligations: defeasibility in preference-based deontic logic. PhD thesis, Erasmus University Rotterdam (1997)

Detecting Deontic Conflicts in Dynamic Settings

Silvano Colombo Tosatto[1,3], Guido Governatori[2], and Pierre Kelsen[1]

[1] University of Luxembourg, Luxembourg
{silvano.colombotosatto,pierre.kelsen}@uni.lu
[2] NICTA and Queensland University of Technology, Australia
guido.governatori@nicta.com.au
[3] Università di Torino, Italy

Abstract. Regulations, through the use of obligations and permissions, are widely used in modern society to define acceptable behaviours. Thus it is indeed important that these regulations do not conflict with each other and contain contradicting obligations. In the present paper we focus on identifying conflicts between obligations in dynamic settings. We first show the need of an alternative semantics rather than the more classic modelled by standard deontic logic. Second we introduce a new semantics for the obligations capable of representing and reasoning about them in these dynamic settings, and lastly we use it to identify the necessary and sufficient conditions to identify conflicting obligations.

Keywords: Normative Reasoning, Conflicts, Time.

1 Introduction

Nowadays, wherever we may go, there are always regulations influencing what we can and cannot do. The modern society makes a heavy use of regulations to define which are the desirable behaviours in almost any foreseeable scenario.

We argue that a clear understanding about how obligations interact is imperative to avoid situations where the obligations contradict each other, turning into dilemmas [16], where desirable behaviours are not discernible anymore.

Example 1. The "working week" defines that workers have to work from monday to friday. Islam defines that friday is an holy day and it is forbidden to work.

The example describes a conflicting situation resulting from merging different regulations, religious and business. The issue of conflicting regulations has been already studied in *normative reasoning*, like by Elhag et al. [6], Beirlaen and Straßer [2], and Sartor [21]. In particular, since regulations define what is obligatory, prohibited and permitted, *deontic logic* [14] and its variants have been extensively used to reason about them. For instance Hansen [12] studies the conflicts between obligations using *dyadic deontic logic*.

The deficiency of standard deontic logic to deal with conflicts has been already studied by Beirlaen et al. [3]. Whereas Beirlaen et al. focus on identifying conflicts between both permissions and obligations in single time instants, in this paper we

F. Cariani et al. (Eds.): DEON 2014, LNAI 8554, pp. 65–80, 2014.

study conflicts in a dynamic setting, consisting of scenarios evolving through time and we refer to them as *traces*. We show that standard deontic logic appears to be too restrictive while reasoning about normative conflicts in dynamic settings. Therefore we propose an alternative formalisation capable of reasoning about the obligations and detecting conflicts in these settings.

The paper is structured as follows: Section 2 introduces standard deontic logic. Section 3 introduces the traces. Section 4 introduces an alternative semantics to reason about obligations using the traces. Section 5 redefines the concept of deontic conflicts according to the alternative semantics. Section 6 describes how conflicts can be detected between the obligations using the alternative semantics. Sections 7 and 8 extend the alternative semantics introducing respectively preemptive and compensable obligations, and study how conflicts are detected given these additional semantics. Section 9 concludes the paper.

2   Standard Deontic Logic

Firstly introduced in 1951 by von Wright [26] as a system for reasoning about what is necessary or allowed, *Standard Deontic Logic* is one of the successors of this system.

The syntax of this logic is composed of an infinite set of propositional variables, the classical logical operators $(\neg, \wedge, \vee, \rightarrow)$ and two modal operators $\mathcal{O}$ and $\mathcal{P}$ used respectively to identify what is obligatory and what is permitted.

2.1   Consistency

Standard deontic logic is a normal KD logic where the axioms: $\mathcal{P}\top$ and $\mathcal{O}\alpha \rightarrow \neg\mathcal{O}\neg\alpha$, and the equivalence $\mathcal{O}\alpha \equiv \neg\mathcal{P}\neg\alpha$ hold. The equivalence expresses a relation between obligations and permissions, in other words it states that if something is obligatory, then the opposite is not permitted. The first axiom: $\mathcal{P}\top$, states that tautologies are always permitted and the second axiom: $\mathcal{O}\alpha \rightarrow \neg\mathcal{O}\neg\alpha$, states that if something is obligatory then its complement must not be obligatory.

We define internal consistency and external consistency using the two axioms and the equivalence. Internal consistency expresses the fact that something contradictory, like a proposition and its negation, cannot be obligatory.

Definition 1 (Internal Consistency). *A set of norms is internally consistent iff there is no formula α such that $\mathcal{O}(\alpha \wedge \neg\alpha)$ is entailed by the set of norms.*

Accordingly internal consistency corresponds to axiom $\mathcal{P}\top$:

$$\neg\mathcal{O}(\alpha \wedge \neg\alpha) \equiv \neg\mathcal{O}\bot \equiv \mathcal{P}\top$$

External consistency expresses that two contradictory obligations cannot coexist, like for instance the obligation of performing an action along with the prohibition of performing it.

Definition 2 (External Consistency). *A set of norms is externally consistent iff there are no formulae α such that $\mathcal{O}\alpha \wedge \mathcal{O}\neg\alpha$ is entailed by the set of norms.*

Accordingly internal consistency corresponds to axiom $\mathcal{O}\alpha \rightarrow \neg\mathcal{O}\neg\alpha$:

$$\neg(\mathcal{O}\alpha \wedge \mathcal{O}\neg\alpha) \equiv \mathcal{O}\alpha \rightarrow \neg\mathcal{O}\neg\alpha \equiv \mathcal{O}\alpha \rightarrow \mathcal{P}\alpha$$

In standard deontic logic the two axioms $\mathcal{P}\top$ and $\mathcal{O}\alpha \rightarrow \neg\mathcal{O}\neg\alpha$ are equivalent. The two consistency measures defined in the present section are used in standard deontic logic to identify inconsistencies. Although inconsistencies involving permissions are also possible, in this paper we focus on inconsistencies between obligations.

3 Traces: A Dynamic Setting

The information contained in single time instants is often not sufficient to decide whether a real obligation has been fulfilled or violated. This problem has been previously approached by Segerberg [22] using dynamic deontic logic.

Example 2. The authors of this paper must submit it to Δeon before the deadline, which is set on Sunday. This also means that the paper has to be finished before the submission deadline.

The scenario contained in Example 2 illustrates a situation comprising an obligation for which considering unique time instants to decide whether it is violated or not is often not sufficient. Because even when considering a time instant where the submission is executed, in order to evaluate the obligation we also need the information regarding whether the submission has been executed before or after the deadline.

To evaluate the obligation in Example 2 we need to consider a time interval. More precisely, we consider the time instants occurring between the event triggering the obligation (being an author of this paper) and the deadline terminating it (Sunday). Considering these time instants between the trigger and the deadline allows to evaluate the obligation by verifying if the paper is submitted in one of them.

3.1 Traces

Each time instant is associated to a state describing the world at that precise point in time. We use finite sets of literals to describe the situation holding in a point in time.

Definition 3 (Universe $\mathcal{L}$). *Given a finite set of atomic elements E, the universe $\mathcal{L}$ is $E \cup \{\neg e \mid e \in E\}$. For $e \in E$, let $\bar{a} = \neg e$ iff $a = e$ and $\bar{a} = e$ iff $a = \neg e$.*

Definition 4 (Consistent Set). *A set of literals L is* consistent *if and only if* $\forall l \in L, \neg \exists \neg l : \neg l \in L$.

Definition 5 (State). *Let $\mathcal{I} = (t_1, t_2, \dots)$ be a discrete linear order of instants of time and L a consistent finite set of literals. A* state *is a tuple $\sigma = (t_i, L)$.*

The sequence of states contained in a *trace* describes the evolution of the world during that time interval.

Definition 6 (Trace). *Given a potentially infinite discrete linear order $\mathcal{I} = (t_1, t_2, \dots)$, a* trace *$\theta$ is a sequence of states: $(\sigma_1, \dots, \sigma_n, \dots)$, such that for each $\sigma_i = (t_i, L_i)$ and $\sigma_j = (t_j, L_j)$, $\sigma_i \prec \sigma_j$ if and only if $t_i \prec t_j$ in $\mathcal{I}$.*

Example 3 (Trace). Considering again Example 2, a hypothetical trace on which is enforced the obligation of submitting the paper can be the following:

$$\theta = ((t_1, \{a\}), \dots, (t_i, \{s\}), \dots, (t_k, \{d\}), \dots)$$

The state $(t_1, \{a\})$ represents the trigger of the obligation, where the authors are acknowledged, The state $(t_i, \{s\})$ represents where submission of the paper is executed and finally $(t_k, \{d\}))$ when the deadline is reached.

The trace illustrated in Example 3 does not violate the obligation of submitting the paper before the deadline. A violating trace can be constructed from it by exchanging the state at time t_k with the one at time t_i.

4   Obligations' Semantics

In this paper we adopt a simpler semantics than Segerberg's [22] to describe and reason about the obligations. We use linear time models, avoiding branching time, and our semantics focuses on obligations leaving permissions out of the picture.

From Example 2 and as previously pointed out by Governatori et al. [7], an obligation requires a lifeline (a trigger), a deadline (determining the obligation termination) and a condition determining what is required from the obligation. In this paper we disregard the first two elements and adopt a more abstract approach by using a function that given a state of a trace, returns the set of obligations holding in that state.

Definition 7 (Obligation in force). *Given a state σ, we define a function*

$$Force : 2^{\mathcal{I}} \times 2^{\mathcal{L}} \mapsto 2^{\circledcirc}$$

where $\circledcirc$ is a set of obligations.

Definition 8 (Obligation). *An* obligation *is a structure $\langle t, c \rangle$, where $t \in \{s, a, m\}$ represents the type of the obligation. The element c is a propositional formula composed by elements in $\mathcal{L}$ and represents the content of the obligation. We use $\mathcal{O}^t \langle c \rangle$ to represent an obligation.*

The content c of an obligation $\mathcal{O}^t \langle c \rangle$ is obligatory in the deontic sense. The formal semantics of how the content is obligatory depends on the type of the obligation t considered. We distinguish three types of obligations: standard ($\mathcal{O}^s$) which replicates the semantics of the obligations of standard deontic logic; achievement ($\mathcal{O}^a$), which captures the semantics of the obligations like the one in Example 2; and maintenance ($\mathcal{O}^m$), which captures the semantics of obligations similar to the one in the following example.

Example 4. To access secure data, the proper credentials must be retained for the whole access period.

Considering the trace illustrated in Example 3 and the obligation in Example 2, we can use s to represent the condition. The obligation, according to Definition 8, is represented as follows: $\mathcal{O}^a \langle s \rangle$.

4.1 Evaluating the Formulae

The states are not necessarily complete, meaning that given a proposition α, a state can either contain such proposition, its negation ($\neg \alpha$) or neither of them.

Definition 9 (Formula Entailment). *Let $\models$ be the standard propositional entailment. Given a state $\sigma = (t, L)$ and a formula α, $\sigma \models \alpha$ if and only if $\bigwedge x \models \alpha$, where each $x \in L$.*

4.2 Standard Obligations

Obligations in standard deontic logic are evaluated in a single state. We can mirror these obligations by forcing the instances of these obligations to be in force for exactly one state. We refer to the mirrored obligations as standard obligations. A standard obligation is represented as follows: $\mathcal{O}^s \langle c \rangle$.

Definition 10 (Comply with Standard). *Given a standard obligation $\mathcal{O}^s \langle c \rangle$ and a trace θ, θ is compliant with $\mathcal{O}^s$ if and only if: $\forall \sigma_i \in \theta$ such that $\mathcal{O}^s \langle c \rangle \in Force(\sigma_i), \sigma_i \models c$*

Consistency of Standard Obligations. We expect that both internal and external consistency measures (Definitions 1 and 2) still apply to standard obligations.

Proposition 1. $\neg \exists \theta | \theta$ *complies with* $\mathcal{O}^s \langle \alpha \wedge \neg \alpha \rangle$.

Proof (Sketch). If we assume an obligation $\mathcal{O}(\alpha \wedge \neg \alpha)$ to be possible, then the translated standard obligation would be the following: $\mathcal{O}^s \langle \alpha \wedge \neg \alpha \rangle$.

From Definition 10 it follows that a trace must contain a state σ_i such that $\sigma_i \models \alpha \wedge \neg \alpha$ in order to comply with the standard obligation unless *Force* of each state of θ is empty (eg. $\forall \sigma \in \theta, Force(\sigma) = \emptyset$). However such state could not exist according to Definition 5 since each state must be consistent. Thus a standard obligation whose condition is a contradiction would never be complied by any trace.

Therefore internal consistency applies to standard obligations. $\Box$

Proposition 2. $\neg \exists \theta | \theta$ *complies with* $\mathcal{O}^s \langle \alpha \rangle$ *and* θ *complies with* $\mathcal{O}^s \langle \neg \alpha \rangle$.

Proof (Sketch). Assume a trace containing the state σ_i, where $\{ \mathcal{O}^s \langle \alpha \rangle, \mathcal{O}^s \langle \neg \alpha \rangle \} \in Force(\sigma_i)$.

From Definition 2, $\mathcal{O}\alpha \wedge \mathcal{O}\neg \alpha$ is translated in standard obligations as follows: $\mathcal{O}^s \langle \alpha \rangle$ and $\mathcal{O}^s \langle \neg \alpha \rangle$ both belonging to the same set returned by applying *Force* the a given state σ_i of a given trace. According to Definition 10, since both standard obligations are in force in σ_i, then both conditions have to be verified in the same state.

A state σ_i, in order to fulfil both obligations, needs to contain in its state both α and $\neg \alpha$, however this is in contradiction with Definition 5, stating that a state has to be consistent. Thus it follows that a state σ_i satisfying both α and $\neg \alpha$ cannot exist.

Therefore a trace compliant with both standard obligations $\mathcal{O}^s \langle \alpha \rangle$ and $\mathcal{O}^s \langle \neg \alpha \rangle$ cannot exist. Thus no solution can exist when such pair of obligations is considered. $\Box$

4.3   Non-standard Obligations

We define the semantics of the additional types of obligations in a similar way as already defined by Governatori et al. [7].

Achievement obligations require that at least a state included in their in force interval satisfies the condition.

Definition 11 (Comply with Achievement). *Given an achievement obligation* $\mathcal{O}^a \langle c \rangle$ *and a trace* θ, θ *is compliant with* $\mathcal{O}^a \langle c \rangle$ *if and only if:*

$\forall$ *maximal subsequences* $\theta_s \in \theta$ *such that* $\forall \sigma_i \in \theta_s, \mathcal{O}^a \langle c \rangle \in Force(\sigma_i), \exists \sigma_h \in \theta_s$ *such that* $\sigma_h \models c$.

An operator with a similar semantics to the one just presented has been defined and analysed by Broersen et al. [5], the operator designed combines the semantics of computation tree logic and standard deontic logic.

Similarly, a maintenance obligation also requires to verify the condition when they are in force. However as we can see from Example 4, for each state where a maintenance obligation is in force, the state needs to satisfy the obligation's condition.

Definition 12 (Comply with Maintenance). *Given a maintenance obligation* $\mathcal{O}^m \langle c \rangle$ *and a trace* θ, θ *is compliant with* $\mathcal{O}^m \langle c \rangle$ *if and only if:*

$\forall \sigma_i \in \theta$ *such that* $\mathcal{O}^m \langle c \rangle \in Force(\sigma_i), \sigma_i \models c$.

Relations with Standard Obligations. Standard obligations are a particular case of both achievement and maintenance obligations. If we constrain the activation period of an achievement obligation to a single state, then such state must satisfy the condition. The same applies to maintenance obligations, if the activation period is limited to only one state, then such state has to fulfil the condition. Therefore if the activation is limited to a single state, then the semantics of both achievement and maintenance obligations collapse in the semantics of standard obligations.

5 Deontic Conflicts

We show here that the external consistency measure of standard deontic logic is too restrictive when used in a dynamic setting. The following example extends Example 2.

Example 5. The authors of this paper must submit it to Δeon before the deadline, which is set on Sunday. This also means that the paper has to be finished before the submission deadline. However, as usual on the weekends, the authors must go to the pub to meet their friends on Saturday or Sunday.

Example 5 contains two obligations: submitting the paper and going to the pub. We assume that the authors cannot finish and submit the paper while at the pub, hence we consider these obligations to be complementary. Thus if the proposition α represents "finishing and submitting the paper", then we can use $\neg\alpha$ to represent "going to the pub".

To formalise the example we discretise time in days. We use the propositions *sat* and *sun* to represent Saturday and Sunday respectively. Lastly we use the proposition *aut* to represent being an author of the present paper. We formalise the obligation of going to the pub: $\mathcal{O}^a\langle\neg\alpha\rangle$ and the obligation of submitting the paper as: $\mathcal{O}^a\langle\alpha\rangle$.

Both obligations are of type achievement. Despite the conditions of the obligations being complementary, it is still possible to provide a trace complying with both obligations.

$$\theta = (\ldots, (t_i, \{aut\}), \ldots, (t_j, \{sat, \neg\alpha\}), (t_k, \{sun, \alpha\}))$$

Assuming that $\mathcal{O}^a\langle\neg\alpha\rangle$ is in force in both $(t_j, \{sat, \neg\alpha\})$ and $(t_k, \{sun, \alpha\})$ and $\mathcal{O}^a\langle\alpha\rangle$ is in force from $(t_i, \{aut\})$ till the end of θ.

Example 5 describes a situation where two complementary obligations coexist during the weekend, but can be both fulfilled. According to the consistency measures provided by standard deontic logic, this situation would result in a conflict since it violates the external consistency measure. From the present analysis it follows that standard deontic logic is ill suited to reason about dynamic settings, more precisely the external consistency measure is too restrictive.

5.1 Redefining Conflicts

We now propose a new definition of inconsistent obligations, suited to be used in dynamic settings.

Definition 13 (Dynamic Conflict). *A set of obligations, written* $\circledcirc$, *is conflicting if and only if it is not possible to construct a trace in such a way that it is compliant with each obligation belonging to the set,* $\neg\exists\theta|\theta$ *compliant with* $\mathcal{O}, \forall\mathcal{O} \in \circledcirc$

The necessary conditions for two obligations to be conflicting is that their fulfilment conditions are complementary and their activation periods need to overlap. Depending on the type of obligations considered, the necessary condition may not be sufficient to determine whether they are conflicting. To focus on this aspect of the problem we introduce a function *Interval*, which when applied to an obligation and a trace returns the sub-intervals of the trace in which the obligation is activated.

Definition 14 (Interval). *Given a trace* θ, *let* θ_p *be the complete set of the sub-intervals of* θ. *Given an obligation* $\mathcal{O}$, *the partial function* Interval *is defined as follows:*

$$Interval : 2^{\mathcal{O}} \times 2^{\theta} \mapsto 2^{\theta_p} \text{ such that } \forall\varphi \in Interval(\mathcal{O}, \theta), \forall\sigma \in \varphi, \mathcal{O} \in Force(\sigma)$$

The function *Interval* returns all the intervals of a trace in which an obligation is active. *Interval* is defined as a partial function since it can be the case that an obligation is never activated in a trace, hence the set of intervals determining the activation period would be represented by the empty set.

5.2   Pair-Wise Conflicts

In Definition 13, conflicts are defined for sets of obligations. The following example illustrates a case where a conflict arises from a set of obligations and, when any proper subset of the obligations is considered a conflict does not arise.

Example 6 (Conflicting Set). Assume a trace θ and a set of obligations composed of a single achievement obligation $\mathcal{O}^a\langle\alpha\rangle$ and k standard obligations $\mathcal{O}^s\langle\neg\alpha\rangle$ such that $Interval(\mathcal{O}^a\langle\alpha\rangle, \theta) \equiv \bigcup I \in Interval(\mathcal{O}^s\langle\neg\alpha\rangle, \theta)$ and $\bigcap I \in Interval(\mathcal{O}^s\langle\neg\alpha\rangle, \theta) = \emptyset$. In other words the activation periods of the standard obligations are all distinct and entirely cover the activation period of the achievement obligation.

From Example 6, we can see that a trace compliant with all the obligations belonging to the set proposed cannot exist because it would require a state containing both α and $\neg\alpha$.

The behaviour of the standard obligations in Example 6 can be simulated using a single maintenance obligation. The behaviour required from a trace to be compliant with the set of standard obligations $(\mathcal{O}^s\langle\neg\alpha\rangle)$ is that in such trace $\neg\alpha$ holds for the interval determined by the obligations. The same result can be obtained by using a single maintenance obligation requiring $\neg\alpha$ to hold for the same interval. Thus the set of standard obligations can be substituted with a single maintenance obligation satisfying the following condition on the activation period:

$$\bigcup \varphi \in Interval(\mathcal{O}^s\langle \neg \alpha \rangle, \theta) \equiv Interval(\mathcal{O}^m\langle \neg \alpha \rangle, \theta)$$

Therefore we focus on analysing pair-wise conflicts between obligations.

6   Conflict Detection

The two necessary conditions to detect whether two obligations conflict are the following:

1. Their fulfilment conditions have to be complementary: $\mathcal{O}_1\langle \alpha \rangle$ and $\mathcal{O}_2\langle \beta \rangle$, such that $\alpha \wedge \beta \to \bot$.
2. The intersection of their activation periods must be not empty: $\exists x, y | x \in Interval(\mathcal{O}_1\langle \alpha \rangle, \theta), y \in Interval(\mathcal{O}_2\langle \beta \langle, \theta)$ and $x \cap y \neq \emptyset$.

We identify here the sufficient conditions to decide whether two obligations are conflicting. Being standard obligations a special case of both achievement and maintenance, it is sufficient to analyse the three combinations involving these types ($\mathcal{O}^m - \mathcal{O}^m$, $\mathcal{O}^m - \mathcal{O}^a$ and $\mathcal{O}^a - \mathcal{O}^a$). To do so we introduce two auxiliary functions, which applied to an interval or a trace, returns the first state belonging to them: min, or the last state: max.

6.1   Maintenance - Maintenance

We consider here two maintenance obligations.

Definition 15 ($\mathcal{O}^m - \mathcal{O}^m$ Conflict). *Let $\mathcal{O}^m\langle \alpha \rangle$ and $\mathcal{O}^m\langle \beta \rangle$ be two complementary maintenance obligations. $\mathcal{O}^m\langle \alpha \rangle$ and $\mathcal{O}^m\langle \beta \rangle$ are conflicting if and only if:*

$$\exists I \in Interval(\mathcal{O}^m\langle \alpha \rangle, \theta) \text{ and } \exists I' \in Interval(\mathcal{O}^m\langle \beta \rangle, \theta) : I \cap I' \neq \emptyset$$

Proposition 3 ($\mathcal{O}^m - \mathcal{O}^m$ Conflict). *Let $\mathcal{O}^m = \langle \alpha \rangle$ and $\mathcal{O}'^m = \langle \beta \rangle$ be conflicting maintenance obligations, then there does not exist a trace complying with both obligations.*

Proof ($\mathcal{O}^m - \mathcal{O}^m$ Conflict). We prove that the condition provided in Definition 15 is sufficient to identify whether two maintenance obligations are conflicting.

1. Let $\mathcal{O}^m = \langle \alpha \rangle$ and $\mathcal{O}'^m = \langle \beta \rangle$ be two complementary maintenance obligations, meaning that $\alpha \wedge \beta \to \bot$.
2. From the hypothesis we know that $\exists I, I'$ such that $I \in Interval(\mathcal{O}^m, \theta), I' \in Interval(\mathcal{O}'^m, \theta)$ and $\exists \sigma$ such that $\sigma \in I$ and $\sigma \in I'$.
3. From Definition 12 and 2. it follows that $\forall \sigma \in I, \sigma \models \alpha$ and $\forall \sigma' \in I' \sigma' \models \beta$.
4. Assume that there exists a trace θ such that θ is compliant with $\mathcal{O}^m$ and $\mathcal{O}'^m$.
5. From 4. it follows that $\forall I, I'$ such that $I \in Interval(\mathcal{O}^m, \theta)$ and $I' \in Interval(\mathcal{O}'^m, \theta), I \subseteq \theta$ and $I' \subseteq \theta$ and $\forall \sigma \in I, \sigma \models \alpha$ and $\forall \sigma' \in I', \sigma' \models \beta$.

6. From 2. and 5. it follows that $\exists \sigma \theta$ such that $\sigma \models \alpha$ and $\sigma \models \beta$.
7. From Definition 9 and 6. it follows that $\exists \sigma \in \theta$ such that $\{\alpha, \beta\} \in \sigma$.
8. From 1. we know that $\alpha \wedge \beta \rightarrow \bot$, hence from 7. and Definition 5 it follows that a state σ is inconsistent and a trace containing such state cannot exists.

Therefore we have proven that the condition provided in Proposition 3 is sufficient to identify to conflicting complementary maintenance obligations. $\square$

Two maintenance obligations are conflicting as soon as they are complementary and their activation periods overlap. In this case the sufficient condition is also the necessary condition previously introduced.

We do not provide propositions and formal proofs for the following definitions since they are analogous of the one provided for Definition 15.

6.2 Maintenance - Achievement

We consider here a maintenance and an achievement obligation.

Definition 16 ($\mathcal{O}^m - \mathcal{O}^a$ Conflict). *Let $\mathcal{O}^m\langle \alpha \rangle$ be a maintenance obligation and $\mathcal{O}^a\langle \beta \rangle$ be a complementary achievement obligation. $\mathcal{O}^m\langle \alpha \rangle$ and $\mathcal{O}^a\langle \beta \rangle$ are conflicting if and only if:*

$$\exists I \in Interval(\mathcal{O}^a\langle \beta \rangle, \theta) \ and \ \exists I' \in Interval(\mathcal{O}^m\langle \alpha \rangle, \theta) : I \subseteq I'$$

The sufficient condition captures the fact that an achievement obligation requires be fulfilled in a single state, hence a conflict arise only if the activation period of the maintenance obligation is a superset of the activation period of the achievement obligation.

6.3 Achievement - Achievement

We consider here two achievement obligations.

Definition 17 ($\mathcal{O}^a - \mathcal{O}^a$ Conflict). *Let $\mathcal{O}^a\langle \alpha \rangle$ and $\mathcal{O}^a\langle \beta \rangle$ be two conflicting achievement obligations. $\mathcal{O}^a\langle \alpha \rangle$ and $\mathcal{O}^a\langle \beta \rangle$ are conflicting if and only if:*

$$\exists I \in Interval(\mathcal{O}^a\langle \alpha \rangle, \theta) : I \in Interval(\mathcal{O}^a\langle \beta \rangle, \theta) \ and \ ||I|| = 1$$

The sufficient condition requires that there exists an activation period common to the two complementary achievement obligations and that such activation period is of length one. These restrictive conditions are necessary due to the flexibility allowed to comply with achievement obligations. Two achievement obligations are actually conflicting if and only if both behave as standard obligations in at least a shared activation period.

The sufficient condition required to identify conflicting standard obligations (Definition 10) is the following:

$$\exists I \in Interval(\mathcal{O}^s\langle \alpha \rangle, \theta) : I \in Interval(\mathcal{O}^s\langle \beta \rangle, \theta)$$

As it is expected, this sufficient condition is a particular case of all the other conditions identified in the present section.

7   Preemptive Obligations

Achievement and maintenance are capable of representing a good deal of obligations used in real world scenarios. However, there are still some which could be not translated using their semantics.

Example 7. Anti-Money Laundering and Counter-Terrorism Financing Act 2006. Clause 54 (Timing of reports about physical currency movements).

1. A report under Section 53 must be given:
    (a) if the movement of the physical currency is to be effected by a person bringing the physical currency into Australia with the person-at the time worked out under subsection (2); or
    . . .
    (d) in any other case-at any time before the movement of the physical currency takes place.

Example 7 illustrates an Australian regulation aimed at monitoring physical currency movements. The obligation states that a report must be provided when transaction occurs, however clause **(d)** states that this report can be provided before the transaction takes place. This obligation is still an achievement obligation, however due to clause **(d)**, this obligation can be preemptively achieved as it has been defined by Governatori and Rotolo [9].

We introduce a sub-type of achievement obligations called *preemptive achievement obligation*, denoted $\mathcal{O}^{-a}$, which allow to be fulfilled in states preceding their triggering state.

Definition 18 (Comply with Preemptive Achievement). *Given a preemptive achievement obligation $\mathcal{O}^{-a}\langle c\rangle$ and a trace θ, θ is compliant with $\mathcal{O}^{-a}$ if and only if:*

$\forall$ *maximal subsequences $\theta_s \in \theta$ such that $\forall \sigma_i \in \theta_s, \mathcal{O}^{-a}\langle c\rangle \in Force(\sigma_i), \exists \sigma_h \in \theta_s$ and $\exists \sigma_j \in \theta$ such that $\sigma_j \models c$ and $\sigma_j \preceq \sigma_h$.*

7.1   Conflicts for Preemptive Achievement Obligations

As it has been done previously for the two main types of obligations, we define the sufficient conditions to identify conflicts involving a preemptive achievement obligation.

Maintenance - Preemptive Achievement. We now consider a maintenance and a preemptive achievement obligation.

Definition 19 ($\mathcal{O}^m - \mathcal{O}^{-a}$ Conflict). *Let $\mathcal{O}^m\langle\alpha\rangle$ be a maintenance obligation and $\mathcal{O}^{-a}\langle\beta\rangle$ be a complementary preemptive achievement obligation. $\mathcal{O}^m\langle\alpha\rangle$ and $\mathcal{O}^{-a}\langle\beta\rangle$ are conflicting if and only if:*

$\exists I \in Interval(\mathcal{O}^{-a}\langle\beta\rangle, \theta), \exists I' \in Interval(\mathcal{O}^m\langle\alpha\rangle, \theta): I \subseteq I'$ and $min(I') = min(\theta)$

The sufficient condition is an extension of Definition 16, to which has been added the additional condition, requiring that the activation period of the maintenance obligation contains the first state of the trace. The stricter sufficient condition follows from the less strict fulfilment condition for preemptive achievement obligations (Definition 18) with respect to achievement obligations (Definition 11).

Preemptive Achievement - Achievement. We now consider a preemptive achievement and an achievement obligation.

Definition 20 ($\mathcal{O}^a - \mathcal{O}^{-a}$ Conflict). *Let $\mathcal{O}^a\langle\alpha\rangle$ be an achievement obligation and $\mathcal{O}^{-a}\langle\beta\rangle$ be a preemptive achievement obligation. $\mathcal{O}^a\langle\alpha\rangle$ and $\mathcal{O}^{-a}\langle\beta\rangle$ are conflicting if and only if:*

$$\exists I \in Interval(\mathcal{O}^{-a}\langle\beta\rangle, \theta)\colon I \in Interval(\mathcal{O}^a\langle\alpha\rangle, \theta), ||I|| = 1 \text{ and } min(I) = min(\theta)$$

The sufficient condition is an extension of Definition 17. The additional constraint: "and $min(I) = min(\theta)$" requires that an activation period for both obligation to include the first sate of the trace and be of length 1.

Preemptive Achievement - Preemptive Achievement. The sufficient condition to identify whether two preemptive achievement obligations are conflicting is the same as the one identified between an achievement and a preemptive achievement obligation in Definition 20.

8   Compensable Obligations

In complex systems, the possibility that regulations may not be followed has to be taken into account. Lomuscio and Sergot [17] studied this in the context of multi-agent systems. Compensable obligations define in addition to their obligation, which we call primary from now on, also what needs to be done when they are violated through secondary obligations as defined by Governatori and Rotolo [11]. Secondary obligations are a particular type of obligation whose activation depends on the violations of the primary obligation they try to compensate.

Definition 21 (Activation). *An activation of an obligation $\mathcal{O}^t\langle c\rangle$ in a trace θ consists of a maximal subsequence θ_s of θ where $\forall\sigma_i \in \theta_s, \mathcal{O}^t\langle c\rangle \in Force(\sigma_i)$.*

A violation can raise for each activation in which a primary obligation is not complied with. This means that if there is no state satisfying the condition, then an achievement obligation (for both types of obligations, Definitions 11 and 18) is raised in the last state belonging to the activation. For maintenance obligations (Definition 12), if exists a state in the activation which does not satisfy the condition, then a violation is raised in the earliest[1] state of the activation which does not satisfy the condition.

---

[1] We consider the earliest to be the one not satisfying the condition and not preceded by any other which does not satisfy the condition.

Definition 22 (Violations). *Given an activation θ_s of an obligation $\mathcal{O}^t\langle c\rangle$, a violation v of $\mathcal{O}^t\langle c\rangle$ is identified a function $V(\theta, \mathcal{O}^t\langle c\rangle)$ which identifies the earliest state of θ_s where $\mathcal{O}^t\langle c\rangle$ is not complied with.*

Compensable obligations are composed of two separate components: a primary obligation which describe the obligation which has to be complied with for each activation of the compensable obligation, and a secondary obligation that needs to be complied with for each violation of the primary obligation.

Definition 23 (Compensable Obligation). *A compensable obligation, written $\mathfrak{O} = \mathcal{O} \otimes \mathcal{O}_c$, is composed of a primary obligation $\mathcal{O}$ and a compensation $\mathcal{O}_c$.*

The relations between the activation periods of $\mathcal{O}$ and $\mathcal{O}_c$ are the following: $\forall I \in Interval(\mathcal{O}_c, \theta), \exists v \in V(\mathcal{O}, \theta) : min(I) = v$. Moreover $\forall v \in V(\mathcal{O}, \theta), \exists I \in Interval(\mathcal{O}, \theta) : max(I) = v$.

The compensation $\mathcal{O}_c$ can be as well a compensable obligation.

Compensable obligations can be seen as sequences of obligations connected by the operator $\otimes$.

Definition 24 (Comply with Compensable Obligations). *Given a trace θ and a compensable obligation $\mathfrak{O} = \mathcal{O} \otimes \mathcal{O}_c$. θ is compliant with $\mathfrak{O}$ if and only if θ is compliant with $\mathcal{O}_c$.*

A trace is compliant with a compensable obligation if it is compliant with its secondary obligation. This follows from Definitions 11, 12 and 18, where a trace θ is always considered to be compliant with an obligation $\mathcal{O}$ if $Interval(\mathcal{O}, \theta) = \emptyset$. This means in this case that either the primary obligation is not violated or if it is violated, then each violation has been compensated.

Example 8. An example of compensable obligations is the following: When you dine at a restaurant you have to pay for your meal. If you don't, then you have to wash the dishes.

This compensable obligation can be formalised as follows: $\mathcal{O}^a\langle\alpha\rangle \otimes \mathcal{O}^a\langle\beta\rangle$, where α represents "paying the bill" and β represents "washing the dishes".

8.1   Conflicts for Compensable Obligations

We define now the sufficient conditions to identify pair-wise conflicts involving compensable obligations. A compensable obligation is not a new type of obligation, but rather a way of structuring the existing types of obligations. A non compensable obligation is a special case of compensable which compensable obligation cannot be fulfilled if triggered. Therefore we analyse the more general case of deciding which are the sufficient conditions to determine whether two compensable obligations are conflicting.

Definition 25 ($\odot$ - $\odot$ Conflict). *Let $\odot = \mathcal{O} \otimes \mathcal{O}_c$ and $\odot' = \mathcal{O}' \otimes \mathcal{O}'_c$ be two compensable obligations. $\odot$ and $\odot'$ are conflicting if and only if:*

$$\mathcal{O}_c \text{ is conflicting with } \mathcal{O}'_c$$

To determine whether two obligation "conflict" we reuse the sufficient conditions from Definitions 15, 16, 17, 19 and 20. The sufficient condition expressed in Definition 25 requires that the compensations of the two compensable obligations are conflicting. A compensation $\mathcal{O}_c$ is triggered by a violation of the primary obligation $\mathcal{O}$, hence $||Interval(\mathcal{O}_c, \theta)|| = ||V(\mathcal{O}, \theta)||$. If the two secondary obligation are conflicting, it means that both $V(\mathcal{O}, \theta)$ and $V(\mathcal{O}', \theta)$ are not empty due existing conflicts between $\mathcal{O}$ and an obligation in $\odot'$ and vice versa.

9   Conclusion

In the present paper we show that standard deontic logic is not well suited to reason about conflicting obligations in a dynamic setting. Therefore we first provide an alternative semantics more suited to reason about obligations in such a setting and second we show how the newly defined semantics can be used to detect conflicting obligations in this type of dynamic setting. The sufficient and necessary condition for identifying conflicting obligations can be used as constraints while designing regulations for normative systems, in order to avoid systems where any behaviour would violate the regulations proposed.

The conditions identified in the present paper can be also used to detect conflicts in existing systems. Resolving these conflicts is also an important part as has been shown by Prakken and Sartor [20]. Another work relative to conflict detection and resolution by Vasconcelos et al. [25], proposes a similar approach as the one in this paper by considering overlapping periods of the obligations for conflict detection. The work of Vasconcelos et al. includes also conflict resolution techniques to solve the conflicts detected, however in the present paper we focus on explicitly identifying and highlighting the sufficient and necessary conditions to detect conflicts between different types of obligations, which we are capable of achieving using a simpler semantics for the obligations. Additionally we claim that a further utility of the conditions identified in the present paper is that they can be used as constraints to design *conflict free* normative systems.

The most closely related work is [10] presenting a temporal version of deontic defeasible logic equipped with deontic operators corresponding to all classes of obligations discussed in the paper (excluding preemptive obligations) and supplemented with an operator for compensatory obligations [11]. The conflicts are not explicitly given but are embedded in the various proof conditions.

Another important element in normative reasoning is constituted by permissions, which, as described by Boella and van der Torre [4], and Makinson and van der Torre [18], can be used as a mean to limit the applicability of obligations and prohibitions, as already has been studied by Stolpe [23] where the semantics is defined using AGM belief revision [1], Input/Output logic [19] and Defeasible Logic [8]. Conflict detection involving permission has already been studied by Hansen [13], however we plan to study it in a dynamic setting as future work.

Acknowledgements. NICTA is funded by the Australian Government as represented by the Department of Broadband, Communications and the Digital Economy and the Australian Research Council through the ICT Centre of Excellence program.

Silvano Colombo Tosatto is supported by the National Research Fund, Luxembourg.

References

1. Alchourrón, C.E., Gärdenfors, P., Makinson, D.: On the logic of theory change: Partial meet contraction and revision functions. J. Symb. Log. 50(2), 510–530 (1985)
2. Beirlaen, M., Straßer, C.: A paraconsistent multi-agent framework for dealing with normative conflicts. In: Leite, et al. (eds.) [15], pp. 312–329
3. Beirlaen, M., Straßer, C., Meheus, J.: An inconsistency-adaptive deontic logic for normative conflicts. J. Philosophical Logic 42(2), 285–315 (2013)
4. Boella, G., van der Torre, L.W.N.: Permissions and obligations in hierarchical normative systems. In: ICAIL, pp. 109–118 (2003)
5. Broersen, J., Dignum, F., Dignum, V., Meyer, J.-J.: Designing a deontic logic of deadlines. In: Lomuscio, A., Nute, D. (eds.) DEON 2004. LNCS (LNAI), vol. 3065, pp. 43–56. Springer, Heidelberg (2004), http://dx.doi.org/10.1007/978-3-540-25927-5_5
6. Elhag, A.A., Breuker, J.A., Brouwer, B.W.: On the formal analysis of normative conflicts. In: van den Herik, H., et al. (eds.) JURIX 1999: The Twelfth Annual Conference, GNI, Nijmegen. Frontiers in Artificial Intelligence and Applications, pp. 35–46 (1999)
7. Governatori, G., Hulstijn, J., Riveret, R., Rotolo, A.: Characterising deadlines in temporal modal defeasible logic. In: Orgun, M.A., Thornton, J. (eds.) AI 2007. LNCS (LNAI), vol. 4830, pp. 486–496. Springer, Heidelberg (2007)
8. Governatori, G., Olivieri, F., Rotolo, A., Scannapieco, S.: Computing strong and weak permissions in defeasible logic. Journal of Philosophical Logic 42(6), 799–829 (2013)
9. Governatori, G., Rotolo, A.: A conceptually rich model of business process compliance. In: Link, S., Ghose, A. (eds.) 7th Asia-Pacific Conference on Conceptual Modelling (APCCM 2010), January 18-21. CRPIT, vol. 110, pp. 3–12. ACS (2010), http://crpit.com/confpapers/CRPITV110Governatori.pdf
10. Governatori, G., Rotolo, A.: Justice delayed is justice denied: Logics for a temporal account of reparations and legal compliance. In: Leite, et al. (eds.) [15], pp. 364–382
11. Governatori, G., Rotolo, A.: Logic of violations: A gentzen system for reasoning with contrary-to-duty obligations. The Australasian Journal of Logic 4, 193–215 (2006), http://www.philosophy.unimelb.edu.au/ajl/2006/
12. Hansen, J.: Conflicting imperatives and dyadic deontic logic. In: Lomuscio, A., Nute, D. (eds.) DEON 2004. LNCS (LNAI), vol. 3065, pp. 146–164. Springer, Heidelberg (2004)
13. Hansen, J.: Reasoning About Permission and Obligation. In: Hansson, S.O., Wansing (eds.) [24] (2014)
14. Hilpinen, R., McNamara, P.: Deontic logic: a historical survey and introduction. In: Gabbay, D., Horty, J., van der Meyden, R., Parent, X., van der Torre, L. (eds.) Handbook of Deontic Logic and Normative Systems. College Publications (2013)

15. Leite, J., Torroni, P., Ågotnes, T., Boella, G., van der Torre, L. (eds.): CLIMA XII 2011. LNCS, vol. 6814. Springer, Heidelberg (2011)
16. Lemmon, E.J.: Moral dilemmas. The Philosophical Review 71(2), 139–158 (1962)
17. Lomuscio, A., Sergot, M.: Violation, error recovery, and enforcement in the bit transmission problem. In: Horty, J., Jones, A.J.I. (eds.) Proceedings of the 6th International Workshop on Deontic Logic in Computer Science (DEON 2002), London, England, May 22-24, pp. 181–202. Imperial College London, Informal Proceedings (2002)
18. Makinson, D., van der Torre, L.: Permission from an input/output perspective. J. Philosophical Logic 32(4), 391–416 (2003)
19. Makinson, D., van der Torre, L.W.N.: What is input/output logic? input/output logic, constraints, permissions. In: Boella, G., van der Torre, L.W.N., Verhagen, H. (eds.) Normative Multi-agent Systems. Dagstuhl Seminar Proceedings, vol. 07122. Internationales Begegnungs- und Forschungszentrum für Informatik (IBFI), Schloss Dagstuhl, Germany (2007)
20. Prakken, H., Sartor, G.: A dialectical model of assessing conflicting arguments in legal reasoning. Artif. Intell. Law 4(3-4), 331–368 (1996)
21. Sartor, G.: Normative conflicts in legal reasoning. Artificial Intelligence and Law 1(2-3), 209–235 (1992)
22. Segerberg, K.: DΔL: a dynamic deontic logic. Synthese 185, 1–17 (2012)
23. Stolpe, A.: Abstract Interfaces of Input/Output Logic. In: Hansson, S.O., Wansing (eds.) [24] (2014)
24. Hansson, S.O., Kracht, M., Smets, L.M.S.: Outstanding Contributions to Logic. Springer, Dordrecht (2014)
25. Vasconcelos, W., Kollingbaum, M., Norman, T.: Normative conflict resolution in multi-agent systems. Autonomous Agents and Multi-Agent Systems 19(2), 124–152 (2009), http://dx.doi.org/10.1007/s10458-008-9070-9
26. Von Wright, G.H.: Deontic logic. Mind 60(237), 1–15 (1951)

Denial of Responsibility and Normative Negation

Federico L.G. Faroldi

University of Pisa, Pisa, Italy
University of Florence, Florence, Italy
`federico.faroldi@for.unipi.it`

Abstract. In this paper I provide some linguistic evidence to the thesis that responsibility judgments are normative.

I present an argument from negation, since the negation of descriptive judgments is structurally different from the negation of normative judgments. In particular, the negation of responsibility judgments seem to conform to the pattern of the negation of normative judgments, thus being a *prima facie* evidence for the normativity of responsibility judgments. I assume — for the argument's sake — Austin's distinction between justification and excuse, and I sketch how to accommodate the distinction between internal (justification) and external (excuse) negation of responsibility within a language with a second-order analogous of existential generalization and λ operator.

In the end I confront with and refute some objections against this argument.

Keywords: Responsibility, Negation, Excuses, Justifications.

1 Introduction

In this paper I suggest that negations of responsibility judgments are isomorphic to the negation of normative sentences. It is only external negation that inverts the value of responsibility judgments, thus providing a *prima facie* evidence to consider responsibility judgments non-descriptive and normative-like.

First, I contrast *two* sorts of negation: negation of *normative* sentences and negation of *descriptive* sentences, pointing out where they differ. My provisional hypothesis is that *internal* negation and *external* negation work in opposite ways for descriptive sentences and normative sentences. (i) In descriptive sentences internal negation inverts their (truth) value;[1] whereas (ii) in normative sentences it is external negation that changes their (normative) value.

Second, I consider denials of responsibility. I show that negation of responsibility judgments falls under case (ii). It is only external negation that inverts the value of responsibility judgments, thus suggesting at least an analogy between responsibility judgments and normative judgments.

---

[1] Of course the value inversion occurs only in classical two-valued logic. In multivalued logics, it assigns its complement. This observation applies every time I mention truth-values.

F. Cariani et al. (Eds.): DEON 2014, LNAI 8554, pp. 81–94, 2014.

In the end of this paper I confront with *three* apparently possible objections to my argument: *first*, I have begged the question in the definition of responsibility judgments; *second*, all I have shown is that external negation inverts the value of sentences if they are not descriptive, but this tells nothing about the exact nature of those sentences; *third*, that these features of negation hold for other modalities, so there is nothing special about normativity. This paper aims at clarifying the various kinds of negation in logic and natural language (in §2). It then advances an interpretation of normative negation (§3) and considers how this model might shed light on responsibility judgments and in particular on negative responsibility judgments (§4).

2   Negation, Negations

I shall now briefly introduce some concepts I use in this paper, namely: (i) the difference between negation, denial and rejection; (ii) the difference between logical negation and natural language negations, including internal *vs* external negation and metalinguistic negation.[2]

2.1   Negation, Denial, Rejection

For the purposes of this chapter, I shall adopt the now common distinction among *negation, denial* and *rejection*.[3] While these definitions are apodictically stated, nothing significant for my arguments relies on them.

Very roughly, *negation* acts on *contents*. For instance, '*un*happy' is the *negation* of 'happy'.[4]

Denial is, instead, an *act*. It can be either a linguistic act, or a non-linguistic act (for instance: shaking one's head).

Rejection is, instead, a mental *attitude*.[5]

2.2   Internal *vs.* External Negation

Due to a felicitous intuition in [26],[6] the well-known sentence:

(1) The King of France is *not* bald
can be given two readings, usually paraphrased as follows:[7]

---

[2] For an engaging yet theory-driven introduction to negation, see [17].

[3] For a survey on the matter, see [24]. The paper discusses even some theories about the respective relationships among *negation, denial* and *rejection.*

[4] We got 'un', in English, from a reconstructed *en-, from Proto-Indoeuropean *n- (probably zero grade of *ne-), prefix usually found in most Indo-European tongues, cf. at least [6,23,38].

[5] On rejection, see [12,19,31,34].

[6] As far as I am aware, [10] (for instance in [10]) did not notice this phenomenon or shunned it.

[7] For instance by [17, §6].

(1a) INTERNAL: The King of France is *not*-bald (is *un*-bald);[8]

(1b) EXTERNAL: It is *not* the case (true)[9] that the King of France is bald.[10]

The former (1a) is usually read as an example of *internal* negation; whereas the latter (1b) is usually read as an example of *external* negation.[11]

In propositional logic internal negation and external negation are equivalent, that is, they both equally invert the logical value of a given sentence.[12]

So, for instance:

(2) Maria is brunette

changes its truth-value both in (2a) and (2b), examples of internal negation and external negation, respectively:

(2a) INTERNAL: Maria is *not* brunette;

(2b) EXTERNAL: It is *not* the case (true) that Maria is brunette.

Please keep this point in mind because it will become handy *infra* at §3, when we shall see that internal negation and external negation are not equivalent in normative sentences.[13]

2.3   Metalinguistic Negation

Metalinguistic negation is defined[14] as a formally negative utterance used to object to a previous utterance on any grounds (even of intonations, assertability, and so on).

---

[8] $\exists x(\forall y(Kxf \leftrightarrow y = x) \land \neg Bx)$

[9] 'True' was proposed by [20].

[10] $\neg \exists x(\forall y(Kxf \leftrightarrow y = x) \land Bx)$

[11] [17, §6] questions the use of 'true' and underlines how *no* known natural language employ two distinct negative operators corresponding directly to internal and external negation, even if a given language employs two (or more) negative operators, for instance (former: *declarative* negation; latter: *emphatic* negation): Ancient Greek: 'ou' *vs.* 'mē'; Modern Greek: 'den' *vs.* 'me'; Hungarian: 'nem' *vs.* 'ne'; Latin: 'non' *vs.* 'nē'; Irish: 'nach' *vs.* 'gan'; Sanskrit: 'na' *vs.* 'mā'. There is another 'un-' in English which is not a negative operator, but it is analogous to German 'ent-' as in '*un*-fold', '*ent*-falten'. See Horn's interesting list of languages with distinct negative operators at p. 366.

[12] But please keep in mind that *duplex negatio affirmat* only in propositional logic and some natural languages, for instance contemporary standard English. Both in Old and Middle English, along with contemporary languages such as Italian, Portuguese and many others, *duplex negatio n e g a t*.

[13] This point was noticed also by St. [2]: "dicimus etiam nos "non debere peccare" pro "debere non peccare". Non enim omnis, qui facit, quod non debet, peccat, si proprie consideretur." Cf. [28, p. 36]. For an interesting survey of modal logics in Anselm, see [15] and [36,37].

[14] For instance by [18,17].

Here is an example of metalinguistic negation:

(3) John didn't *manage* to pass his viva — it was quite easy for him. (Emphasis signals stressed intonation here.)

(4) Ben is meeting a man this evening. No, he's not — he's meeting his brother.

So one does not object to the truth of a sentence, but to its (felicitous, appropriate) assertability.

Another interesting feature of metalinguistic negation is its inability to be incorporated prefixally:

(5) The King of France is not happy (*unhappy) — in fact there isn't any king of France.[15]

2.4 Illocutionary or Neustic Negation

Introduced as "neustic" negation by Hare ([14, p. 21] [13, p. 35]) and later called "illocutionary" negation (originally by Searle, cf. [22,29]), it should apply to what expresses illocutive force in a sentence or the neustic.

Here it is an example.

(8) I promise to come.

(9a) I promise not to come.

(9b) I don't promise to come.

According to Searle, (9a) is simply a propositional (or internal) negation, whereas (9b) is an example of illocutionary negation: one denies the very linguistic act, not its content. (9a) and (9b) are not equivalent.

Illocutionary negation, if it exists, seems non-truth conditional. Is it assimilable to metalinguistic negation? As [21] maintains, not always: in fact metalinguistic negation need not to be expressed linguistically, whereas illocutionary negation is necessarily linguistic.

Some doubts about the very existence of illocutionary (or neustic) negation are expressed by [7,11,16] and [21].

[16] has proposed a very interesting reading of illocutionary negation not as external or metalinguistic negation (ie, a negation of the whole speech-act), but simply as an internal negation.

According to him,

(10) It is not the case (that) I promise to come

it is not equivalent to (9b).

But (9b) must be read not as the internal negation of the coming, but as the negation of promise (as in not-promise):

(9b) I don't *promise* to come.

(9b) would be — at most — the negation of a preceding speech-act, rather the negation of that very speech-act produced by uttering (9b).

---

[15] [17, p. 392].

To Sum Up. *First*, there is *logical negation*. Logical negation is a logical operator (for instance: '¬') which is unambiguous: it always inverts the truth-value of a given sentence p.[16] Moreover, internal (logical) negation and external (logical) negation are functionally equivalent.[17]

Second, there is *natural negation*, ie negation in natural languages. As we have seen *supra*, negation in natural languages is much more complex a phenomenon than logical negation. *Firstly*, it may be pragmatically ambiguous (as [17, §6] and [32] masterly argued); *secondly*, other than descriptive negation, natural negation can be realized externally or metalinguistically, and it is not the case that it be always used to act on the truth of a given sentence; *thirdly*, non-descriptive negation cannot always be semantically analyzed in terms of external or metalinguistic negation, because there are pragmatic phenomena (intonation, phonetics, etc.) involved: external or metalinguistic negation can be realized implicitly, without fixed semantic features ('it is not the case that', 'it is not true that', etc.). *Fourthly*, not all (negated) sentences in natural language are truth-functional, but they may be commands, prayers, wishes or insults.

Third, natural negation, for instance via metalinguistic negation, can be used not only to invert the truth-value of a sentence, but also to reject or question its assertability.

3   Normative Negation

Last section was devoted to analyze different kinds of negation in logic and natural languages.

In this section I try to give an account of normative negation. I maintain that it can be differentiated from non-normative negation because normative negation cancels (at least) one of its presuppositions, whereas non-normative negation preserves the presuppositions of the negated sentence.

I have argued elsewhere that is not possible to have distinct species of negation for descriptive and normative language, but only different realizations of a single attitude.[18] I therefore propose to extend the model we have sketched in the preceding sections to normative language.

We have seen that logical negation, although unambiguous, is quite limited. Natural negation is instead a complex phenomenon, it does not always act on truth-values and it can be pragmatically ambiguous, divided among at least internal and external or metalinguistic negation.

Moreover, following [17, §6], we have noticed that at least metalinguistic negation is a formally negative utterance used to object to a previous utterance on any grounds, especially its assertability.

---

[16] In many-valued logics, it assigns p's truth-value complement. In logics with more than one negation, they are nonetheless unambigous.

[17] Of course I am referring here to classical propositional logic. Intuitionistic logics do not accept the equivalence of internal and external negation, nor the law of double negation: $\neg\neg B \neq B$.

[18] See [9, §5].

I propose to extend this model also to normative language. To stick to a logical level, even [25, §§31-2] noticed that while external and internal negation are functionally equivalent in propositional logic, internal negation and external negation differ quite radically in deontic logic: the fact you are under an obligation not to teach deontic logic ($O\neg\delta$), for instance, it is quite different from the fact you are not under an obligation to teach deontic logic ($\neg O\delta$).

In an analogous fashion, I maintain that internal normative negation keeps the sentence binding or, so to speak, normative, only to invert its deonticity: from obligatory to forbidden, and so on.[19] (Please note that I am not forced to assign normative sentences truth-aptness, because truth does not tell us the whole story even when (non-normative) natural language is concerned.)

External or metalinguistic negation is a rejection of the assertability (*lato sensu*) of a *prima facie*, allegedly normative sentence. Specifically, though, rejection of the assertability of a normative sentence is (implicity, I maintain) not a normative judgment, but a judgment on its normativity (or bindingness, or you name it). If a speaker feels[20] a given (non-normative) proposition unassertable, he rejects it metalinguistically; if he feels a given (*prima facie* normative) proposition not binding or not normative, he rejects it metalinguistically or externally, canceling its presupposition of normativity.[21]

Consider the following normative sentence:

(1) Abortion is wrong

and its *prima facie* negation:

(2) Abortion is not wrong.

Both are moral (normative) judgments, and share — among others — the following presupposition:

(0) Abortion can be an object of a genuine moral judgment.

Now consider external negation of (1):[22]

(3) It is not the case that abortion is wrong.

Now, while (2) is still a normative judgment, (3) seems intuitively a judgment on the normativity of (1).

(3) cancels (1)'s and (2)'s presupposition (0), because it simply rejects that abortion can be object of (that) moral judgment.

Let's now make a comparison with internal and external negation of nonnormative sentences.

Let's consider

(4) He stopped beating his wife

---

[19] I am well aware that not all normative sentences (or propositions) are in deontic terms. This was only an example to illustrate the general principle I want to bring forth.

[20] Please note that 'to feel' here is used generally has no intended reference to emotivism or expressivism.

[21] I am using this as a sort of a term of art, in order to make a general point without supporting a substantive theory of normativity either in terms of reasons (cf. for instance [27,30]), good (cf. [35]) or oughts.

[22] Of course it can be realized also metalinguistically.

its internal negation:

(5) He didn't stop beating his wife

and its external negation:

(6) It's not true that he stopped beating his wife.

Neither (5) nor (6) modify the ("factive") presuppositions of (4) such as that he has a wife, and he used to beat her.

Since — as we have seen — not every instance of external or metalinguistic negation is analyzable with distinct semantic (or, for that matter, syntactic) features, I assume a paraphrase in terms of external negation will account for the phenomenon, at least for our present purposes.

Considering the problem with normative negation only from the point of view of truth is quite limited, because truth does not tell the whole story even in non-normative negation, as I pointed out in the case of metalinguistic negation. This turns out to be a plus, because normative sentences are usually not considered truth-apt.[23]

In this section I contrasted descriptive and normative sentence by considering negation. I showed that normative negation, usually realized externally or metalinguistically, cancels its presupposition(s) of normativity.

Next section applies this conclusion to judgments of responsibility, showing that their structure with respect to negation is akin to normative sentences.

4   Denial of Responsibility

In last section I contrasted descriptive and normative sentence by considering negation. I suggested that normative negation, usually realized externally or metalinguistically, cancels its presupposition(s) of normativity.

In this section, I apply these results to judgments of responsibility, in order to provide an argument to the thesis that responsibility judgments are normative,

---

[23] And consequently one may maintain that (a) what you negate is not their truth; or that (b) norms cannot be negated. (a) was the position of the very first philosopher known to have written on this topic: Jerzy Sztykgold. In [33], he argued that you cannot negate the truth of norms, but only their righteousness [słuszność] in terms of non-righteousness [niesłuszność]. (Righteousness and unrighteousness are, for Sztykgold, the strict análogon of truth and falseness.)

(b) was instead the position of Karel Engliš ([8]), according to which:

(i) logical operations are possible only for "descriptive judgments" [soudy]);

(ii) negation [popření] is a logical operation;

(iii) norms [normy] and postulates [postuláty], although sentential, are not "descriptive judgments";

and therefore

(iv) logical operations don't apply to norms and postulates.

In particular:

(v) norms cannot be negated.

Of course Engliš' argument shows — at most, if premise (i) holds — that negation *as* a logical operator doesn't apply to norms. But negation is not exclusively a logical operator. Negation exists outside logic, in natural language, with different characteristics.

and namely an argument "from negation". I shall show that when one denies responsibility, what happens is (a) what happens when one denies normative statements; (b) what happens is the case *only* when normative entities are concerned. This might show that judgments of responsibility are normative.[24]

Here is a more schematic version of my fourth argument:

1. when you deny a responsibility judgment, what happens (what obtains) is a cancelation of its presuppositions;
2. canceling of presuppositions obtains only when normative judgments are negated;
3. Therefore, responsibility judgements are normative judgments.

Let's begin. I shall use negation a test to isolate a normative entity. We have seen back in §3 that negation of descriptive and normative entities differs in at least one substantial point: internal and external negation work in opposite ways.

Here is an example for descriptive statements:

Internal negation (1) "John isn't tall'

vs.

External negation (2) "It's not the case that John is tall"

Now, let's take a normative statement (for simplicity's sake, I shall consider an imperative):

Internal negation $O(\neg W)$ (3): "Don't shut the window!" (that is: "Shut not the window").

Note that (3) and its "positive"
(3a) Shut the window
share a presupposition of normativity.
Now, (3a)'s external negation:

External negation $\neg O(W)$ (4a) "I do not accept that is the case of shutting the window"/ (4b) "I do not accept the command 'Shut the window'''/ (4c) "I don't care".[25]

instead, rejects (cancels) the presupposition of normativity that both (3) and (3a) shared.

---

[24] This is by no means the standard theory. When judgments of responsibility are kept separate from responsibility or concepts of responsibility, they are usually considered *non*-normative; for example, judgments of responsibility are considered *explanatory* by [5,4]. Anderson ([1, §3.1] and p.c.) considers responsibility judgments to be normative, even though he does not provide any arguments for this thesis.

[25] Of course I am aware these are only some possible paraphrases — there might be many more. The most important fact is that internal and external negation can be consistently kept separable.

As I explained in §2, for descriptive sentences it is *internal* negation that might change their truth-value (from truth to false and viceversa); vice versa, for normative (imperative, in this case) sentences, it is *external* negation that changes their normativity-value, by rejecting the presupposition of normativity.

Now, let's apply this test to responsibility.

Internal negation (5) "He is not responsible for killing A, because..."

vs.

External negation (6) "It is not the case that he is responsible for killing A, because..."/

Now, if (5) stands to (1) as (6) stands to (2), we can confidently conclude that (5) and (6) are statements analogous to (1) and (2), that is, non-normative.

Quite on the contrary, if (5) stands to (3) as (6) stands to (4), we can confidently conclude that (5) and (6) are statements analogous to (3) and (4), that is, broadly normative.

It turns out, unfortunately, that you cannot really tell if (5) — internal negation of responsibility — tells us something of significance, for the very simple reason that its interpretation requires an understanding of responsibility. If you think responsibility is an objective state-of-affairs, that can be somehow empirically ascertained, then you would interpret (5) as a descriptive statement, whose truth-value is to be checked against the world; and vice versa.

Therefore, let's turn to (6) to seek some clarification of the matter.

My hypothesis is that a statement such as (5) stands for a justification; while (6) stands for an excuse. I take advantage of the paradigm excuse *vs.* justification developed in [3].

With a justification, I maintain, we remain in the domain of the normative: we accept A, and even add some *reasons* for it. The presupposition of normativity is kept.

Quite on the contrary, an excuse, in a way, suspends what was going on, it makes "normativity freeze" because it refers to conditions other than the very act A, conditions that (by definition) rule out responsibility (duress, infancy, mental incapacity, maybe psychopathy for moral responsibility). The presupposition of normativity is canceled.

In the words of Austin:

> [i]n the one defence [= justification], briefly, we accept responsibility but deny that it was bad: in the other [= excuse], we admit that it was bad but don't accept full, or even any, responsibility ([3]).

> it is not quite fair or correct to say baldly "X did A". We may say it isn't fair just to say X did it; perhaps he was under somebody's influence, or was nudged. Or, it isn't fair to say baldly he *did* A; it may have been partly accidental, or an unintentional slip. Or, it isn't fair to say he did simply *A* – he was really doing something quite different and A was only incidental, or he was looking at the whole thing quite differently ([3, p.2]).

First, excuses are *denial* of responsibility because, in giving excuses, a person contests or opposes a previously ascribed responsibility, by rejecting constitutive elements of the accusation: for instance, by denying having committed anything. He simply denies that the previous ascription of responsibility is sound.

Second, excuses are *rhetic* (and not thetic) negations (denials) of responsibility because they do not seek to cancel or nullify responsibility, since they assume that there is *no* responsibility whatsoever. Absence of responsibility is constitutive of excuses: if there were responsibility, they would not be excuses but — at most — justifications. Excuses do not presuppose responsibility, but only ascription of responsibility.[26]

Justifications, instead, are not at all negations of responsibility because justifications presuppose responsibility: justifications affirm responsibility, but deny it is responsibility for something bad. (A paradigmatic example seems to me "self defense": a admits to having killed b, but b was assaulting him with a knife, for instance.)

Negative Properties and Existential Generalization. A possible way to account for the difference between internal and external negation, and the existence of a given property is to consider a plausible analogous of Existential Generalization at the second order (I am not arguing for it at this point; I shall only make my point with a somewhat sloppy notation).

(EG1) $Fa \models \exists x F x$

(1) $\neg Fa \not\models \exists x F x$

But with a λ operator we can gain negative properties:

(2) $\lambda x(\neg F x)a \models \exists x(\neg F x \wedge x = a)$

Likewise, it is plausible to hold the following:

(EG2) $Fa \models \exists P \exists x(Px \wedge x = a \wedge P = F)$[27]

(3) $\neg Fa \not\models \exists P \exists x(Px \wedge x = a \wedge P = F)$ but

(4) $\lambda P \lambda x(\neg F x)a \models \exists P \exists x(Px \wedge x = a \wedge P = \lambda x(\neg F x))$

While both (1) and (3) are plain external negations and don't license any inference to the existence of either something or some property; (2) and (4) can, with the use of λ-abstraction, represent internal negation. Internal negation seems to license an inference to the existence of some property of sort.

The connection with internal and external negation of responsibility, while stretched, is significant: in fact, we suggested that with external negation of responsibility (excuse) there is no more responsibility (and normativity) involved, whereas with internal negation of responsibility (justification) the normativity is kept.

---

[26] As I noted with accusations, not all excuses are pled using a verb like 'to excuse' or 'scusare'; in an analogous fashion, it is not only the use of 'to excuse' or 'scusare' that can make an excuse.

[27] Of course one needs to explain what '=' among P and F means. I thank Tim Williamson for discussion on this point.

Two Examples: Excuses *vs.* Justifications. I am going to illustrate the difference between justification and excuses. I ask the reader to imagine two fictional criminal cases (both involve a death), and to abstract from particular legal systems in order to focus on the general point.

In the first, let's call it WIFE, a man comes back home and sees an intruder trying to rape or kill his wife. By chance, there is the intruder's loaded gun at hand. The man takes it up, aims and finally shoots the intruder down — killing him. In court, he admits the murder and puts forward his reasons. His lawyer says: "Look, *he is* not *responsible* for the killing, because that was self-defence: he was trying to defend and save his own wife." This is a *justification*: you admit your deed (there are all the relevant required elements: *actus reus, mens rea,* volition, intention, knowledge and so on to make that killing a murder) but you have a (good) reason for you action.

In the second, let's call it MAD, a mentally-ill man escapes from a psychiatric hospital, manages to get a gun, and shoots down a random passer-by. His lawyer says: "Look, he is *not* responsible for the killing, because *it is* not *the case he is (= can be) responsible at all*: he is mad (under duress, in infancy...)." This is an *excuse*: you may admit the deed, but it was done without the relevant required conditions: without *mens rea*, for instance, or without those capacities required for a death or a killing to be a murder.

To sum up, with a justification you deny your responsibility for that deed *qua* a particular action (but you admit, nonetheless, that you are under the domain of responsibility, that you can be responsible); with an excuse you deny your responsibility tout court, you deny that you are under the very domain of responsibility.

The lawyer's sentence in WIFE: "*he is* not *responsible* for the killing" is comparable to (3): "Don't shut the window" and (5): "He was not responsible", inasmuch as they are internal negations.

On the contrary, the lawyer's sentence in MAD: "*it is* not *the case he is (= can be) responsible at all*" seems to me analogous to (4): "It is not the case you order me to shut the window" and (6): "It is not the case that he is responsible for A, because..."

As (3) conserved the imperative nature of the sentence, so WIFE conserved the domain of responsibility. As (4) instead went out the domain of the imperative, to make a non-imperative claim, in the same way MAD appealed to a condition — in a way a non-normative, even factual condition — to be excluded from the domain of responsibility.

This linguistic evidence is consistent with the conceptual arguments I put forward earlier in this section: while justifications aren't at all denial of responsibility because they presupposes responsibility, excuses are in fact denial of responsibility, because they reject it.

With justifications and excuses, negation of responsibility coincides both with a linguistic act (denial) and a mental state (rejection).

We suggested that

- (i) when we deny responsibility, we have (at least) two cases: internal negation (which stands for a justification) and external negation (standing for an
  excuse). Then, we have seen that
- (ii) internal negations of responsibility do not exit the domain of responsibility (they presuppose responsibility); whereas external negations do (they
  reject the presupposition of responsibility). But this was exactly what happened with normative sentences (as I showed in §3): internal negation keeps
  the sentence normative (it keeps the presupposition of normativity), whereas
  external negation rejects it (it cancels the presupposition of normativity).

If we suppose that this kind of negation is at work only with non-descriptive
(and namely, normative statements), we can therefore conclude that

- (iii) since judgments denying responsibility are structurally akin to normative sentences, responsibility judgments are akin to normative sentences.

Caveats and Assumptions. Now, some caveats. I have limited my discussion
to the word (and the concept) of responsibility in the proper, fuller sense. I am
very well aware that there may be pragmatical ways to express a responsibility
judgment without mentioning the word 'responsibility' or any related. I am also
aware that we may get indicative (or descriptive) sentences (to express/ascribe
responsibility). For this (and for other) reasons linguistic arguments are interesting but not conclusive. I offer more (non linguistic) arguments for the thesis
that responsibility is normative in [9].

Last not least, my argument makes the following assumption: there are only
two kinds of language relevant to our investigation here: descriptive and normative language. This may not be the case: there are several other language
domains I am not considering: prayers, exclamations, insults, whose "status"
with regard to negation is unclear. Therefore, it might be the case that the
different ways negation works (in descriptive and normative domains) is not exclusive: negation might work in prayers as in normativity, and the second premise
of my argument would be factually undermined. Assuming the *prima facie* evidence I discussed as conclusive might be too strong, and other interpretations
are certainly possible depending on substantive theories of normativity, modality, and responsibility. But even if in general this argument does not prove to be
conceptually unassailable, I think it is still very telling.

Objections. I consider here *three* possible objections to my argument: *first*, I
have begged the question in the definition of responsibility judgments; *second*, all
I have shown is that external negation inverts the value of sentences if they are
not descriptive, but this tells nothing about the exact nature of those sentences;
third, these features of negation may be shared by other kinds of modality, so
there is not special about normativity.

To the *first* objection, I put forward a twofold reply: *first*, there is no shared
consensus either on what responsibility is or on what responsibility judgments

are: a degree of arbitrariness is needed anyway; *second,* there is no conceptual reason precluding my analysis to be extended further, given the right premisses.

To the *second* objection, I reply that I have at least shown that responsibility judgments are not descriptive; nonetheless I believe a linguistic test such as mine cannot exhaust the richness of human practices — in other words, normativity is not a sheer linguistic notion.

To the *third* objection, I reply that, examples with "oughts" notwithstanding, it is not clear whether normativity is a modality or not (it may be a property, for one). Moreover, other modalities may be normative as well (recently [30] so argued for necessity, the *a priori,* and other modalities), and thus these features of negation shouldn't come as a surprise.

Acknowledgements. For discussion on various points, suggestions, and critiques I thank Amedeo Giovanni Conte, Mattia Bazzoni, Guglielmo Feis, Sergio Filippo Magni, Timothy Williamson and two anonymous referees.

References

1. Anderson, S.: Coercion. In: Zalta, E.N. (ed.) The Stanford Encyclopedia of Philosophy (Winter 2011)
2. Anselm of Canterbury. "Lambeth's Fragments". In: Schmitt, F.S. (ed.) Ein neues unvollendetes Werk des hl. Anselm von Canterbury. Aschendorff, Münster i. W. (1936)
3. Austin, J.L.: A Plea for Excuses: The Presidential Address. In: Proceedings of the Aristotelian Society, vol. 57, pp. 1–30 (1956)
4. Björnsson, G., Persson, K.: A Unified Empirical Account of Responsibility Judgments. In: Philosophy and Phenomenological Research (forthcoming)
5. Björnsson, G., Persson, K.: The Explanatory Component of Moral Responsibility. Noûs 46(2), 326–354 (2012)
6. Brückner, A.: Słownik etymologiczny języka polskiego. Wiedza Powszechna, Warszawa (1957)
7. Cohen, L.J.: Do Illocutionary Forces Exist? The Philosophical Quarterly 14, 118–137 (1964)
8. Engliš, K.: Postulát a norma nejsou soudy. In: Časopis pro právní a státní vědu XXVIII, pp. 95–113 (1947)
9. Faroldi, F.L.G.: The Normative Structure of Responsibility. Law, Ethics, Neuroscience (forthcoming)
10. Frege, F.L.G.: Der Gedanke: Eine logische Untersuchung. In: Beiträge zur Philosophie des Deutschen Idealismus I, pp. 58–77 (September 1918)
11. Garner, R.T.: Some Doubts about Illocutionary Negation. Analysis 31, 106–112 (1971)
12. Gomolińska, A.: On the Logic of Acceptance and Rejection. Studia Logica 60(2), 233–251 (1998)
13. Hare, R.M.: Some Sub-Atomic Particles of Logic. Mind 98, 23–37 (1989)
14. Hare, R.M.: The Language of Morals. Clarendon Press, Oxford (1952)
15. Henry, D.P.: St. Anselm on the Varieties of 'Doing'. Theoria 19, 178–183 (1953)
16. Hoche, H.-U.: Do Illocutionary, or Neustic, Negations Exist? Erkenntnis 43, 127–136 (1995)

17. Horn, L.R.: A Natural History of Negation. University of Chicago Press, Chicago, Illinois (1989)
18. Horn, L.R.: Metalinguistic Negation and Pragmatic Ambiguity. Language 61(1), 121–174 (1985)
19. Incurvati, L., Smith, P.: Rejection and Valuations. Analysis 70(1), 3–10 (2010)
20. Karttunen, L., Peters, S.: Conventional Implicature. In: Oh, C.-K., Dinneen, D. (eds.) Syntax and Semantics 11: Presupposition, pp. 1–56. Academic Press, New York (1979)
21. Moeschler, J.: Negation, Scope and the Descriptive/Metalinguistic Distinction. Generative Grammar in Geneva 6, 29–48 (2010)
22. Peetz, V.: Illocutionary Negation. Philosophia: Philosophical Quarterly of Israel 8, 639–644 (1979)
23. Pokorny, J.: Indogermanisches etymologisches Wörterbuch. Francke (1994)
24. Ripley, D.: Negation, Denial, and Rejection. Philosophy Compass 6(9), 622–629 (2011)
25. Ross, A.N.C.: Directives and Norms. Humanities Press, New York (1968)
26. Russell, B.: On Denoting. Mind 14(56), 479–493 (1905)
27. Scanlon, T.M.: Being Realistic about Reasons. Oxford University Press, Oxford (2014)
28. Schmitt, F.S.: Ein neues unvollendetes Werk des hl. Anselm von Canterbury. A-schendorff, Münster i. W. (1936)
29. Searle, J.R., Vanderveken, D.: Foundations of Illocutionary Logic. Cambridge University Press, New York (1985)
30. Skorupski, J.: The Domain of Reasons. Oxford University Press, Oxford (2010)
31. Smiley, T.: Rejection. Analysis 56(1), 1–9 (1996)
32. Speranza, J., Horn, L.R.: A Brief History of Negation. Journal of Applied Logic 8(3), 277–301 (2010)
33. Sztykgold, J.: Negacja normy. Przegląd filozoficzny 39, 492–494 (1936)
34. Tamminga, A.: Logics of Rejection: Two Systems of Natural Deduction. Logique et Analyse 37(146), 169–208 (1994)
35. Thomson, J.J.: Normativity. Open Court, Chicago, Illinois (2008)
36. Uckelman, S.L.: Anselm's Logic of Agency, Amsterdam: Institute for Logic, Language and Computation (ILLC). University of Amsterdam (2007)
37. Uckelman, S.L.: Modalities in Medieval Logic. Institute for Logic, Language and Computation, Amsterdam (2009)
38. Vasmer, M.: Russisches etymologisches Wörterbuch. C. Winter, Heidelberg (1958)

Factoring Disjunction Out of Deontic Modal Puzzles

Melissa Fusco

UC Berkeley

Abstract. Ross's puzzle (Ross, 1941) and the paradox of Free Choice Permission (Kamp, 1973), puzzles involving disjunction under deontic operators, have received wide discussion in recent work in natural language semantics. First, I contrast two opposed modal views—call them the "box-diamond" theory and EU theory—that form two poles of the contemporary debate. The opposition between them is underwritten by distinct, well-developed conceptions of what it is for an action to be good. I present an axiomatization of obligation and permissibility—of 'ought' and 'may'—that is *neutral* between the two theories. Adding in the interpretation of 'or' as Boolean union, we get the received dialectic in the literature between the two theories on explaining Ross and FCP. Factoring out this assumption, we get a picture of how far apart the two theories are as theories of value, with no questions begged about the semantics of sentential disjunction.

1 Introduction

In this paper I will discuss two puzzles. The first is Ross's Puzzle [20]: from a premise like

(1)   Alice ought to call her mother.
       $Ought(C)$

one may not, it seems, infer

(2)   Alice ought call her mother or rob the bank.
       $Ought(C \text{ or } R)$

...despite the fact that disjunction introduction in the scope of 'ought' is valid on many semantic theories of 'ought' and 'or'. Call this

(Ross) $Ought(\phi) \nRightarrow Ought(\phi \text{ or } \psi)$

The second puzzle is the corresponding one for 'may' instead of 'ought'. From

(3)   Alice may take the bus.
       $May(B)$

F. Cariani et al. (Eds.): DEON 2014, LNAI 8554, pp. 95–107, 2014.

it seems unreasonable to infer that

(4)   Alice may take the bus or hijack a car.
        $May(B \text{ or } H)$

The failure of the inference from (3) to (4) has a better-known positive half: the paradox Free Choice Permission [23], by which (4) (but not (3)) seems to entail (5):

(5)   Alice may take the bus and Alice may hijack a car.
        $May(B) \wedge May(H)$

Call this:

(FC) $May(\phi \text{ or } \psi) \Rightarrow May\ (\phi) \wedge May\ (\psi)$

In this paper, I shall focus mostly on the weaker, negative datum, which we can call (FC-):

(FC-) $May(\phi) \nRightarrow May(\phi \text{ or } \psi)$

This data has, in the literature, primarily been interpreted as bearing on the interpretation of deontic modals rather than bearing on the interpretation of 'or'.[1] As I will explain in §3, it makes tempting an expected utility (EU) approach to these operators. My goal in this paper is to contrast this strategy with an approach to (Ross) and (FC-) from a less-examined angle: a revisionary semantics for disjunction.

First, I contrast the opposed modal views—call them the "box-diamond" theory and EU theory—that form the two poles of the debate about these natural language modals. The opposition between them is underwritten by distinct, well-developed conceptions of what it is for an action to be permissible. I present an axiomatization of obligation and permissibility—of 'ought' and 'may'—that is *neutral* between EU and box-diamond theories: both assign the same truth-conditions when applied to prejacents that describe actions which are *basic* in the relevant model, but different truth-conditions when applied to prejacents that are multiply realizable in the model.

Adding in the interpretation of 'or' as Boolean union—that is, as the relevant kind of multiple realizability—we get the received dialectic in the literature between the two theories on explaining (Ross) and (FC-). Factoring out this assumption, we get a picture of how far apart the two theories are as theories of value, with no questions begged about disjunction. In the rest of the paper, I take up this position to lay out a theory of 'or' in a 2-dimensional semantic framework that both camps should be able to agree to; these are conditions under which a semantics for 'or' *could* block disjunction introduction by the lights of

---

[1] See, for example, [18,2,7], and von Wright himself [23].

either side, since it is framed from a neutral standpoint. Finally, I respond to an argument from [2] that raises a challenge for revisionary theories of disjunction.

2   The Standard Modal Theory versus Disjunction

Let us begin with a standard modal approach to deontic operators. To simplify our demands on the model—to avoid making stipulations, in particular, about accessibility relations—I consider a language without iterated modalities.

Definition 1 (Well-Formed Formulas and Models)
Let prop *be a set of propositional well-formed formulas such that, if* $\phi, \psi \in$ prop, *so is* $\ulcorner \phi$ *or* $\psi \urcorner$. *Any* $\phi \in$ prop *is a well-formed formula. For any* $\phi \in$ prop, $\ulcorner Ought(\phi) \urcorner$, $\ulcorner May(\phi) \urcorner$ *are also well-formed formulas.*

A model $\mathcal{M}$ *is a tuple* $\langle W, OPT, V \rangle$, *where* W *is a nonempty set of possible worlds,* $OPT : W \to \mathcal{P}(W)$ *is a function from a given world* w *to a sets of worlds (worlds 'deontically ideal' from the point of view of* w), *and* $V : $ wff $\times W \to \{0, 1\}$ *is a recursive valuation function on well-formed formulas in* prop.

It will also be convenient to speak of the intension $I(\phi)$ of a sentence ϕ, defined in terms of the extensional valuation function, V:

Definition 2 (Intensions)
$I(\phi) = \{w' : V_{\mathcal{M}}(\phi, w') = 1\}$.

The standard deontic modals quantify existentially and universally, respectively, over worlds in $OPT(w)$.

Definition 3 (Quantificational Modals)
$\mathcal{M}, w \vDash May(\phi)$ *iff* $\exists w' \in OPT(w): w' \in I(\phi)$.
$\mathcal{M}, w \vDash Ought(\phi)$ *iff* $\forall w' \in OPT(w): w' \in I(\phi)$.

What counts as deontically ideal relative to a world w—which worlds are in $OPT(w)$—may be context-sensitive; this sensitivity may include, but perhaps not be limited to, what is *known* at w.[2]

It is a result of the semantic entries for the quantificational modals that they are *Upward Closed*:

---

[2] The premise semantics of Angelika Kratzer [12] is a generalization of this theory, according to which a set of premises determines what counts as good, where these premises may be inconsistent. The result is that worlds may be *ordered* by context, according to how many premises they satisfy. Modulo the Limit Assumption [15], it will still be the case that 'ought' is a univeral quantifier, and 'may' is an existential quantifier, over a modal base, which can be characterized as follows: any world in the modal base satisfies more premises than any world outside the modal base. For Kratzer's discussion of the Limit Assumption, see [12], §3.

(Consequence) $\phi \vDash \psi$ iff, for any model $\mathcal{M}$ and any $w \in W_{\mathcal{M}}$,
                                 if $\mathcal{M}, w \vDash \phi$, then $\mathcal{M}, w \vDash \psi$.

(Upward Closure (UC))   An operator O is upward-closed just in case,
                                 if $\phi \vDash \psi$, then $O(\phi) \vDash O(\psi)$.

((UC) for Deontic Modals)   If $\phi \vDash \psi$, then $Ought(\phi) \vDash Ought(\psi)$ and
                                 $May(\phi) \vDash May(\psi)$.

This result is discomfited by (FC-) and (Ross), where embedded 'or' introduction seems to be blocked. For on a Boolean 'or', $\phi \vDash (\phi$ or $\psi)$. In terms
of I, the intension $I(\phi$ or $\psi)$ of a disjunction, relative to any world w, is just
$I(\phi) \cup I(\psi)$; it is a *multiply realizable* outcome that obtains in any ϕ-world and in
any ψ-world. To the extent to which (Ross) and (FC-) strike us as problematic
inferences, whatever is wrong with them must be explained in the pragmatics,
rather than in the semantics. At first blush, this doesn't look hard to do for
'ought' (see, for example, [22], [8]). [13] take a similar approach to the prima
facie the more complex case of 'may': disjunction introduction under 'may' gives
rise via a *second-order* implicature to the effect that both are permissible. My
purpose here is not to weigh in on these projects, but simply to point out that
such views must appeal to pragmatic resources to explain the failure of embedded
disjunction introduction, given this consequence in the semantics.

3   EU to the Rescue?

A much different reaction to the data in (FC-) and (Ross) is to use it to overturn
the standard modal operator semantics, and to give new entries for 'ought' and
'may' that respects these inferences as semantic.

This route models 'ought' and 'may' as reflecting the notions of obligatoriness
and permissibility that are found in Expected Utility Theory. Expected Utility
Theory enjoins an agent perform the act with the highest expected utility, or
one of these options, when there are ties.

Definition 4 (EU Models and Expected Utility)
*An EU-model[3] $\mathcal{M}$ is a tuple $\langle W, Pr, Val, Act, V \rangle$ such that W is a nonempty set
of possible worlds and Pr is a probability function on $\mathcal{P}(W)$; for any $w \in W_{\mathcal{M}}$,
Val_w is a function $\mathcal{P}(W) \to \mathbb{N}$ which, at a world w, takes a proposition p to
a natural number (the utility of p, relative to w);[4] $Act \subseteq \mathcal{P}(W)$ is a set of
available acts (closed under union), and V is a valuation function on well-formed
formulas.*

---

[3] There are many expected utility models in the literature; the simplified one I present
  here most closely follows [9].

[4] It is point familiar from decision theory that an individual's preferences should be
  modeled by a *family* of such functions, unique only up to positive affine transformation [16]. I abstract from this detail here.

Where $I(\phi) \in Act_{\mathcal{M}}$, $EU_w(I(\phi))$ is the expected utility of $I(\phi)$: $\sum(Pr(w_j|I(\phi)) \cdot Val_w(w_j)$ for all $w_j \in W_{\mathcal{M}}$.

The expected utility of ϕ is maximal relative to an EU-model $\mathcal{M}$ and world $w \in W_{\mathcal{M}}$ iff $I(\phi) \in Act_{\mathcal{M}}$ and $\neg \exists q \in Act_{\mathcal{M}} : EU_w(q) > EU_w(I(\phi))$.

Definition 5 (EU Modality)
$\mathcal{M}, w \vDash May(\phi)$ *iff* $I(\phi) \in Act_{\mathcal{M}}$ *and* $EU_w(I(\phi))$ *is maximal.*
$\mathcal{M}, w \vDash Ought(\phi)$ *is true iff* $I(\phi) \in Act_{\mathcal{M}}$ *and* $I(\phi) = \bigcup\{p \in Act_{\mathcal{M}} : EU_w(p)$ *is maximal*$\}$.

EU modals have a swift take on the negative data in (Ross) and (FC-): the problematic inferences are not semantically valid. Whereas the quantificational modals are upward-closed, the EU notion of permissibility is *downward closed*: if $\phi \vDash \psi$, then $May(\psi) \vDash May(\phi)$. Since the expected utility of a multiply realizable option p is the (probability-weighted) average of its realizations, p's *EU* will be maximal only if the *EU* of all its realizations is also maximal.[5] Interpreting Boolean 'or' as multiple realizability, we get the result that, for example, if it is EU-permissible to have coffee or tea, then both the coffee option and the tea option must be EU-permissible.

$$\text{if } \phi \vDash (\phi \text{ or } \psi), \text{ then } May(\phi \text{ or } \psi) \vDash May(\phi)$$

Because EU permissibility is downward entailing, and EU optimality entails the EU permissibility of any option, Boolean disjunction introduction is also blocked in the scope of EU-'Ought.' From the EU point of view, given Boolean disjunction, we get (Ross), (FC-) and the positive datum (FC) all in one go.

4   Does Natural Language Semantics Reflect EU Permissibility?

The ease with which the EU modals account for the puzzles of disjunction under modals raises a natural question: has anyone ever embraced these views? To my knowledge, no one has embraced both EU modals as a package, but they have appeared individually in the literature as a response to our puzzles.

(EU-'May') is EU permissibility imported directly into the object language: if a proposition p is permissible and multiply realizable in context, then every realization, or every *way*, of doing p must be permissible. Such a notion of permissibility—*strong permissibility*—was proposed by [23], who, in turn, was originally motivated by the Free Choice puzzle.[6] von Wright argued that sometimes, what it means to say "you may ϕ" to someone is to give him or her permission to ϕ "in every way." In this vein, EU theory can be seen as an extensive exploration and formal development of von Wright's notion of strong

---

[5] I ignore zero-probability options here.
[6] See, for example, [23, pg. 26].

permissibility—the notion of permissibility which, to von Wright's ear, was simply manifest in (some) natural language uses of 'may.' Von Wright did not, however, have anything like this to say about 'ought.'

For hints of an inclination towards (EU-'Ought'), we can look to [9], [14] and [2]. Lassiter notes simply that $EU(\phi) \geq \theta$ does not imply $EU(\phi \vee \psi) \geq \theta$, where θ is some threshold for expected utility (26). Cariani's semantics for 'Ought(ϕ)' requires that $I(\phi)$ be an option in context and that every atomic act in $I(\phi)$ be above some 'benchmark' of permissibility that is accessible in the metalanguage. This is enough to block disjunction introduction in the scope of 'ought,' on roughly the same grounds as a more straightforward EU semantics would: the failure of the inference is explained by way of holding that the introduced disjunct is not (strongly) permissible. The main difference is whether this type of permissibility *is* (von Wright) or *isn't* (Cariani) identified explicitly with same brand of permissibility that provides the semantics for the object-language 'May.'

Is it true that there is a downward-entailing notion of permissibility that is active in the semantics of our deontic talk? The theory has some drawbacks, which I'll canvas here, first for an object-language theory of 'ought' (for Goble, Lassiter, and Cariani) and then for an object-language theory of 'May' (von Wright).

4.1 'Ought' as Requiring Strong EU-Permissibility

If the truth of $\ulcorner Ought(\phi) \urcorner$ at a model requires the EU-permissibility if $I(\phi)$, ϕ cannot be the prejacent of a true "ought" claim unless every more fine-grained act which is a way of carrying out ϕ is EU-permissible. Disjunctive cases aside, is this claim plausible?

The first drawback concerns an analogy with decision-making—it doesn't seem like we use a principle like this in deciding what to do. But this raises doubts about whether it could really be a hidden feature of what we *ought* to do. Call this problem (Means-Ends); we do not limit ourselves to actions such that *every* way of carrying them out is permissible.

To illustrate this, consider the case of

(PROFESSOR PUNCTUAL.) Professor Punctual is invited to review a book on whose subject matter he is the world's foremost expert. If Punctual accepts the invitation and writes the review, the book will receive a high-quality assessment—this is the best possible outcome. If Punctual accepts and does not write, the delay will constitute an injustice to the author and an embarrassment for the journal. If Punctual declines the invitation, another, less-qualified person will write a mediocre review. Finally, Professor Punctual is dutiful. He indefatigably fulfills his commitments in a timely manner.

It seems perfectly normal for Professor Punctual to accept, and overwhelmingly natural to say that he *ought* to accept. However, there is a salient way of accepting the invitation to write the review that would bring about the worst possible outcome (this is obviously a feature Punctual's case shares with the case of his better-known co-author, Professor Procrastinate [10].) If Strong Permissibility is really a necessary condition on the truth of "ought" claims, "Professor Punctual ought to accept" is false. But this doesn't seem right; it doesn't seem to be a necessary condition on the truth of $\ulcorner Ought(\phi) \urcorner$ that *every* way of ϕ-ing is permissible.

Is it possible that, because of Punctual's punctuality in w, the option of accepting and failing to write is not represented in the model's set of acts at w? Consider dialogues with *fronted* alternatives:

(6)   a.   May I bring some wine to the party?

       b.   No—the host is allergic. But you ought to bring something.

On a straightforward application of (EU-'Ought'), this dialogue is inconsistent. In (6), the possibility of bringing wine to the party is explicitly raised and classified as impermissible. But then it seems that it cannot be true that *every visible way* of bringing something to the party is above benchmark. Yet by (6-b), "You ought to bring something to the party" is true.

4.2   Strong 'May'

The problems for (EU-'May') mirror the problems for the strong permission theory of 'Ought.' It seems we can construct Professor cases in which it is true that

(7)   Punctual may accept.

but it is false that

(8)   Punctual may accept and fail to write.

So it seems, as much as in the 'ought' case, that the requirement *that every way of accepting be permissible* is too strong. Even when we temper this claim with the proviso that it is only the *represented* or *salient* ways of accepting that must be permissible, we can generate cases with fronted alternatives:

(9)   a.   May I bring some wine to the party?

       b.   No—the host is allergic. But you may bring *something*.

Deliberatively, as well, (Means-Ends) resurfaces for the 'may' case: it is implausible that we take this piecemeal approach to action, at each earlier moment minimizing the harm we can do at some later moment: rather, we often undertake actions which will make things go much worse, if we fail to follow through. Since this is a pervasive feature of the kinds of actions we *do* undertake, it is hard to believe that the model for our deontic talk would tell us that we *may* not do such things.

My interest, in the rest of this paper, is in isolating an argument for blocking embedded disjunction introduction that doesn't rely on 'Ought' and 'May' being downward entailing—in fact, is compatible with their being *upward*-entailing. The EU theorist has a shorter way home, of course. But if I can do this, I can offer someone tempted by the EU modals a way to get the data without having to bite the bullets in (Punctual), (Fronted Alternatives), and (Means-Ends). (FC-) and (Ross) can, perhaps, be had for less.

5 Another Route

The first thing to do is to isolate what the two competing theories of 'Ought' and 'May' have in common. Both begin with the notion of a fine-grained possibility: fine-grained, that is, with respect to their relevant models. For modal logic, a fine-grained possibility is a *possible world*; for EU theory, this is an atomic act in the set *Act*. Some of these possibilities are good, according to the model, and some are not; call the good ones *P*-states ('*P*' for *permissible*.) The two theories are different in how they interpret the normative status of multiply realizable possibilities—how they interpret the information that one's action will place one within a *set* of fine-grained options, some which are *P*-states and some of which are not-P ($\overline{P}$) states.

Definition 6 (*P*-States in EU Theory and Deontic Logic)
EU Theory.
(Base Case). Any P_w-state p is such that $EU_w(p)$ is maximal in $\mathcal{M}$.
(Recursive Clause). Any union of P_w-states and $\bar{P}_w$-states is an $\bar{P}_w$-state.

Classic Deontic Logic.
(Base Case). Any P_w-state p is a subset of $OPT(w)$ in $\mathcal{M}$.
(Recursive Clause). Any union of P_w-states and $\bar{P}_w$-states is a P_w-state.

Visually, under union, an *EU* theory sees the $\overline{P}$ status as infective: it takes any multiply realizable option to $\overline{P}$, since averaging maximal and non-maximal expected utilities will always result in a lower-than-maximal expected utility.[7] The modal theory is more forgiving: it interprets the P status as *modal compatibility* with the best outcome(s), and if a proposition p is modally compatible with the best outcome(s), then so is any superset of p.

$$
\begin{array}{c|c}
p & q \\
\hline
P & \overline{P}
\end{array}
\quad\Rightarrow\quad
\begin{array}{c}
p \cup q \\
\hline
\overline{P}
\end{array}
$$

EU Theory

$$
\begin{array}{c|c}
p & q \\
\hline
P & \overline{P}
\end{array}
\quad\Rightarrow\quad
\begin{array}{c}
p \cup q \\
\hline
P
\end{array}
$$

Modal Theory

---

[7] Once again, I ignore the case of zero-probability propositions.

From this perspective, both semantic theories endorse the following semantics for 'Ought' and 'May':

Observation 1 (Common Core). *For any EU model or modal model $\mathcal{M}$ and any $w \in W_{\mathcal{M}}$,*
$\mathcal{M}, w \vDash May(\phi)$ iff $I(\phi)$ is a P_w-state in $\mathcal{M}$.
$\mathcal{M}, w \vDash Ought(\phi)$ iff (i) $I(\phi)$ is a P_w-state in $\mathcal{M}$, and (ii) no proposition p disjoint from $I(\phi)$ is such that p is a P_w-state in $\mathcal{M}$.

This factorization of the views is convenient, because it divides them into a shared starting point (the basic notion of a P-state and its relation to the object language) and two nonequivalent notions of how the status of being a P-state propagates up under propositional union. According to the Boolean 'or', disjunction *just is* propositional union. So if we add Boolean 'or' to this picture, we get the received dialectic: (Punctual), (Fronted Alternatives) and (Means-Ends) on one side of a sharp divide, and (FC-) and (Ross) on the other—such that we cannot interpret both sets of inferences in terms of semantic consequence.

6   Disjunction in 2 Dimensions

Let us (i) keep the common core of 'ought' and 'may' axiomatized according to P-states, and (ii) reject the Boolean idea that

$$\phi \vDash (\phi \text{ or } \psi).$$

There are many frameworks which reject unrestricted disjunction introduction (for example, linear logic and relevance logic). What I propose to explore here, though, is fleshing out (ii) by going to a 2-dimensional semantics, as in [11,5].

According to a 2-dimensional semantics, the interpretation function V on sentences ϕ in the language must be evaluated relative to *two* worlds in $W_{\mathcal{M}}$, a world-as-actual (call this 'y') and an evaluation world (call this 'x'). So instead of

$$V_{\mathcal{M}}(\phi, w) \in \{0, 1\}$$

we have

$$V_{\mathcal{M}}(\phi, x, y) \in \{0, 1\}$$

Now, the intension of a sentence ϕ is once again a set of worlds, but this set must be relativized to y, the world-as-actual; instead of

$$I(\phi) = \{w' : V_{\mathcal{M}}(\phi, w') = 1\}$$

we have

$$I(\phi, y) = \{w' : V_{\mathcal{M}}(\phi, w', y) = 1\}$$

for a well-formed formula ϕ and $x, y, w \in W$.[8]

The relativity of $I(\phi)$ to a world $y \in W$ allows us to model the idea that ϕ might express different intensions at different possible worlds. For example, if, in w_1, Alice called her mother, but in w_2, Alice forgot to call her mother, we might like to say that "It ought to be that Otto does what Alice actually did" is true in w_1 and false in w_2, in virtue of the fact that "Otto does what Alice actually did" expresses a different intension in w_1 than it does in w_2. Intuitively, w_1 and w_2 differ, not in respect of what is morally required at each, but in virtue of what is expressed by "what Alice actually did" in each.

The simplest upgrade of our deontic modals to a two-dimensional system will reflect the sensitivity of intensions to the world-as-actual.

Definition 7 (2D Modals)
$\mathcal{M}, x, y \vDash May(\phi)$ iff $\{w' : V(\phi, w', y) = 1\}$ is a P_x-state.
$\mathcal{M}, x, y \vDash Ought(\phi)$ iff (i) $\{w' : V(\phi, w', y) = 1\}$ is a P_x-state, and (ii) no proposition p disjoint from $\{w' : V(\phi, w', y) = 1\}$ is such that p is a P_x-state.

With these new points of evaluation, we distinguish two relevant notions of consequence, which I will call *diagonal* ($\vDash_D$) and *unrestricted* ($\vDash$), respectively:

Definition 8 (Notions of Consequence)
For any well-formed formulas ϕ and ψ:
$\phi \vDash_D \psi$ *iff for all $w \in W_\mathcal{M}$, if $\mathcal{M}, w, w \vDash \phi$, then $\mathcal{M}, w, w \vDash \psi$.*
$\phi \vDash \psi$ *iff for all $x, y \in W_\mathcal{M}$, if $\mathcal{M}, x, y \vDash \phi$, then $\mathcal{M}, x, y \vDash \psi$.*

Following a common strain in 2D semantics, let us assume that it is *diagonal* that most closely approximates intuitive consequence relations between natural language sentences.[9]

With all this on board, the non-Boolean 'or' we need, I suggest, is just an 'or' such that Disjunction Introduction is valid at diagonal points, but not at nondiagonal points.

Proposal 1 (A Non-Boolean 'or') A Non-Boolean 'or'
$\phi \vDash_D (\phi$ or $\psi)$, *but* $\phi \nvDash (\phi$ or $\psi)$.

6.1   Putting It All Together

I claimed above that a non-Boolean semantics for 'or' could offer an explanation of (FC-) and (Ross) that both theories of $\ulcorner Ought(\phi) \urcorner$ and $\ulcorner May(\phi) \urcorner$ could accept. The relevant feature of both theories is that, in 2 dimensions, each requires

---

[8] Going forward, I implicitly retain the idea that the intension of ϕ relative to a point of evaluation in the model is both (i) a set of possible worlds and (ii) the only notion of compositional semantic value that embeds under deontic modals. This contrasts with an *inquisitive semantics* approach to Free Choice Permission and Ross's Paradox in the vein of [1,3,4], and [19].

[9] See, for example, the corresponding notion of validity in [11, pg. 547], and the notion of *real world validity* in [5].

the semantic value of the embedded formula ϕ to be evaluated at nondiagonal points, but only requires the modalized sentences $\ulcorner Ought(\phi)\urcorner$ and $\ulcorner May(\phi)\urcorner$ to be evaluated at diagonal points.

$\phi \vDash_D (\phi$ or $\psi)$ iff, for any $\mathcal{M}$ and $w \in W_{\mathcal{M}}$,
if $V_{\mathcal{M}}(\phi, w, w) = 1$, then $V(\ulcorner \phi$ or $\psi \urcorner, w, w) = 1$.

$May(\phi) \vDash_D May(\phi$ or $\psi)$ iff, for any $\mathcal{M}$ and $w \in W_{\mathcal{M}}$,
if $\{w' : V_{\mathcal{M}}(\phi, w, w') = 1\}$ is a P_w-state in $\mathcal{M}$, then $\{w' : V_{\mathcal{M}}(\ulcorner \phi$ or $\psi \urcorner, w, w') = 1\}$ is a P_w-state in $\mathcal{M}$.

On the non-Boolean 'or', the inference from $\phi \vDash_D (\phi$ or $\psi)$ to $May(\phi) \vDash_D May(\phi$ or $\psi)$ fails, since $V(\phi, w, w') = 1$ does not entail $V(\ulcorner \phi$ or $\psi \urcorner, w, w') = 1$. The inference to $Ought(\phi) \vDash_D Ought(\phi$ or $\psi)$ fails for the same reason. The 2D deontic modals are upward closed—but only when we consider prejacents ϕ and ψ such that ψ is a *general* consequence, and not merely a *diagonal* consequence, of ϕ. We can preserve upward closure and still block disjunction introduction; we just have to flesh out a 2-dimensional non-Boolean 'or,' which coincides with Boolean 'or' at diagonal points, but departs from it off the diagonal.

6.2   A Comparison: "I am Here Now"

What would it look like to have a logic in which $(\phi$ or $\psi)$ is a diagonal, but not an unrestricted, consequence of ϕ? Disjunction introduction will pattern with cases in which it is valid to introduce a disjunct *outside* the scope of an upward-entailing intensional operator O, but not *inside* its scope. The status of disjunction introduction—the inference from ϕ to $(\phi$ or $\psi)$—will be an *a priori contingent* inference, in the sense of [6]. It is like one's knowledge of the truth of the sentence

(10)   I am here now.
        IHN

Since (10) is true at all diagonal points, conjoining it with any sentence will preserve truth at a diagonal point. We might call an inference rule that reflects this fact '$\wedge IHN$'-Introduction: from any ϕ, conclude $(\phi \wedge IHN)$.

'$\wedge IHN$' Introduction	$\dfrac{\phi}{\phi \wedge IHN}$	$\dfrac{O(\phi)}{O(\phi \wedge IHN)}$
	valid	invalid

For example, if $2+2 = 4$, then $2+2=4$ and I am here now; but from the fact that it is (metaphysically) necessary that $2+2=4$, it does not follow that it is (metaphysically) necessary that $(2+2=4$ and I am here now), since it is not metaphysically necessary that I am here now.

Disjunction introduction—an 'or ψ' rule—works the same way:

'or ψ' Introduction	$\dfrac{\phi}{\phi \text{ or } \psi}$	$\dfrac{O(\phi)}{O(\phi \text{ or } \psi)}$
	valid	invalid

For example, if I am mailing the letter, it follows that I am mailing it or burning it; but from the fact that I *ought* to mail the letter, it does not follow that I ought to mail it or burn it. This is the perspective we can begin to get from the semantics of the non-Boolean 'or.'

7   Coda: Is "Blaming Disjunction" Too General?

In this paper, I've given an overview of the debate over disjunction within the scope of deontic modals, and sketched the ground for a semantic explanation of the data which jettisons the Boolean 'or.' I've merely laid a groundwork, of course, for I still haven't even begun to offer an explanation of (FC)—the positive inference for which the failure of in-scope disjunction introduction is merely the negative half. However, what we've done already accomplishes something: it is compatible with upward closure for the modals, and it begins to explain how it is that disjunction introduction might be *unimpeachable*, but also *unembeddable*.

In closing, I'd like to consider an objection, advanced by [2], to my approach to (FC-) and (Ross) via disjunction (an approach Cariani calls a "BD" approach, for "blame disjunction.") Cariani's claim bears direct quotation: BD accounts are too general, because they

> do not predict that deontic modals and epistemic modals should give rise to disanalogous predictions. In fact they naturally predict the opposite— that an epistemic 'must' taking scope over a disjunction should pattern in the relevant respects with a deontic 'ought' in the same position. (21)

It would be bad, I think, if this outcome were predicted by the approach I just sketched. But it isn't predicted, as should by now be clear. Epistemic modals, whatever their precise semantics is, should generate a logic in which sentences true at all diagonal points—the *a priori truths*—are axioms. This is just to say that, for example,

(11) $\Box_e$(I am here now)

should be true at any diagonal point (with '$\Box_e$' marking that the relevant necessity is epistemic) just as its unembedded prejacent should be.

It is a point familiar from Kaplan's own remarks that we can capture what is distinctive about a priori truths by looking at what is true at every diagonal point (see, for example, [11, pg. 509].) The most natural way of marking these a priori truths in the object language is with epistemic necessity operators, and indeed a 'monstrous' approach to them—where one quantifies over diagonal points, rather than points that are constant in one of the two dimensions—has been proposed as the distinctive feature of epistemic operators by [17,21], and others. It is epistemically necessary that I am here now, but it is not deontically necessary; it could well be permissible for me to be elsewhere. That is just the pattern we recapitulate on our nascent semantics for 'or': 'or'-introduction is predicted to be valid in the scope of upward-entailing epistemic operators—including, of course, epistemic 'must'—but not valid in the scope of our deontic operators, 'ought' and 'may.'

References

1. Aher, M.: Free choice in deontic inquisitive semantics (DIS). In: Aloni, M., Kimmelman, V., Roelofsen, F., Sassoon, G.W., Schulz, K., Westera, M. (eds.) Amsterdam Colloquium 2011. LNCS, vol. 7218, pp. 22–31. Springer, Heidelberg (2012)
2. Cariani, F.: 'ought' and resolution semantics. Noûs, 1–28 (2011)
3. Ciardelli, I., Aloni, M.: A logical account of free choice imperatives. In: Aloni, M., Franke, M., Roelofsen, F. (eds.) The Dynamic, Inquisitive, and Visionary Life of ϕ, $?\phi$, and $\Diamond\phi$: a festschrift for Jeroen Groenendijk, Martin Stokjof, and Frank Veltman (2012)
4. Ciardelli, I., Groenendijk, J., Roelofsen, F.: Inquisitive semantics: a new notion of meaning. Language and Linguistics Compass 7(9), 459–476 (2013)
5. Davies, M., Humberstone, L.: Two notions of necessity. Philosophical Studies 38(1), 1–30 (1980)
6. Evans, G.: Reference and contingency. The Monist 62, 161–189 (1977)
7. von Fintel, K.: The best we can (expect to) get?: Challenges to the classic semantics for deontic modals (February 2012), Central APA session on Deontic Modals
8. Follesdal, D., Hilpinen, R.: Deontic logic: An introduction. In: Hilpinen, R. (ed.) Deontic Logic: Introductory and Systematic Readings. Dordrecht Reidel (1971)
9. Goble, L.: Utilitarian deontic logic. Philosophical Studies 82(3), 317–357 (1996)
10. Jackson, F., Pargetter, R.: Oughts, options and actualism. The Philosophical Review 95(2), 233–255 (1986)
11. Kaplan, D.: Demonstratives. In: Almog, J., Perry, J., Wettstein, H. (eds.) Themes from Kaplan. Oxford University Press (1989)
12. Kratzer, A.: The notional category of modality. In: Words, Worlds, and Context. de Gruyter (1981)
13. Kratzer, A., Shimoyama, J.: Indeterminate pronouns: The view from japanese. In: Proceedings of the Third Tokyo Conference on Psycholinguistics (2002)
14. Lassiter, D.: Nouwen's puzzle and a scalar semantics for obligations, needs, and desires. In: Proceedings of SALT 21, pp. 694–711 (2011)
15. Lewis, D.: Counterfactuals. Blackwell, Oxford (1973)
16. von Neumann, J., Morgenstern, O.: Theory of Games and Economic Behavior. Princeton University Press (1944)
17. Perry, J., Israel, D.: Where monsters dwell. Logic, Language and Computation 1, 303–316 (1996)
18. Portner, P.: Permission and choice. Georgetown University (2010) (manuscript)
19. Roelofsen, F.: Algebraic foundations for the semantic treatment of inquisitive content. Synthese 190(1), 79–102 (2013)
20. Ross, A.: Imperatives and logic. Theoria 7(1), 53–71 (1941)
21. Weatherson, B.: Indicative and subjunctive conditionals. The Philosophical Quarterly 51(203), 200–216 (2001)
22. Wedgwood, R.: The meaning of "ought". Oxford Studies in Metaethics 1, 12–60 (2006)
23. von Wright, G.H.: An Essay on Deontic Logic and the General Theory of Action. North Holland, Amsterdam (1969)

Toward a Linguistic Interpretation of Deontic Paradoxes

Beth-Reichenbach Semantics Approach for a New Analysis of the Miners Scenario

Dov Gabbay[1,2], Livio Robaldo[3], Xin Sun[2],
Leendert van der Torre[2], and Zohreh Baniasadi[2]

[1] Department of Computer Science, King's College London
[2] Faculty of Science, Technology and Communication, University of Luxembourg
[3] Department of Computer Science, University of Turin
dov.gabbay@kcl.ac.uk, robaldo@di.unito.it,
{xin.sun,leon.vandertorre}@uni.lu,
zohreh.baniasadi.001@student.uni.lu

Abstract. A linguistic analysis of deontic paradoxes can be used to further develop deontic logic. In this paper we provide a Beth-Reichenbach semantics to analyze deontic paradoxes, and we illustrate it on the single agent decision problem of the miners scenario. We also introduce extensions with reactive arrows and actions, which can be used to give a linguistic interpretation of multi-agent dialogues.

1 Introduction

Consider the following discussion by Condoravdi and van der Torre [6].

"**Example (Linguistic interpretation of Chisholm's paradox)** *The most notorious story from the deontic logic literature is known as Chisholm's paradox:*
1. *a certain man ought to go to the assistance of his neighbours,*
2. *if he goes, he ought to tell them he is coming,*
3. *if he does not go, he ought not to tell them he is coming,*
4. *he does not go.*
Analyses of the three conditional obligations have led to preference-based deontic logic, temporal deontic logic, action deontic logic, non-monotonic deontic logic, and more. A more general linguistic analysis would also question the fourth sentence: what does it mean that the man does not go? Does it mean that he cannot go, that he intends not to go, or that he did not go? Taking into account the temporal perspective of the fourth premise and, more generally, the context in which the reasoning takes place constitutes new challenges for the logical analysis of the paradox." [6]

The example of Condoravdi and van der Torre suggests that a linguistic analysis of deontic paradoxes can be used to further develop the logic of obligations and permissions. In this paper, we take up their challenge, and we start the development of a *semantics*

F. Cariani et al. (Eds.): DEON 2014, LNAI 8554, pp. 108–123, 2014.

for such a linguistic interpretation. We use the following fragment of the miners scenario[1] to motivate, develop and validate our approach. As in the analysis of Chisholm's paradox above, we consider not only the obligations, but also the temporal perspective of the factual premise and, more generally, the context in which the reasoning takes place.

Example 1 (Miners scenario, single agent decision problem). The miners scenario is introduced by Kolodny and MacFarlane [12]. Ten miners are trapped either in shaft A or in shaft B, but we do not know which one. Water threatens to flood the shafts. We only have enough sandbags to block one shaft but not both. If one shaft is blocked, all of the water will go into the other shaft, killing every miner inside. If we block neither shaft, both will be partially flooded, killing one miner. The decision problem is summarised in the following table:

Action	if miners in A	if miners in B
block shaft A	all saved	all drowned
block shaft B	all drowned	all saved
block neither shaft	one drowned	one drowned

Lacking any information about the miners' exact whereabouts, and without the possibility to obtain such information, it seems acceptable to say that:

(1)
   a. We ought to block neither shaft.
   b. If the miners are in shaft A, we ought to block shaft A.
   c. If the miners are in shaft B, we ought to block shaft B.
   d. Either the miners are in shaft A or they are in shaft B.

However, (1.a-d). seem to entail (2), contradicting (1.a).

(2) Either we ought to block shaft A or we ought to block shaft B.

Various consistent representations of the scenario have been given [12,3,4,16,5]. Moreover, Kolodny and MacFarlane [12] extend the single agent decision problem above to multi-agent dialogues, leading to additional logical developments. While these representations focus on the interpretation of deontic modality and conditionals in the first three sentences, leading to new developments in deontic logic like information sensitivity, decision rules, and dynamic semantics, we use in this paper ideas from intuitionistic logic [15] and reactive Kripke semantics [11] to form a new analysis focussing on the fourth sentence.

Research Question. Which semantics can be used to give a linguistic analysis of paradoxes and use of normative language, and thus to further develop deontic logic?

---

[1] Kolodny and MacFarlane [12] and Willer [16] call it a paradox, while Cariani et al. [3] call it a puzzle. In this paper, we do not consider the question whether it is a paradox, and call it "the miners scenario."

This general objective breaks down into the following three subquestions:

1. How to define a special Beth-Reichenbach semantics capable of analyzing the miners scenario?
2. How to augment the Beth-Reichenbach semantics with reactive arrows and sharpen our analysis of the miners scenario?
3. How to further extend the Beth-Reichenbach semantics to obtain a logic capable of modeling actions in the miners scenario?

Our linguistic analysis questions the disjunction in the fourth sentence. For example, is this information relevant at the current moment or in a reference time later in the narrative of the story? If we read the wording of the miners scenario and the natural flow of events involved in the situation described by the scenario, we have a story about what we know at the beginning (namely, we do not know where the miners are), we have actions we want to take (block the shafts) which intuitively we should not be taking until we know where the miners are, and when we know where the miners are we immediately have the obligation to take the proper action. Our use of the Beth-Reichenbach semantics starts by observing that, on a temporal perspective, disjunctions represent limited information. We *do not know* where the miners are, so we are only able to state a disjunction that "enumerates" the places where they could be: shaft A and shaft B. Therefore, we need a logical account where disjunctions are interpreted in that way, regardless of the actions that the agent will decide to take.

Classical logic does not have components to model the desired semantics at the object level. We need to somehow add to classical logic, at the object level, a component of knowledge, time, and actions in a natural way, where by "natural" we mean a way which mirrors our human perception of the story. Classical logic can describe the above flow of knowledge, time and actions only by acting as a meta-language, but when it is used as a meta-language, it can equally describe the cooking of an omelette. This is not what we mean by a natural logic to represent the miners scenario. We therefore do not move from classical logic to the machinery of the temporal modal action logic [2], as a sort of meta-language to describe the miners scenario [12,3,4]. Instead, we modify the traditional semantics for classical logic by moving to the Beth-Reichenbach semantics.

In this paper we do not introduce a full-fledged deontic logic, as there are various ways to use the Beth-Reichenbach semantics for normative reasoning. For example, we can add a modal operator for obligation to the semantics, to obtain a kind of intuitionistic standard deontic logic, or we can use the intuitionistic logic as the base logic in the input/output logic framework [8]. We leave these developments for further research, and focus in this paper on the linguistic interpretation of the miners scenario.

The paper is structured as follows. In section 2, we present the Beth-Reichenbach semantics for classical logic. In section 3, we present our case study of the miners scenario. In section 4 we introduce reactive arrows and in section 5, we introduce actions. Section 6 formalizes the miners scenario and section 7 compares our representation of the miners scenario with the literature. We conclude in section 8.

2   The Beth-Reichenbach Semantics for Classical Logic

In this paper our starting point is propositional logic. Therefore for brevity we only introduce the propositional fragment of Beth-Reichenbach semantics. Drawing from Reichenbach [14], Beth [1] introduced his semantics in 1956 as a candidate semantics for intuitionistic logic, and it is combined with Kripke semantics to form the Beth-Kripke semantics [9]. It became popular as the semantics of intuitionistic logic. Finite Beth-Reichenbach models comprize semantics for classical logic. The basic idea of finite Beth-Reichenbach models can be described by the following example.

Example 2 (Police scenario). Imagine a police officer collecting evidence on a murder case and preparing a file for the prosecution of a certain suspect for the murder. At any given moment of time the police officer can go in different directions collecting evidence and according to what he finds, different statements can be verified to be true. There are three options for a statement A:

(a) There is enough evidence now to prove A.
(b) It is clear now that no matter how our investigations will proceed, there will not be enough evidence to prove that A is true; therefore for the purpose of prosecution it is acceptable that $\neg A$ is true.
(c) Although there is not enough evidence to establish that A is true, it may be possible in the future that some new evidence will be uncovered that will establish the truth of A. Therefore neither A nor $\neg A$ are established as true now.

The police has a deadline by which time the investigation and the prosecution file has to be prepared, and therefore the model is finite.

   A Beth-Reichenbach model is an overview of the different states of evidence in the investigation. It is a finite ordered set with the relation '$\leq$' such that '$t \leq s$' means that s has more established evidence than t. Thus if at t statement A can be proven true, it could be also be proven true at s. Such a finite Beth-Reichenbach model provides semantics for classical logic, because if we look at the endpoints of the process, namely all the possible files where no further investigation and collection of evidence is performed, we get a classical model. What is true in that final node is what can be proven and what is false in that final node is what cannot be proven (which may be seen as a kind of close world assumption). There are some immediate properties of this mental picture:

(a) The nodes together with $\leq$ relation forms an acyclic order (for example a tree-structure) with finite depth.
(b) If a statement is proven at moment t, then it will remain proven later than t.
(c) To prove a statement of the form "not φ" at moment t, the policeman must be certain that no matter what further investigation is done: φ will *never* be proven. In other words, "$\neg\varphi$" is proven at moment t iff φ is not proven later than t.
(d) A statement of the form "φ or ψ" is proven by the policeman at moment t iff no matter how he stops his investigation, at least one of φ and ψ will have been proven when he stops.

The Beth-Reichenbach semantics for classical logic contains a component of progressive knowledge which is compatible with the progression of knowledge aspects we find in the miners scenario. Thus we use the Beth-Reichenbach semantics to model the miners scenario. But first we have to give formal definitions.

The language we use contains atomic formulas q, negations of atoms $\neg q$, conjunctions and disjunctions. Every well-formed formula is a disjunction of conjunctions of atoms or their negations, and we do not have implication in our language.

Definition 1 (Beth-Reichenbach Semantics for Classical Logic). *Consider the language* L_C *of classical propositional logic with atoms* $Q = \{p, q, \ldots\}$ *and the connectives* $\neg, \wedge, \vee$. *Well-formed formulae (wff) be of the following form* $\bigvee_i \bigwedge_j \pm q_{i,j}$ *where* $q_{i,j}$ *is atomic,* $+q$ *is* q *and* $-q$ *is* $\neg q$. *A Beth-Reichenbach model for* L_C *has the form* (T, R, h), *where* T *is a set of reference points (worlds, information states),* $R \subseteq T \times T$, (T, R) *is a finite tree.* h *is an assignment giving a subset* $h(q) \subseteq T$ *to each atomic* q . *Furthermore, let* $<$ *be the transitive closure of* R, *and* $x \leq y$ *be* $x = y$ *or* $x < y$. *We require the following to hold:*

(a) For each $t \in T$, *say that* t *is an endpoint iff* $\neg \exists_x (t < x)$. *Let* $E_t = \{x \in T \mid t \leq x$ *and* $\neg \exists_{y \in T}(x < y)\}$. *Intuitively* E_t *is the set of all endpoints of* t. *We require that for each* $t \in T$, $E_t \neq \emptyset$. *This means that:* $\forall_t \exists_s (t \leq s$ *and* s *is an endpoint).*
(b) $E_t \subseteq h(q)$ *iff* $t \in h(q)$.

We now define satisfaction of a formula A in a model. We write $t \models_{index} A$, where the *index* gives the type of satisfaction we are defining. In our formalization, there will be two possible values for *index*: bs and br. In BS-semantics, we only need satisfaction on bs, while satisfaction on br will be used in our extended BR-semantics. Satisfaction on bs in t depends on the truth values at the endpoints and the ones assigned by h in t.

Definition 2 (Satisfaction in bs in Beth-Reichenbach model (T, R, h)).

1. $t \models_{bs} q$ *iff* $t \in h(q)$, *where* $t \in T$ *and* q *atomic*
2. $t \models_{bs} \neg A$ *iff for all* t', $t \leq t'$ *implies* $t' \not\models_{bs} A$
3. $t \models_{bs} A \wedge B$ *iff* $t \models_{bs} A$ *and* $t \models_{bs} B$.
4. $t \models_{bs} A \vee B$ *iff* $t' \models_{bs} A$ *or* $t' \models_{bs} B$, *for all* $t' \in E_t$.

Remark 1. The interpretation of the atoms, $\neg$ and $\vee$ can be understood as modal S4 interpretation. The atoms q are understood as $\square q$, and $\neg$ and $\vee$ are understood as $\square \neg$ and $\square \vee$.

We analyze the miners scenario by dropping condition (b) of Definition 1 to obtain our BR-semantics. A BR model is defined as an BS model as in Definition 1, but without condition (b), i.e. there is no connection between $E_t \subseteq h(q)$ and $t \in h(q)$.

Definition 3 (BR-model). *A BR-model has the form of a model of Definition 1 without requirement (b).*

Satisfaction on br is the same as satisfaction on bs, except the condition on negation. To highlight the difference we use a different symbol for negation, namely '$\sim$'. The second interpretation rule in definition 4 implements close world assumption, and evaluate a negative atomic formula without looking at the endpoints. In case h does not assign a positive value to an atom q, then q is asserted as false.

Definition 4 (Satisfaction $\models_{br}$ in Beth-Reichenbach model (T, R, h)).

1. $t \models_{br} q$ *iff* $t \in h(q)$, *where* $t \in T$ *and* q *atomic*
2. $t \models_{br} \sim A$ *iff* $t \not\models_{br} A$.
3. $t \models_{br} A \wedge B$ *iff* $t \models_{br} A$ *and* $t \models_{br} B$.
4. $t \models_{br} A \vee B$ *iff* $t' \models_{br} A$ *or* $t' \models_{br} B$, *for all* $t' \in E_t$.

Remark 2. When we evaluate atoms we do not look at the endpoints. On the other hand, for evaluating disjunction, we look at the endpoints. For example, let $T=\{r, t, s\}$, $r < t$, $r < s$, and $h(Z)=\{t, s\}$, $h(X)=\{t\}$, $h(Y)=\{s\}$. In this example, the reference point "r" does not belong to $h(Z)$. This is to show that even if Z may be true at all endpoints, in *br* it is false r'. On the other hand, in *br* $(Z \vee Z)$ holds at now, according to the interpretation rule of disjunction. This represents a big difference with respect to our logic and classical logic. In our logic, it is not true that $(Z \vee Z) \models_{br} Z$.

Disjunction constitutes a "connection" between the two kinds of satisfactions we are going to use ($\models_{br}$ and $\models_{bs}$). The way formulae are satisfied is different for both atomic formulae and boolean operators, except the disjunction where we look at the endpoints in both kinds of satisfaction.

In BR semantics we read disjunction as modal, namely we read $A \vee B$ as $\Box(A \vee B)$. So our semantics is modal logic in disguise. The advantage is that we are adding to classical logic just enough modal properties to address the miners scenario in a natural way, without having to bring in and commit to a lot of unnecessary modal machinery.

3   A Case Study: The Miners Scenario

Several authors provide consistent representations of the miners scenario. Kolodny and MacFarlane [12] give a detailed discussion of various consistent representations, but they conclude that the only satisfactory representation of the scenario is to invalidate the argument from (1.b-d) to (2) by rejecting modus ponens. Willer [16] argues that there are good reasons to preserve modus ponens and develops another consistent representation by falsifying monotonicity. Charlow [5] proposes a comprehensive representation which requires rethinking the relationship between relevant information (what we know) and practical rankings of possibilities and actions (what to do). Cariani et al. [3] argue that the traditional Kratzer's semantics [13] of deontic conditionals is not capable of representing the scenario satisfactorily. They propose to extend Kratzer's standard account by adding a parameter representing a "decision problem" to solve the scenario. Finally, Carr [4] argues that the proposal of Cariani et al. is still problematic, in that it packs decision theory in the semantics of modals.

We choose an approach different from the above mentioned treatment. In a nutshell, instead of invalidating the argument from (1.b-d) to (2), we address the scenario by making (1.a-d) and (2) compatible. According to our BR-semantics, the problem with the miners scenario is that we have three reference points. See Figure 1. Each reference point represents an information state. At now we do not have information where the miners are. Later, at point l_1, we know in which shaft the miners are, and later still (point l_2) we block the correct shaft. The meaning of "(Miners in A $\vee$ Miners in B)" is

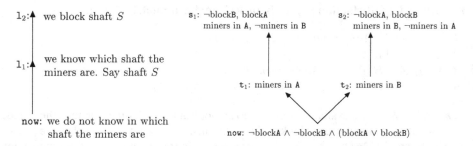

Fig. 1. Reference points in the miners scenario

Fig. 2. Items in our solution to the miners scenario

that no matter how information evolves, either we have that the assertion "Miners in A" holds or that "Miners in B" holds.

In a sense, we are operating in modal logic without bringing an explicit modality into the language, and we read "Either the miners are in shaft A or they are in shaft B" as □ (Either the miners are in shaft A or they are in shaft B.) and we read "we ought to block neither shaft" as ¬□ block A ∧¬□ block B. If we regard the worlds in the Beth-Reichenbach model as possible worlds with a reflexive and transitive accessibility relation, then the semantic condition we gave to formulas of the form $x \wedge y$ is as if we have a necessity operator □ in front of it. We can thus have that $(X \vee Y) \wedge \neg X \wedge \neg Y$ can be consistent:

(3.a) now$\models_{br} (X \vee Y) \wedge \neg X \wedge \neg Y$

By instantiating (3.a), we can get (3). We thus have that at reference point now, "(block A ∨ block B)" is true but "block A" and "block B" are both false.

(3) now$\models_{br}$ (block A ∨ block B) ∧ ¬block A ∧ ¬block B

Moreover, if we add the deontic modal operator '○' to the language, we can instantiate (3.a) differently as in (4) (we will formally define deontic modality below in Section 5.1).

(4) now$\models_{br}$ (○ block A ∨ ○ block B) ∧ ¬ ○ block A ∧ ¬ ○ block B

This means that at now "we ought to block A or we ought to block B" is true but "we ought to block A " is false and "we ought to block B " are false. Therefore (1.a-d) and (2) are compatible. (4) moreover gives the right prediction to the miners scenario: the prediction given by (4) is "not block A" and "not block B" at now, although given more information "we will eventually either block A or block B".

Some readers may think that we can represent the miners scenario in standard deontic logic (SDL), augmented with a K (Knowledge) modality:

1. $\bigcirc(\neg A \wedge \neg B)$
2. $KA \rightarrow \bigcirc A$

3. $KB \to \bigcirc B$
4. $K(A \vee B)$

Now $\bigcirc A \vee \bigcirc B$, which would conflict with 1, is not derivable. This could be a simple and effective solution based on recognizing the epistemic context. However, SDL has its problems and of course we need to give problems free axioms for the $\bigcirc$, K logic.

The perceptive reader might think that the Beth-Reichenbach semantics approach seems to provide a solution by coincidence. This is not the case. Indeed that there is an epistemic reading of the Beth-Reichenbach semantics itself enabling it to give a solution. Given that SDL has its problems we believe that it is better to offer a simple well known semantics , namely the Beth-Reichenbach semantics. The question to ask now is whether we can use the Beth-Reichenbach semantics to solve the difficulties of SDL. This could be a future research, where SDL is based on intuitionistic logic.

4   Reactive Semantics

We now describe the ReBR-semantics. This is an intermediate semantics where we augment BR models with reactive double arrows, as defined in Gabbay [11], but no actions. It is only a temporary step, for reasons of exposition, to lead the reader towards the final ReBRA semantics with actions. The reactive arrows are not needed to solve the miners scenario but it is compatible with it and can expand and solve other problems related to multi-agent and their respective progression of knowledge in time.

So we explain the idea with a diagram. Consider Figure 3. As the agents traverse an arc, if there is a double arrow emanating from the arc to another arc, the double arrow will disconnect the target arc.

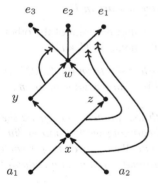
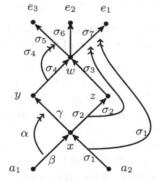

Fig. 3. Double arrow model

Fig. 4. Model for Figure 3 enriched with actions and reactivity

So if an agent passes through the path $a_2 \to x \to z \to w$, passing through the arc $x \to z$, the double arrow $(x \to z) \twoheadrightarrow (w \to e_3)$ gets active and disconnects (blocks) the arc $(w \to e_3)$. Similarly, if the agent moves to node w along the path $a_1 \to x \to y \to w$, then he cannot go to e_1 because by passing through $y \rightarrow w$ there is a double arrow disconnects the arow from w to e_1. We identify an agent with the path of the following form:

$$\Pi = (x_1, x_2, \ldots, x_n), \text{ such that:}$$
$$x_1 \rightarrow x_2 \rightarrow x_3 \rightarrow \ldots \rightarrow x_n$$

x_1 is the starting point of the agent and x_n is the current point of the agent. Mathematically, if $((x_i, x_{i+1}), (u, v))$ is a double arrow, then $u \rightarrow v$ is blocked, in case $x_i \rightarrow x_{i+1}$ is in Π. We say in this case that $u \rightarrow v$ is blocked by Π. In particular, we may have that $(u \rightarrow v) \in \Pi$.

Now, we continue to explain the diagram in Figure 3. e_1, e_2, and e_3 are endpoints, while x, y, and z are reference points without being endpoints. In the BS-semantics for this figure, *atomic* formulae may be true or false at the reference points. Their value is *given* in terms of an assignment h. On the other hand, complex formulae (negation, disjunctions, and conjunctions) are *evaluated* in terms of the truth values of the atoms at the endpoints above them according with the interpretation rules given in Definition 2 for the BS-semantics. In other words, atomic values are always given, while for evaluating complex formulae we must look at the endpoints.

The role of reactivity is in the notion of what it means to be above a point. For an endpoint e to be above a point t we need to have a path from t leading to e such that none of its arcs are blocked by the path itself. For example, the path $y \rightarrow w \rightarrow e_1$ blocks its own arc $w \rightarrow e_1$. If the agent goes to w through y, he can reach the endpoints $\{e_2, e_3\}$, while if he goes from x to w through z he can reach $\{e_1, e_2\}$. This means that passing the path from x to z does not let the agent reach the endpoint e_3.

We define now a legitimate path.

Definition 5 (Legitimate Path). *Given a set of point S and a relation $R \subseteq S \times S$ and a relation $R^* \subseteq R \times R$, a path Π is legitimate if it does not block itself, i.e. if there is not an arc in Π that activates a double arrow in R^* that blocks another arc in Π.*

Two paths can be concatenated if the last point of one is the starting point of the other. We refer to the concatenation of two paths Π and Π' via the notation $\Pi * \Pi'$

Definition 6. *A reactive Beth-Reichenbach model for L_C has the form (T, R, R^*, h), where T is a set of worlds, $R \subseteq T \times T$, $R^* \subseteq R \times R$, and h is an assignment giving a subset $h(q) \subseteq T$ for each atomic q.*

We require the following to hold: Given a path Π with last point t, let E_Π be the set of all endpoints x such that there exists a path Π' beginning with t and ending with x such that the concatenation of Π with Π' is legitimate. We require that for each legitimate Π, $E_\Pi \neq \emptyset$. Note that this means that every legitimate path has an endpoint above it that can be legitimetely reached.

Definition 7. *Satisfaction $\models_{br}$ in reactive Beth-Reichenbach model (T, R, R^*, h), with respect to a legitimate path Π.*

1. *$(\Pi \models_{br} q)$ iff t is in $h(q)$, where q is atomic and t is the last point of Π.*
2. *$(\Pi \models_{br} \neg A)$ iff $(\Pi \not\models_{br} A)$.*
3. *$(\Pi \models_{br} A \wedge B)$ iff $(\Pi \models_{br} A) \wedge (\Pi \models_{br} B)$*
4. *$(\Pi \models_{br} X \vee Y)$ iff in any endpoint $t \in E_\Pi$ we have: either $(t \models_{br} X)$ or $(t \models_{br} Y)$*

5   Action Involved Model

In this section we add action to our semantics. We begin with an explanation and then give a formal definition. Consider Figure 5. In the state a_1 Mary has the laptop. John wants the laptop and in state x John has the laptop. On our previous model there is nothing else to say. We do not know *in what way* John has become the owner of the laptop. In our new model we want to add actions and specify that the move from a_1 to x was the result of an action. Let us list (all) the actions available to John.

- Action α: steal the laptop
- Action β: buy the laptop
- Action γ: buy insurance for the laptop

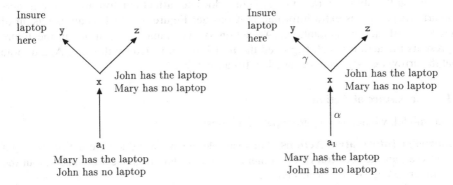

Fig. 5. A model explaining the need of actions **Fig. 6.** Model of Figure 5 enriched with actions

In our new model we need to write annotation as to which action was used in the transitions. Therefore if John stole the laptop then he cannot insure it. So we have Figure 7. But if he bought the laptop then he can insure it. So we have Figure 8.

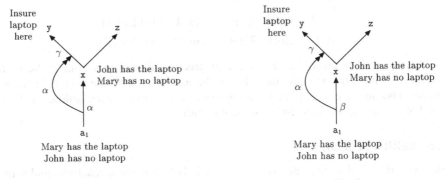

Fig. 7.   **Fig. 8.**

Therefore, in our modelization, reactive arrows define a *dependency* among actions. Action α can block the execution of action γ. This is like a deontic rule on action: $\alpha \rightarrow O(\neg \gamma)$. Indeed, Gabbay [10] uses reactive arrows to give semantics to contrary-to-duty obligations.

In other words, we have an opportunity to use the reactive arrows, in the following sense: if the laptop is stolen it *ought* not to be insured. Therefore we model this deontic/legal rule in Figure 8 using annotated arrows with actions to indicate which action was taken to move across the arrows and we also annotate double arrows by actions to indicate what actions activate the double arrow.

The double arrow $(a_1 \rightarrow_\alpha x) \rightarrowtail_\alpha (x \rightarrow_\gamma y)$ cancels $(x \rightarrow_\gamma y)$, but the double arrow $(a_1 \rightarrow_\beta x) \rightarrowtail_\alpha (x \rightarrow_\gamma y)$ does not cancel $(x \rightarrow_\gamma y)$, because α is different from β (the laptop was bought, not stolen). So for a reactive cancellation to work $(t_1 \rightarrow_\gamma t_2) \rightarrowtail_\varepsilon (s_1 \rightarrow_\eta s_2)$, we must have $\gamma = \varepsilon$.

Note that the above extra action structure does not affect our solution of the miners scenario. It just gives extra information. Consider Figure 4, which is an extension of Figure 3 with action annotations. An action symbol annotating an arc in the figure represents the action which triggered the transition indicated by this arc. Again, note that the arrow $(x \rightarrow y)$ is not cancelled, because $\beta \neq \alpha$,

5.1 The Nature of Action

In our model, we need two pure types of actions:

Knowledge Information Actions. For example, let us introduce the action $\delta = $ "get info about the location of the miners". δ is non-deterministic we can find out the miners are in shaft A or in shaft B.

Facts Actions. For example, let us introduce the action $\varepsilon = $ "kill all the miners". ε also yields information, e.g. that all miners are dead.

In order to model actions, it is necessary to add a definition of how actions are to be executed, and what is the form of actions. We adopt a traditional AI view:

$$\text{Action } \alpha: (\text{precondition, postcondition})$$

Both precondition and postcondition are represented by wff of our language L_C. In the miners scenario we have:

$$\text{Action block-}A: (\text{miners in } A, A \text{ is blocked})$$
$$\text{Action block-}B: (\text{miners in } B, B \text{ is blocked})$$

The two actions are blocked when their preconditions do not hold at the current reference point. For example, most likely the action "kill all the miners" cannot be executed because there could be severe preconditions to execute the action, e.g. finding out that the miners are terrorist hiding in the shaft.

5.2 ReBRA-Semantics

Having introduced all ingredients, namely basic BR-Semantics, reactivity and actions, we are ready now to define our final semantics, which we call ReBRA-Semantics.

Definition 8. *ReBRA-Semantics model.*
Let (S, R), $R \subseteq S \times S$ be a finite network. This is our basic set of states/worlds/reference points and the transition relation R. Let E be the set of endpoints of (S, R), i.e. $e \in E$ iff $\neg \exists_y (eRy)$. Let $<$ be the transitive closure of R and let $x \leq y$ be $x = y$ or $x < y$. Let $E_t = \{set\ of\ all\ endpoints\ s\ such\ that\ t \leq s\ \}$. We also have a stock of actions A of the form $\alpha = (Pre_\alpha, Post_\alpha)$, where Pre_α is a statement being the precondition of α and $Post_\alpha$ is the postcondition. Then:

A model is a system with (S, R, R^*, A, h), where S is the set of states, A is the set of actions, $R \subseteq S \times S$, and $R^* \subseteq A \times R \times R$, and for each atomic q, $h(q) \subseteq S$.

Satisfaction of a model is given with respect to an annotated path:

Definition 9. *Annotated paths.*
Taken S, R, and A of definition 8, let $\Pi = (s_0, s_1, \ldots, s_k)$ to denote a path. We have $s_i R s_{i+1}$, $0 \leq i \leq k$. We also write $s_0 \to s_1 \to \ldots \to s_k$. A path is the history of the agent as he moves around the network from one state to the next. s_0 is the beginning state of the path and s_k is the last state of the path.

An annotated path Π has the form $(s_0 \to_{\alpha_0} s_1 \to_{\alpha_1} \ldots \to_{\alpha_{k-1}} s_k)$. This is a path where each transition $s_i R s_{i+1}$ is labelled by actions. We imagine an agent moving along the path $(s_0 \to s_1 \to \ldots \to s_k)$. Each move from state s_i to state s_{i+1} is done by action α_i. s_0 is where the agent started and s_k is where the agent is currently situated. Arrows annotated by actions are elements of $A \times S \times S$.

We also have annotated reactive double arrows:

Annotated reactive double arrows have the form $(t \to s) \twoheadrightarrow_\alpha (x \to y)$, where t, $s \in S$, tRs, $x,y \in S$, xRy. The annotated double arrows are elements in $A \times R \times R$.

We need now to define satisfaction in a model.

Definition 10. *Satisfaction, Legitimacy, and Coherence with respect to actions. Let Π be an annotated path. Π is a legitimate annotated path iff both (a) and (b) hold.*

(a) *There does not exist two arcs of the form $x \to_\alpha y$ and $u \to_\beta v$ in Π such that $(x \to_\alpha y) \twoheadrightarrow_\alpha (u \to_\beta v)$ is in R^*. I.e. $(\alpha, (\alpha, x, y), (\beta, u, v)) \in R^*$.*
(b) *Whenever $x \to_\alpha y$ is in the path then $\Pi_x \models_{br} A$, where Π_x is the initial path of Π up to node x and A is the precondition of α and $\models_{br}$ is the br-satisfaction defined in the next item 3.*

Two paths can be concatenated if the last point of one is the starting point of the other. We refer to the concatenation of two paths Π and Π' via the notation $\Pi * \Pi'$.

Let Π be a legitimate path. We define E_Π, the set of legitimate endpoints of Π, is defined as follows: a point t is in E_Π iff t is an endpoint and for some Π' such that the first element of Π' is equal to the last element of Π and the last element of Π' is t and $\Pi * \Pi'$ is legitimate.

Finally, let Π be a legitimate path. We define satisfaction $\models_{br}$ as follows:

(a) $(\Pi \models_{br} q)$ iff t is in $h(q)$, where q is atomic and t is the last point of Π.

(b) $(\Pi \models_{br} \neg A)$ iff $(\Pi \not\models_{br} A)$.

(c) $(\Pi \models_{br} A \wedge B)$ iff $(\Pi \models_{br} A) \wedge (\Pi \models_{br} B)$

(d) $(\Pi \models_{br} X \vee Y)$ iff in any endpoint $t \in E_\Pi$ we have: either $(t \models_{br} X)$ or $(t \models_{br} Y)$

The last component we need to complete our formal framework is a mechanism that allows agents to choose the actions they will execute, among those available.

In the present paper, given an action "α: $(prec(\alpha), post(\alpha))$" and a path Π whose last point is t, we say that α can be executed given Π iff $\Pi \models_{br} prec(\alpha)$ holds. In other words, we understand every action as a conditional norm. There we define an action to be obligatory iff the precondition of the action is satisfied. Formally,

Definition 11. *Obligatory actions.*

$$\Pi \models_{br} \bigcirc \alpha \text{ iff } \Pi \models_{br} prec(\alpha)$$

Agents must execute all actions whose precondition holds.

Of course, actions could be selected in other different ways. According to [4], actions should be ranked according to probability, expected values, goals of the agents, etc. The present account only uses agents in deontic mode, i.e. agents are always obligated to perform an action when the preconditions hold, while an extension to ordinary action is left as future work.

Given the definition of obligation in def.11, we deduce the following:

(5) if $\Pi \not\models_{br} prec(\alpha)$, then $\Pi \models_{br} \neg \bigcirc \alpha$.

In the miners scenario, (5) can be used to derive $\neg \bigcirc block\ A$ and $\neg \bigcirc block\ B$ as long as the precondition of $block\ A$ and $block\ B$ are not satisfied in the reference point now.

6 Using ReBRA-Semantics for the Miners Scenario

We now illustrate the use of actions in the context of the miners scenario. We are not going to use reactive arrows for the moment. Figure 2 becomes now Figure 9. Let us summarize figure 9. At now we have true that: "A is not blocked", "B is not blocked", "A is blocked $\vee$ B is blocked". We do not know where the miners are. We take action "get-info" and get two non-deterministic results: "t_1: miners in A" and "t_2: miners in B".

The preconditions of the action "get-info" is $\top$. Therefore, we can get-info at any point. The postcondition of "get-info" is that we know where the miners are and so we reduce the number of endpoints. After the execution of get-info, for instance, we could move to either t_1 or t_2 so that we will see a single endpoint.

We take action "block-A" at t_1. We take action "block-B" at t_2. We get s_1 and s_2 respectively. However, at now, actions "block-A" cannot be executed because we require the precondition of "block-A" to be "miners in A", and similarly for the action "block-B". We therefore accept "not ought to block A" at reference point now by referring the Kantian law "ought implies can". Moreover, we take for granted that at t_1

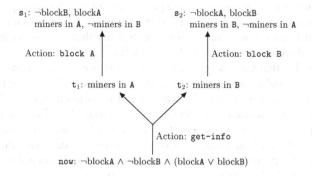

Fig. 9. Items in the miners scenario with actions

"ought to block A" is true and at t_2 "ought to block B" is true. Then by the semantics of disjunction we have "ought to block A or ought to block B" is true at now. That is to say, we have (4) in the miners scenario. As we state at the end of Section 3, (4) is logically consistent in our semantics and gives the right prediction.

7   Related Work

We have proposed a new approach based on reactive semantics that appears to be promising for handling normative multi-agents systems [7] and their respective progression of knowledge in time. We used our new logic in this paper to specifically solve a well-known puzzle that recently gained popularity in the scientific community: the miners scenario. This section highlights the differences between our approach and some recent solutions to that scenario in the literature.

Several authors observe that the paradox arises by applying deduction rules that are commonly assumed to be valid in deontic logic, and they propose a revision of such rules. [12] examine various options, i.e. rejecting either one of the premises or one of the three deduction rules used in the derivation (disjunction introduction, disjunction elimination, or modus ponens). They come to the conclusion that we must reject modus ponens for indicative conditionals. The validity of modus ponens - they argue - must be "information-sensitive", i.e. it must be defined with respect to the knowledge that is at the disposal of the agent at the time of the inference.

Other authors do not accept the solution by [12], focussed on modus ponens, arguing that contextual (pragmatic) preferences ought to be included in the semantics of modal operators. [3] and [5] belong to this school of thought. They modify the semantics of modal operators by including some kinds of decision rules that allow prefererence for one of the available options, so that inconsistency does not arise.

Some have questioned this solution, e.g. Carr and Willer [4], [16]. Carr, in particular, shows that encapsulating pragmatic decision rules within the meaning of modals - a solution that seems at odds with standard literature on modals - makes the formal framework too rigid. It is no longer possible, for instance, to rank the available actions that an agent can perform according to the probability of getting the expected outcome. With respect to the miners scenario, for instance, it is not possible to model a scenario where the agent knows that there is a 99% probability that the miners are in shaft A, and so he could decide to take the risk and block the shaft.

We acknowledge that Carr is on the right track. However, in our view, she has not fully achieved the goal of keeping pragmatic constraints distinct from the semantics. Carr's work started from a question by Krazter and von Fintel: "why pack information about rational decision making into the meaning of modals?". She developed a formal theory where preference rules are asserted in terms of separate (context-dependent) functions that affect the truth conditions of modals and conditionals, but they are not part of the formal representations of their meaning. She states that the decision rules regarding actions only "determine the meanings of modals".

Although this allows for a more expressive and flexible management of pragmatic constraints, modals still need decision rules to be interpreted in a model. In other words, we do not achieve neat independence between semantics and pragmatics if the choice of a certain action is needed to determine the truth values of modals and conditionals.

Our basic BR-semantics is already capable of solving the miners scenario as it adopts a different account of disjunction. Disjunctions, used to express limited knowledge, are interpreted with respect to the endpoints, not the current reference point. In this respect, our approach is more similar to that of [12]. However, rather than rejecting modus ponens, we reject disjunction introduction since in our logic "$A \rightarrow (A \lor B)$" does not hold. We have also shown that our basic BR-semantics may be extended with actions into a new semantics that we call ReBR-semantics, to enable the implementation of all pragmatic preferences and constraints that affect the selection of the proper actions to be taken. A complete and exhaustive formalization of such constraints, however, deserves much further work.

8   Conclusion

This paper presents the following contributions. First, we show how a linguistic interpretation of deontic paradoxes can be used to further develop deontic logic, by introducing a special Beth-Reichenbach semantics and using it to represent the single agent decision problem of the miner's scenario. In further research we will complete this argument, for example by extending the semantics with a deontic modal operator, or by using the language as a base logic in the input/output logic framework.

Second, we give a new analysis of the single agent decision problem of the miners scenario. We bring modal meaning to disjunction which mirror our intuitions and eliminates the cause of the scenario, without bringing in the full machinery of modal or non-classical logic.

Third, we augment the BR-semantics with reactive arrows and actions, obtaining a new semantic that we call ReBR- and ReBRA-semantics. In future work we will illustrate how this extended semantics can be used to represent the multi-agent dialogues of the miners scenario.

References

1. Beth, E.W.: Semantic construction of intuitionistic logic. Kon. Nederlandse Ac. Wetenschappen afd. Letteren: Mededelingen, 357–88 (1956)
2. Broersen, J.M.: Modal Action Logics for Reasoning about Reactive Systems. PhD thesis, Faculteit der Exacte Wetenschappen, Vrije Universiteit Amsterdam (2003)
3. Cariani, F., Kaufmann, M., Kaufmann, S.: Deliberative modality under epistemic uncertainty. Linguistics and Philosophy 36(3), 225–259 (2013)
4. Carr, J.: Deontic modals without decision theory. Proceedings of Sinn und Bedeutung 17, 167–182 (2012)
5. Charlow, N.: What we know and what to do. Synthese 190(12), 2291–2323 (2013)
6. Condoravdi, C., van der Torre, L.: Deontic modality: linguistic and logical perspectives. In: ESSLLI 2014 course descriptions (2014)
7. Gabbay, D., Horty, J., Parent, X., van der Meyden, R, van der Torre, L.: Handbook of Deontic Logic and Normative Systems. College Publications (2013)
8. Gabbay, D., Parent, X., van der Torre, L.: An intuitionistic basis for input/output logic. In: Hansson, S.O. (ed.) David Makinson on Classical Methods for Non-Classical Problems. Series Oustanding Contributions to Logic, vol. 3, pp. 263–286. Springer (2014)
9. Gabbay, D.M.: Semantical Investigations in Heyting's Intuitionistic Logic. D. Reidel Publishing Company (1981)
10. Gabbay, D.M.: Reactive kripke models and contrary to duty obligations. part A: Semantics. Journal of Applied Logic 11(1), 103–136 (2013)
11. Gabbay, D.M.: Reactive Kripke Semantics. Cognitive Technologies. Springer (2013)
12. Kolodny, N., MacFarlane, J.: Iffs and oughts. Journal of Philosophy 107(3), 115–143 (2010)
13. Kratzer, A.: The notional category of modality. In: Eikmeyer, H.J., Rieser, H. (eds.) Words, worlds, and Contexts: New Approaches in World Semantics. de Gruyter, Berlin (1981)
14. Reichenbach, H.: Elements of Symbolic Logic. Macmillan Co., New York (1947)
15. Dalen, D.V.: Intuitionistic logic. In: Handbook of philosophical logic, pp. 225–339. Springer (1986)
16. Willer, M.: A remark on iffy oughts. Journal of Philosophy 109(7), 449–461 (2012)

For a Dynamic Semantics of Necessity Deontic Modals*

Alessandra Marra

Tilburg Center for Logic, General Ethics and Philosophy of Science, Tilburg
University, P.O. Box 90153, 5000 LE Tilburg, The Netherlands
a.marra@uvt.nl

Abstract. Traditional approaches in deontic logic have focused on the
so-called reportative reading of obligation sentences, by providing truth-
functional semantics based on a primitive ideality order between possible
worlds. Those approaches, however, do not take into account that, in natural language, obligation sentences primarily carry a prescriptive effect.
The paper focuses precisely on that prescriptive character, and shows
that the reportative reading can be derived from the prescriptive one.
A dynamic, non truth-functional semantics for necessity deontic modals
is developed, in which the ideality relations among possible worlds can
be updated. Finally, it is proven that the semantics solves several of the
classic deontic paradoxes.

Keywords: deontic logic, update semantics, prescriptive reading, reportative reading, deontic paradoxes.

1 Introduction: Deontics in Everyday Discourse Practice

In philosophy and linguistics, it has become customary to distinguish between
reportative and *prescriptive* readings of obligation sentences:[1]

(1) You must go
   (a) (According to the rules) you must go
   (b) (I command you,) you must go

Under the reportative reading, the sentence *You must go* is interpreted as (1)a:
the speaker is referring to some pre-existing code of norms and rules, and intends
to give a description of that code. Under the reportative reading, therefore, (1)
is a truth-apt sentence. The case of the prescriptive reading is different. When
a speaker utters *You must go* prescriptively, she does not intend to report, but
to bring about an obligation. Under the prescriptive reading, the sentence (1)
can be therefore interpreted as (1)b. In that case the speaker is enacting a

---

* I would like to thank Reinhard Muskens, Alan Thomas, Dominik Klein and the
anonymous referees of DEON 2014 for their suggestions and comments on the issues
discussed in this paper.
[1] See, for instance, [1], [5] and [11].

F. Cariani et al. (Eds.): DEON 2014, LNAI 8554, pp. 124–138, 2014.

new obligation: she is not describing, but rather commanding. Hence, under the prescriptive reading, (1) has no truth-value.

Traditional approaches in deontic logic have focused on the reportative reading of obligation sentences, by providing truth-functional semantics of necessity deontic modals based on a primitive ideality order between possible worlds.[2] Those approaches, however, cannot account for examples like (1)b. The present paper adopts a different perspective. A dynamic deontic semantics is provided, in which the prescriptive reading is taken to be the primary one.[3] The reportative reading can then be derived from the prescriptive. Moreover, it is shown that the proposed dynamic deontic semantics gives also the right predictions with respect to several deontic paradoxes. In the conclusive part of the paper, some conceptual implications of the dynamic deontic semantics are mentioned.

2   The Semantics

2.1   A Dynamic Setting

The present paper focuses on the prescriptive reading of obligation sentences. As noticed by Austin [3], in everyday discourse practice the prescriptive reading takes priority over the reportative one: obligation sentences are typically used to bring about new duties and obligations, rather than to give a true or false description of the existing ones.[4] They are intrinsically prescriptive, and they are used to *create* and *change* an aspect of reality, i.e., its normative configuration. They have, therefore, a dynamic character. That is what we want to model in our semantics.

We develop here a framework from the dynamic logic tradition, especially Veltman's [21] work on Update Semantics. Given that we aim at representing the dynamics of discourse practice, especially concerning obligation sentences, the semantic framework interprets "meanings" as the change that every sentence induces in the context in which it is uttered and accepted by the agents involved in the conversation. As it is said, the slogan "You know the meaning of a sentence if you know the conditions under which it is true" is replaced by the following one: "You know the meaning of a sentence if you know the change it brings about in the information state of anyone who accepts the news conveyed by it."[5]

In a well-run conversation, speaker and listener share a certain background information, or conversational context, which gets updated during the dialogue.[6]

---

[2] See [6], [9], [10] and [15].

[3] As Condoravdi [5] notices, the relation between reportative reading and prescriptive reading is still debated. Some authors, like [7] and [14], have indeed suggested that the prescriptive reading only pertains to the pragmatic domain. In the current paper, however, we adopt a different standpoint, by modeling the prescriptive reading in the semantics. What we mean by semantics, and some advantages of the position adopted here are discussed in Sections 2.1, 2.4 and 4.

[4] On this point, see also [20].

[5] See [21], p. 221.

[6] See [16].

When a sentence is uttered and accepted by the participants in the conversation, the information the sentence carries (i.e., its meaning) modifies and strengthens that context, by eliminating all possibilities that are incompatible with it. Every sentence affects the conversational context: during a conversation, agents can acquire factual knowledge and new obligations. In what follows, we consider the conversational exchange between a moral authority (the speaker) and a listener.

2.2 Van der Torre and Tan's Deontic Update Semantics

The idea of using Veltman's [21] Update Semantics to model the prescriptive reading of obligation sentences was first proposed by van der Torre and Tan [19].[7] The semantics that van der Torre and Tan provide is based on two different deontic operators, **oblige** and **oblige*** which range respectively over W, a set of possible worlds, and $W^* \subseteq W$, a set of epistemically possible worlds. Worlds are ordered according to their deontic ideality: a reflexive $\preceq$-relation on W indicates which worlds are deontically preferable ($w_1 \preceq w_2$ is read as: w_1 is at least as ideal as w_2). The two different sets, W and W^*, are intended to represent what van der Torre and Tan call *context of justification* and *context of deliberation*. The latter, as the name itself suggests, is meant to denote the possibilities that are taken into consideration while agents deliberate about what ought to be done. Clearly, the context of deliberation contains only possibilities that are still open according to agents' knowledge. The context of justification, on the other hand, gives a broader perspective: it is used to indicate that, despite the fact that a certain world w results to be the most ideal in W^*, it might not be among the most ideal worlds if the entire set W is considered. In that case, it is said that a violation occurred, since the most ideal worlds do not belong to W^* anymore.

The deontic semantics of van der Torre and Tan is dynamic, since the ideality relation can be changed by updating a state S with **oblige** and **oblige***. In particular, they define the following update rule for the conditional obligation **oblige**(α/β):

Definition 1 (van der Torre and Tan's update). $S[\textbf{oblige}(\alpha/\beta)]$ is defined as follows:

- if $\textbf{pref}(S\bot\textbf{oblige}(\alpha/\beta) \mid \beta) \models \alpha$ then $S[\textbf{oblige}(\alpha/\beta)]=S\bot\textbf{oblige}(\alpha/\beta)$
- otherwise $S[\textbf{oblige}(\alpha/\beta)]$ results in the absurd state **1**

where $S\bot\textbf{oblige}(\alpha/\beta)$ changes $\preceq$ by eliminating pairs (w_1, w_2) where $w_1 \models \neg\alpha \wedge \beta$ and $w_2 \models \alpha \wedge \beta$, and the clause $\textbf{pref}(S\bot\textbf{oblige}(\alpha/\beta) \mid \beta) \models \alpha$ checks whether, after that change in $\preceq$, the best β-worlds in W satisfy α.[8]

Van der Torre and Tan's semantics constitutes a very important example of how to use Veltman's update semantics to model prescriptive obligation sentences. However, their semantics seem to give rise to some unwelcome predictions. Denote with **0** the *minimal state*, where $W^* = W$ and $\preceq = W \times W$. Given Definition 1, we can get that:

---

[7] For other dynamic (but not update semantics-style) approaches in deontic logic, see [2], [12] and [17].

[8] The case of **oblige***(α/β) is analogous, just replace W with W^*.

(2) **0**[**oblige**(c/k)][**oblige**($\neg c/\top$)]$\neq$ **1**

while:

(3) **0**[**oblige**(c/k)][**oblige**($\neg c/\top$)][**oblige**(c/k)]=**1**

Two reasons why the above results are problematic. The first one is that the updates in (2) would be intuitively expected to give rise to the absurd state: a conflict
emerges if it is commanded that *If k is done, then c ought to be done* and then that
c ought not to be done.[9] Secondly, from (3), it emerges that van der Torre and
Tan's semantics is not suitable to model iterated updates. In particular, it does
not seem to provide an adequate framework to model the acquisition of obligations: the obligation **oblige**(c/k), whose update was successful in (2), is indeed
lost in (3). In what follows, therefore, we will modify the semantics proposed by
van der Torre and Tan in order to overcome those difficulties.

2.3 Our Approach: Information States and Updates

In the present section, we define the basic notions of information state and
updates. We take an information state to represent the conversational context
that speaker and listener share during a conversation. The basic elements of
information states are *possible worlds*, which represent the epistemic possibilities
considered open by the agents involved in the conversation, and *deontic ideality
relations* of possible worlds. Just as it happens during a dialogue, information
states can be modified and evolve.

Definition 2 (Language L_0). Language L_0 is built from a countable set A of
atoms according to the following BNF:

$$\phi := p \mid \top \mid \neg\psi \mid \psi_1 \wedge \psi_2 \mid \psi_1 \vee \psi_2$$

where $p \in A$.

Definition 3 (Valuation Function V). Let A be the set of atoms of L_0, W a
set of epistemically possible worlds, and let $v: A \longrightarrow \mathcal{P}(W)$ be an interpretation
function. A formula $\phi \in L_0$ is true at w under the interpretation v (written
$w \in V_v(\phi)$) iff:

- if ϕ=p, then $w \in v(p)$
- if $\phi = \neg\psi$, then $w \notin V_v(\psi)$
- if $\phi = \psi_1 \wedge \psi_2$, then $w \in V_v(\psi_1)$ and $w \in V_v(\psi_2)$
- if $\phi = \psi_1 \vee \psi_2$, then $w \in V_v(\psi_1)$ or $w \in V_v(\psi_2)$

Language L_0 consists only of factual sentences, i.e., sentences which provide a
true or false description of the facts of the world. In order to model prescriptive
necessity deontic modals, we enrich the language L_0 with the operator **Oblige**.[10]

---

[9] It is the famous Considerate Assassin example: take c as *You offer a cigarette* and k
as *You kill someone*. See, e.g., [13].

[10] The operator **Oblige** is meant to represent necessity deontic modals. In the present
paper, we abstract from the difference between weak and strong necessity deontic
modals.

Definition 4 (Language L_1). Language L_1 is built from a countable set A of atoms according to the following BNF:

$$\phi := p \mid \top \mid \neg\psi \mid \psi_1 \wedge \psi_2 \mid \psi_1 \vee \psi_2$$

$$\pi := \mathbf{Oblige}(\phi) \mid \mathbf{IF}\phi_1, \mathbf{Oblige}(\phi_2)$$

where $p \in A$.

For the semantics of **Oblige** we need to define the deontic selection function d:

Definition 5 (Deontic Selection Function). Let W be a set of epistemically possible worlds. The deontic selection function d assigns to every world $w \in W$ a reflexive $\preceq$-relation on W.[11]

Note that, contrary to van der Torre and Tan's semantics, different worlds may come with different ideality relations. However, obligation sentences are evaluated at the level of the entire information state S. That is because we want our semantics to represent the interplay between factual knowledge and deontic reasoning: obligations are defined with respect to the current knowledge (i.e., at the level of the information state S); and the relation attached to a single world is intended to represent the deontic preferences from the perspective of that world (i.e., if that world turns out to be the actual one, that ideality relation defines an obligation).[12] By doing so, we also move van der Torre and Tan's distinction between context of justification and context of deliberation from the language level to the semantic one. In particular, by attaching ideality relations to worlds, we can express that a world is sub-ideal, i.e., may constitute a violation to an obligation, even before any violation is actually committed. For instance, an ideality relation like $d(w_1) = \{(w_2, w_1), (w_1, w_1), (w_2, w_2)\}$ indicates that, in the very perspective of w_1, w_1 itself is sub-ideal.

Definition 6 (State). An information state is $S = \langle W, v, d \rangle$ where:

- W is a set of epistemically possible worlds
- v is an interpretation function
- d is a deontic selection function

A special class of information states is worth mentioning: the *absurd states* **1**.

Definition 7 (Absurd State). An information state $S = \langle W, v, d \rangle$ is called absurd state **1** iff:

- $W = \emptyset$;

  or:

- for some $w \in W$, $d(w)$ is such that there are $s, t \in W$: $(s, t) \notin d(w)$ and $(t, s) \notin d(w)$

---

[11] So, for every $w \in W$, we represent $d(w)$ as a set of ordered pairs (s, t) such that $s, t \in W$ and s is at least as deontically ideal as t (written: $s \preceq t$). We take $\preceq$ to be reflexive, but we do not assume transitivity. For a counter-example to transitivity, see [19], p.86.

[12] See Definitions 9 and 11.

Intuitively, the absurd state **1** is such that the agents have reached a contradiction either about what they take to be the facts (hence, there are no possibilities left open), or about what ought to be done. By assuming that deontic incompatible worlds lead to the absurd state, we have therefore adopted a notion of absurd state that is, in a certain sense, stricter than the one proposed by van der Torre and Tan.

Every sentence affects the common ground. During a conversation, agents can acquire factual knowledge and new obligations. That amounts to a process of elimination: the acquisition of factual knowledge results in the elimination of worlds in W, while the acceptance of new obligations has the effect of eliminating couples from the relevant ideality relations.

Definition 8 (Update with Factual Sentence). Let $S = \langle W, v, d \rangle$ be an information state and $\phi \in L_0$ a factual sentence. The acceptance of ϕ triggers the following update of S: $S[\phi] = \langle W', v, d' \rangle$, where $W' = W \setminus \{w \in W \mid \phi$ is false at $w\}$, and $d' = d_{\upharpoonright W'}$

Definition 9 (Update with Obligation Sentence). Let $S = \langle W, v, d \rangle$ be an information state and $\phi := \mathbf{Oblige}(\alpha)$ (where $\alpha \in L_0$). The acceptance of ϕ triggers the following update of S: $S[\phi] = \langle W, v, d' \rangle$, where, for every $w \in W$, $d'(w) = d(w) \setminus \{(s, t) \mid \alpha$ is false at s and α is true at $t\}$

Definition 10 (Update with Conditional Obligation). Let $S = \langle W, v, d \rangle$ be an information state and $\phi := \mathbf{IF}\beta, \mathbf{Oblige}(\alpha)$ (where $\alpha, \beta \in L_0$). The acceptance of ϕ triggers the following update of S: $S[\mathbf{IF}\beta, \mathbf{Oblige}(\alpha)] = \langle W, v, d' \rangle$, where,

- for every $w \in W$ such that β is true at w, $d'(w) = d(w) \setminus \{(s, t) \mid \beta \wedge \neg\alpha$ is true at s and $\beta \wedge \alpha$ is true at $t\}$
- for every $w' \in W$ such that β is false at w', $d'(w') = d(w')$

Less formally: Definition 8 is very standard, as it establishes that the acceptance of a factual sentence makes the set of epistemic possibilities W shrink; Definition 9 says that the acceptance of an obligation sentence like $\mathbf{Oblige}(\alpha)$ triggers a change in the deontic ideality relations, by requiring that, from the perspective of every epistemically possible world, $\neg\alpha$-worlds are not deontically preferred over the α-worlds; finally, according to Definition 10, the update with the obligation sentence $\mathbf{Oblige}(\alpha)$ concerns only the ideality relations of the epistemically possible worlds that satisfy the antecedent β. In that sense, our conditional update works as a proper restrictor of possibilities.

We give two last definitions: support and logical consequence. We adopt the corresponding definitions proposed by Veltman [21]:

Definition 11 (Support). An information state S supports a sentence α (written $S \models \alpha$) iff $S[\alpha] = S$

In other words, we say that a sentence α is supported in an information state S if and only if the information the sentence conveys is already subsumed in that information state.[13]

Definition 12 (Logical Consequence). $\alpha_1, ..., \alpha_n \models \beta$ iff for any S: $S[\alpha_1], ..., [\alpha_n]=S'$ such that $S' \models \beta$, i.e., updating any information state S with the premises $\alpha_1, ..., \alpha_n$ in that order yields to new S' which supports β.

It is possible now to see that our semantics gives the desired prediction concerning (2): updating the minimal state **0** with **IF**k,**Oblige**(c) and with **Oblige**$(\neg c)$ makes $c \wedge k$-worlds and $\neg c \wedge k$-worlds incomparable, resulting therefore in the absurd state. Moreover, given the definitions presented above, the semantics results to be suitable to model the process of acquisition of obligations: contrary to what happened in (3), no obligation can be lost during iterated updates.[14]

We conclude the presentation of the dynamic framework by mentioning a feature of our semantics which will become particularly relevant in Section 3.3.

Theorem 1 (Modus Ponens for Conditional Obligations). For every S, $S \models$**IF**β,**Oblige**$(\alpha) \implies S[\beta] \models$ **Oblige**(α).

Proof. By contradiction, assume that, for a certain $S=\langle W, v, d\rangle$, (i) $S \models$**IF**β, **Oblige**(α) and suppose that (ii) $S[\beta] \not\models$ **Oblige**(α). From (i), it follows that, for all $w \in W$ that satisfy β, there is no pair (s, t) in $d(w)$ where $\neg\alpha \wedge \beta$ true at s and $\alpha \wedge \beta$ is true at t. From (ii), it follows that $S[\beta]=\langle W', v, d'\rangle$ where $W' = W \setminus \{z \in W \mid \beta$ is false at $z\} \neq \emptyset$ and, therefore, there are worlds $w \in W'$ which satisfy β and such that there is a pair (s, t) in $d_{\upharpoonright W'}(w)$ where $\neg\alpha \wedge \beta$ true at s and $\alpha \wedge \beta$ is true at t. Contradiction.

---

[13] Given our definitions, it follows that the framework validates the principle "α, therefore α ought to be the case", since for every S, $S[\alpha] \models$ **Oblige**(α). Moreover, it also holds that $S \models$ **Oblige**$(p \vee \neg p)$, and even that $S \models$ **Oblige**$(p \wedge \neg p)$. Those unintuitive predictions have all the same root: using Austin's [3] terminology, we can say that those updates are void speech acts. In fact, such updates cannot trigger any possible change in the ideality relations, since, e.g., there are no $\neg\alpha$-worlds preferred over α-worlds if W contains only α-worlds. One possible solution would consist in imposing the *precondition* that updates with **Oblige**(α) have α as an open possibility in S. Moreover, such a precondition could be motivated by appealing to gricean-style maxims of conversation. A similar solution is also adopted by Condoravdi [4] in her paper on epistemic modals.

[14] No obligation is lost even in the case of updates with factual sentences. Of course, this is not the case for van der Torre and Tan's **oblige***. Consider the state S with $W = W* = \{s, t, k, z\}$ and $\preceq = W \times W$. Let $p \wedge q$ true at s, $p \wedge \neg q$ true at t, $\neg p \wedge q$ true at k and $\neg p \wedge \neg q$ true at z. The updates with **oblige***$(p/\top)$ and **oblige***$(q/\top)$ make s the most ideal world, while t and k become incompatible. But if the new state is now updated with the factual sentence $\neg(p \wedge q)$, s gets eliminated from $W*$ and in the resulting state neither **oblige***$(p/\top)$ nor **oblige***$(q/\top)$ holds anymore. In our framework, we avoid that possible result by anticipating that the sequential update with **Oblige**(p) and **Oblige**(q) results in the absurd state.

2.4   The Reportative Reading, Again

The semantics we have presented is fully dynamic. It permits us to model the prescriptive use of obligation sentences, by allowing the ideality relations to be updated every time a new obligation is uttered and accepted. However, it is worth noticing that we have provided only update rules for obligation sentences of the form **Oblige**(α) and **IF**β,**Oblige**(α), where α, $\beta \in L_0$. As it has been discussed in the literature, under the prescriptive reading not all combinations of deontic modals and connectives are indeed admissible. Consider the following examples: [15]

(4) (a) I don't command you, you have to use your dictionary during the English exam. #

   (b) According to the rules, you don't have to use your dictionary during the English exam.

(5) (a) I command you, you must eat the soup ... or ... I command you, you must eat the salad. #

   (b) According to the rules, you must eat the soup or you must eat the salad.

Under the prescriptive reading (made explicit by the clause *I command you*), the use of necessity deontic modals in (4)a and (5)a is not felicitous, in the sense that an utterance of (4)a or (5)a does not result in a genuine command. However, sentences (4)b and (5)b show that things stand differently if the deontic modals are used reportatively.

While sentences of the form ¬**Oblige**(α) and **Oblige**(α)∨**Oblige**(β) are not felicitous under the prescriptive reading, the case of conjunction scoping over prescriptive deontic necessity modals seems to be less clear:

(6) (a) I command you, you must go to school ... and ... I command you, you must finish your homework. ?

   (b) According to the rules, you must go to school and you must finish your homework.

The reportative (6)b is felicitous, while some authors, e.g., [22], rule out (6)a, by considering it infelicitous on a par with (4)a and (5)a. In the present work we follow that line of analysis, by forbidding that truth-functional connectives take scope over the prescriptive **Oblige**. Moreover, we suggest that if (6)a does not appear to be as infelicitous as (4)a and (5)a, it may be because it can be interpreted as a sequence –rather than a genuine conjunction– of updates, like in $S[\textbf{Oblige}(\alpha)][\textbf{Oblige}(\beta)]$.

The discrepancy between (4)a, (5)a and (6)a, on one side, and (4)b, (5)b and (6)b, on the other, can be taken into account in our dynamic semantics. Thanks to the notion of support, the semantics permits us to consider also the reportative use of obligation sentences. Recall that, according to the reportative

---

[15] For some further examples, see [22]. Even if [22] only takes into account the case of imperatives, his examples can be rephrased using prescriptive necessity deontic modals.

use, an obligation sentence like *You must do* α describes an already existent obligation, and does not trigger any change in the ideality relations. It amounts to saying that the ideality relations already encode the information that the ideal possible worlds are the ones in which you do α, i.e., that the sentence *You must do* α is supported in the information state:

Definition 13 (Reportative Obligations).   The reportative reading of **Oblige** is derived as follows:

- $S \models \textbf{Oblige}(\alpha)$ iff $S[\textbf{Oblige}(\alpha)]=S$
- $S \models \textbf{IF}\beta,\textbf{Oblige}(\alpha)$ iff $S[\textbf{IF}\beta,\textbf{Oblige}(\alpha)]=S$
- $S \models \neg\textbf{Oblige}(\alpha)$ iff $S[\textbf{Oblige}(\alpha)]\neq S$
- $S \models \textbf{Oblige}(\alpha)\wedge \textbf{Oblige}(\beta)$ iff $S[\textbf{Oblige}(\alpha)]=S$ and $S[\textbf{Oblige}(\beta)]=S$
- $S \models \textbf{Oblige}(\alpha)\vee \textbf{Oblige}(\beta)$ iff $S[\textbf{Oblige}(\alpha)]=S$ or $S[\textbf{Oblige}(\beta)]=S$

The reportative use can be, therefore, derived from the prescriptive one. The equivalences above show that a reportative obligation sentence holds in an information state S if an only if certain conditions on the update with the prescriptive **Oblige** apply. The difference between reportative and prescriptive obligations is internal in the semantics, and it can be modeled without enriching the language with a further necessity deontic operator.

3   Deontic Paradoxes

We apply the dynamic deontic semantics presented in the previous section to the analysis of some of the most known deontic paradoxes: Forrester's Paradox, Ross' Paradox and the Miners' Paradox. Particular attention is given to the latter, which represents also an emblematic example of deontic reasoning in case of partial factual knowledge.

3.1   Forrester's Paradox

Conditional obligations are often used to describe so-called "secondary obligations" (of the form $\textbf{IF}\beta,\textbf{Oblige}(\alpha)$) which come into play when "primary obligations" (of the form $\textbf{Oblige}(\neg\beta)$) are violated. It is customary to use the term *contrary-to-duty obligations* (CTDs) to refer to those secondary obligations.[16] It is worth noticing that CTDs do *not* establish *exceptions* to the primary obligations. The two notions, CTDs and exceptions, are indeed different. Consider the primary obligation $\textbf{Oblige}(\neg\beta)$. An exception to the primary obligation may have the form $\textbf{IF}\alpha,\textbf{Oblige}(\beta)$. Being an exception, it leads to a conflict with the primary obligation $\textbf{Oblige}(\neg\beta)$: it establishes, indeed, that if α is the case the primary obligation is canceled.[17] On the contrary, in the case of CTD, the primary obligation is still in force. $\textbf{IF}\beta,\textbf{Oblige}(\alpha)$ establishes a new duty in the case in which the primary obligation is violated, but that does not prevent the primary obligation $\textbf{Oblige}(\neg\beta)$ from holding unconditionally.[18]

---

[16] See, for instance [13].

[17] Therefore, the case of the so-called Considerate Assassin in Section 2.2 is an example of conditional obligation as exception.

[18] See [18].

A typical CTD-scenario is Forrester's Paradox of the Gentle Murder. Imagine that a moral authority utters:

(7) Smith should not murder Jones
(8) If Smith murders Jones, then Smith should murder Jones gently

The paradox emerges in Standard Deontic Logic (SDL) because (7) and (8) cannot be expressed consistently in that framework. However, the problem does not arise in the dynamic deontic semantics we have presented in Section 2.

We adopt the standard formalization of (7) and (8) as:

(9) **Oblige**$(\neg m)$
(10) **IF**m, **Oblige**(g)

where m is *Smith murders Jones*, and g is *Smith murders Jones gently*.

Consider the information state $S = \langle W, v, d \rangle$, where:

- $W = \{s, t, u\}$
- v is such that $s \in V_v(\neg m)$, $s \in V_v(\neg g)$, $t \in V_v(m)$, $t \in V_v(g)$, $u \in V_v(m)$ and $u \in V_v(\neg g)$.
- for every $w \in W$, $d(w) = W \times W$.

Now let us update S with (7) and (8). That amounts to the sequential updates $S[\mathbf{Oblige}(\neg m)][\mathbf{IF}\ m, \mathbf{Oblige}(g)]$. First consider $S[\mathbf{Oblige}(\neg m)] = S' = \langle W, v, d' \rangle$, where:

- $d'(s) = d'(t) = d'(u) = \{(s, t), (s, u), (t, u), (u, t)\}$[19]

Now take $S'[\mathbf{IF}\ m, \mathbf{Oblige}(g)] = S'' = \langle W, v, d'' \rangle$, where:

- $d''(s) = d'(s)$
- $d''(t) = d''(u) = \{(s, t), (s, u), (t, u)\}$

Since updating S with (7) and (8) does not result in the absurd state, no paradox arises. Moreover, the state S'' makes sense of the difference between the primary obligation and the CTD. The primary obligation is supported in S'': from the perspective of every world, the world s, in which Smith does not murder Jones, is better than t and u. Moreover, it is recognized that *if* a violation occurs (i.e., from the perspective of the sub-ideal worlds t and u), the world t in which Smith murders Jones gently, is better than u, in which Smith murders Jones and he does not do it gently.

3.2 Ross' Paradox

The next example we consider is Ross' Paradox:

(11) The letter must be mailed
(12) The letter must be mailed or burned

---

[19] In this case, and in the following ones, we leave the reflexive closure implicit.

Ross' Paradox is a paradox because in SDL (12) follows from (11). The dynamic semantics we have presented in the paper provides, on the other hand, a rather straightforward solution to the paradox. We follow the standard formalization of (11) and (12), and show that (12) is not supported in an information state updated with (11).

Let ma be *The letter is mailed* and bu be *The letter is burned*. Consider an information state $S= \langle W, v, d \rangle$, where:

- $W = \{s, t, u\}$
- v is such that $s \in V_v(ma)$, $s \in V_v(\neg bu), t \in V_v(\neg ma), t \in V_v(bu), u \in V_v(\neg ma)$ and $u \in V_v(\neg bu)$.
- for every $w \in W$, $d(w)= W \times W$.

Moreover, let (13) and (14) formalize (11) and (12), respectively:

(13) **Oblige**(ma)
(14) **Oblige**$(ma \lor bu)$

The update of S with (13) results in the following $S'=\langle W, v, d' \rangle$ where:

- $d'(s)= d'(t)= d'(u)=\{(s,t), (s,u), (t,u), (u,t)\}$

The update with **Oblige**(ma) has the effect that, from the perspective of every world, s is more ideal than t and u. However, nothing is said concerning t and u: they are equally preferred. **Oblige**$(ma \lor bu)$ is indeed not supported in S'. To see that, consider $S'[$**Oblige**$(ma \lor bu)=\langle W, v, d'' \rangle$, where:

- $d''(s)=d''(t)= d''(u)=\{(s,t), (s,u), (t,u)\}$

Clearly $S' \neq S''$, hence (14) does not follow from (13), and Ross' Paradox is blocked.

3.3 The Miners' Paradox

Another well-known deontic paradox is the Miners' Paradox, an example that runs as follows:

> Ten miners are trapped either in shaft A or in shaft B, but we do not know which one. Water threatens to flood the shafts. We only have enough sandbags to block one shaft but not both. If one shaft is blocked, all of the water will go into the other shaft, killing every miner inside. If we block neither shaft, both will be partially flooded, killing one miner.[20]

Why is the Miners' Paradox a paradox? Lacking any information about the miners' position, it seems right to that the outcome should be:

---

[20] See [8].

(15) We ought to block neither shaft.[21]

However, in deliberating about what to do, we accept:

(16) If the miners are in shaft A, we ought to block shaft A.
(17) If the miners are in shaft B, we ought to block shaft B.

We also accept:

(18) Either the miners are in shaft A or they are in shaft B.

But (16)-(18) seem to entail:

(19) Either we ought to block shaft A or we ought to block shaft B.

And this is incompatible with (15). Thus we have a paradox.

In order to block the Miners' Paradox, several escape routes have been proposed.[22] In particular, it has been suggested that two different *oughts* are involved in the paradox: a *subjective ought* and an *objective ought*.

Intuitively, subjectivism and objectivism differ with respect to the body of information in light of which the deontic modal is evaluated. Under the objectivist reading, a sentence like X *ought to do* α indicates that α is the best option available to the agent X in light of all facts, known or unknown; while, under the subjectivist reading, X *ought to do* α indicates that α is the best option available to the agent X in light of what X knows. If one adopts a purely objectivist reading of the deontic *ought*, the paradox does not arise, since the very premise (15) is rejected. On the other hand, if one adopts a purely subjectivist reading of *ought*, premises (16) and (17) are not acceptable.

Both objectivism and subjectivism have, however, some difficulties. As Kolodny and MacFarlane [8] also point out, objectivism seems too strong, since it does not allow deontic reasoning for partially-informed agents.[23] On the other hand, subjectivism seems to be too weak, since it does not validate conditional obligations like *If α, then X ought to do β.*[24]

---

[21] We take (15) as asserting that (i) we do not have the obligation of blocking shaft A and (ii) we do not have the obligation of blocking shaft B. Some authors, e.g., [8], are not explicit about their interpretation of (15), while other authors, e.g.,[23], have argued that (15) asserts that we are obliged not to block A and not to block B (with the negation scoping inside the deontic operator). We think that the last reading, which ranks the worlds in which we do not block A and we do not block B as the most ideal ones, is not satisfactory. Our intuitions indeed are such that the most ideal worlds are the ones in which all the ten miners survive, not the ones in which only nine of them are saved.

[22] See [8].

[23] If the objectivist view had to be adopted, only an omniscient agent with a complete knowledge of all facts would be able to determine whether X *ought to do* α is true. See [8], p.118.

[24] Subjectivism may validate only weaker conditionals such as *If X knows that α, then X ought to do β.* See [8], p.118.

However, that does not imply that some of the claims made by objectivism and subjectivism are not worthy to be considered. In what follows we show that, in the case of the Miners' Paradox, it is possible to provide a solution which takes into account some aspects of the objectivism's and subjectivism's views, but uses one single deontic operator that remains neutral between the two positions.

The Miners' Paradox represents an emblematic example of deontic reasoning in case of partial factual information, as it is about agents who consider what they ought to do in a context where they lack knowledge about the miners' position. We can understand the paradox' scenario in terms of a conversational exchange between a moral authority (the speaker) and a listener. In particular, we can see that even if the listener accepts all the premises of the Miners' Paradox, the paradoxical conclusion (19) is not supported in the updated information state.

We formalize sentences (16)-(18) as follows:

(20) **IFa,Oblige**$(block_a)$
(21) **IFb,Oblige**$(block_b)$
(22) $a \vee b$

where a is *The miners are in shaft A*, b is *The miners are in shaft B*, $block_a$ is *We block shaft A* and $block_b$ is *We block shaft B*.

Consider the following "action"- state, which represents all the possibilities that agents have in the case some action is taken: $S = \langle W, v, d \rangle$, where:[25]

- $W = \{s, t, u, z\}$
- v is such that:
  - $s \in V_v(a)$, $s \in V_v(block_a)$
  - $t \in V_v(b), t \in V_v(block_b)$
  - $u \in V_v(a), u \in V_v(block_b)$
  - $z \in V_v(b), z \in V_v(block_a)$
- for every $w \in W$, $d(w) = W \times W$.

Assume that the moral authority has already uttered *If the miners are in shaft A, we ought to block shaft A* and *If the miners are in shaft B, we ought to block shaft B*, and the listener has already accepted those obligations. Consider the updated information state $S' = \langle W, v, d' \rangle$ such that:

- $d'(s) = d'(u) = \{(s, t), (s, u), (s, z), (t, s), (t, u), (t, z), (u, t), (u, z), (z, s), (z, t), (z, u)\}$
- $d'(t) = d'(z) = \{(s, t), (s, u), (s, z), (t, s), (t, u), (t, z), (u, s), (u, t), (u, z), (z, s), (z, u)\}$

Hence we have that:

- $S' \models$ **IFa,Oblige**$(block_a)$
- $S' \models$ **IFb,Oblige**$(block_b)$
- $S' \models a \vee b$

---

[25] It is worth noticing that the analysis does not change in the case worlds in which both $\neg block_a$ and $\neg block_b$ are added to the state S.

In the state S', all the premises of the paradox are accepted. However, we get that:

- $S' \not\models$ **Oblige**($block_a$)
- $S' \not\models$ **Oblige**($block_b$)

It is indeed the case that $S'[\textbf{Oblige}(block_a)] \neq S'$ and $S'[\textbf{Oblige}(block_b)] \neq S'$.

So S' does not support **Oblige**($block_a$) nor **Oblige**($block_b$). That is, given our current partial factual information, neither we ought to block shaft A nor we ought to block shaft B. Even if all the premises of the Miners' Paradox hold in the information state S', the paradoxical conclusion (19) **Oblige**($block_a$)$\vee$ **Oblige**($block_b$) does not hold. Therefore no paradox arises.[26]

4   Conclusion

The paper aimed at providing a dynamic semantics for necessity deontic modals which could account for the difference between reportative and prescriptive readings of obligation sentences. Contrary to the traditional approaches in deontic logic, we gave the priority to the prescriptive reading. We modeled the dynamic effect that prescriptive obligations carry through a dynamic deontic semantics, and showed that the dynamic semantics can derive the reportative reading and solve some of the main deontic paradoxes which usually affect static, truth-conditional deontic semantics. The reason for giving the priority to the prescriptive reading originally came from everyday discourse practice, in which obligation sentences appear to be primarily used to bring about new obligations and duties, rather than to describe obligations that already exist. The formal results obtained in the dynamic semantics seem to suggest that the linguistic analysis which gives the priority to the prescriptive reading is on a promising track.

References

1. Alchourrón, C.: Philosophical Foundations of Deontic Logic and the Logic of Defeasible Conditionals. In: Meyer, J.-J., Wieringa, R. (eds.) Deontic Logic in Computer Science: Normative System Specification, pp. 43–84. John Wiley & Sons (1993)
2. Aucher, G., Boella, G., van der Torre, L.: Prescriptive and Descriptive Obligations in Dynamic Epistemic Deontic Logic. In: Casanovas, P., Pagallo, U., Sartor, G., Ajani, G. (eds.) AICOL-II/JURIX 2009. LNCS, vol. 6237, pp. 150–161. Springer, Heidelberg (2010)
3. Austin, J.L.: How to do things with words. The William James Lectures delivered at Harvard University in 1955. Oxford University Press, Oxford (1962)
4. Condoravdi, C.: Temporal interpretation of modals: Modals for the present and for the past. In: Beaver, D., Kaufmann, S., Casillas, L. (eds.) The construction of meaning, pp. 59–88. CSLI Publications, Stanford (2002)

---

[26] Moreover, the above solution to the Miners' Paradox differ from [8] and [23]. Contrary to [8], our semantics validates *Modus Ponens* for conditional obligations, and, contrary to [23], the solution to the paradox is provided in a context of a fully dynamic deontic semantics.

5. Condoravdi, C., Lauren, S.: Speaking of Preferences: Imperative and Desiderative Assertions in Context. In: Speaking of Possibility and Time: The 7th Workshop on Inferential Mechanisms and their Linguistic Manifestation, http://www.sven-lauer.net/output/ CondoravdiLauer-SoPaT10-desiderative-imperative-assertions.pdf

6. Hansson, B.: An Analysis of some Deontic Logics. Deontic Logic: Introductory and Systematic Readings, Synthese Library 33, 121–147 (1971)

7. Kaufmann, S., Schwager, M.: A uniform analysis of conditional imperatives. Proceedings of SALT 19 (2009)

8. Kolodny, N., MacFarlane, J.: Ifs and Oughts. Journal of Philosophy 107(3), 115–143 (2010)

9. Kratzer, A.: The notional category of modality. In: Eikmeyer, H.J., Reiser, A. (eds.) Words, Worlds and Contexts. New Approaches in Word Semantics, pp. 38–74. Walter de Gruyter, Berlin (1981)

10. Lewis, D.: Counterfactuals. Oxford, Blackwell (1973)

11. Makinson, D.: On a Fundamental Problem of Deontic Logic. In: McNamara, P., Prakken, H. (eds.) Norms, Logics and Information Systems: New Studies in Deontic Logic and Computer Science. Frontiers in Artificial Intelligence and Application, vol. 49, pp. 29–53. IOS Press (1999)

12. Meyer, J.-J.C.: A Different Approach to Deontic Logic: Deontic Logic Viewed as a Variant of Dynamic Logic. Notre Dame Journal of Formal Logic 29(1), 109–136 (1988)

13. Prakken, H., Sergot, M.: Dyadic Deontic Logic and Contrary-to-duty Obligations. In: Nute, D.N. (ed.) Defeasible Deontic Logic, pp. 223–262. Synthese Library, Kluwer (1997)

14. Schwager, M.: Interpreting Imperatives. PhD thesis, Johann Wolfgang Goethe-Universtät, Frankfurt am Main (2006)

15. Spohn, W.: An analysis of Hansson's dyadic deontic logic. Journal of Philosophical Logic 4, 237–252 (1975)

16. Stalnaker, R.C.: Assertion. In: Cole, P. (ed.) Syntax and Semantics, vol. 9, pp. 315–332 (1978)

17. Yamada, T.: Logical dynamics of some speech acts that affect obligations and preferences. Synthese 165, 295–315 (2008)

18. van der Torre, L.W.N., Tan, Y.: The many faces of defeasibility in defeasible deontic logic. In: Nute, D.N. (ed.) Defeasible Deontic Logic, pp. 79–121. Synthese Library, Kluwer (1997)

19. van der Torre, L.W.N., Tan, Y.: An Update Semantics for Deontic Reasoning. In: McNamara, P., Prakken, H. (eds.) Norms, Logics and Information Systems: New Studies on Deontic Logic and Computer Science, pp. 73–90. IOS Press (1999)

20. van Rooij, R.: Permission to change. Journal of Semantics 17(2), 119–143 (2000)

21. Veltman, F.: Defaults in Update Semantics. Journal of Philosophical Logic 25, 221–261 (1996)

22. Veltman, F.: Or else, what? Imperatives on the borderline of semantics and pragmatics. Lego Seminar presentation, Institute for Logic, Language and Computation, Amsterdam (2012)

23. Willer, M.: A Remark on Iffy Oughts. Journal of Philosophy 109(7), 449–461 (2012)

Proof Analysis in Deontic Logics

Eugenio Orlandelli

Dipartimento di Filosofia e Comunicazione, Università di Bologna, Italy
eugenio.orlandelli2@unibo.it

Abstract. Sequent calculi for normal and non-normal deontic logics are introduced. For these calculi we prove that weakening and contraction are height-preserving admissible, and we give a syntactic proof of the admissibility of cut. This yields that the subformula property holds for them and that they are decidable. Then we show that our calculi are equivalent to the axiomatic ones, and therefore that they are sound and complete w.r.t. neighborhood semantics. This is a major step in the development of the proof theory of deontic logics since our calculi allow for a systematic root-first proof search of formal derivations.

Keywords: Deontic logics, sequent calculi, proof analysis, structural rules, non-normal modal logics.

1 Introduction

In the context of deontic logics it is widely recognized that a logic weaker than the standard deontic logic **KD** should be employed. Deontic paradoxes are one of the main motivations for adopting a non-normal deontic logic.[1] Non-normal logics are quite well understood from a semantic point of view, where they can be studied by means of neighborhood [3] or multi-relational semantics [2]. Their proof theory, nevertheless, is rather limited since it is confined to Hilbert-style axiomatic systems. This situation seems to be rather critical since it is difficult to build countermodels in such semantics and it is difficult to find derivations in axiomatic systems. When the purpose is to find derivations and to analyze their structural properties, sequent calculi are to be preferred to axiomatic systems. However the existing sequent calculi for non-normal logics [6,7] do not allow to eliminate all the structural rules of inference – weakening, contraction and cut – and therefore it is not possible to use them to determine whether a given formula is derivable or not by means of a root-first proof search procedure. Furthermore, although there are rules of inference that capture the deontic axiom $\neg(\mathcal{O}A \wedge \mathcal{O}\neg A)$ [6,13], to our knowledge there is no sequent rule that captures the possibly weaker deontic axiom $\neg\mathcal{O}\bot$. Thus it seems worthwhile to introduce proof-systems for deontic logics that allow for structural proof analysis.

This paper fills this gap by introducing cut- and contraction-free sequent calculi for the deontic logics **ED**, **MD**, **RD** and **KD**, where all the structural

---

[1] See [8] for a survey of deontic paradoxes and proposed solutions, and [3] for naming conventions of deontic logics.

F. Cariani et al. (Eds.): DEON 2014, LNAI 8554, pp. 139–148, 2014.

Table 1. Rules of inference

$$\frac{A \leftrightarrow B}{\mathcal{O}A \leftrightarrow \mathcal{O}B} \, RE \qquad\qquad \frac{A \rightarrow B}{\mathcal{O}A \rightarrow \mathcal{O}B} \, RM$$

$$\frac{(A_1 \wedge \ldots \wedge A_n) \rightarrow B}{(\mathcal{O}A_1 \wedge \ldots \wedge \mathcal{O}A_n) \rightarrow \mathcal{O}B} \, RR \, (n \geq 1) \qquad \frac{(A_1 \wedge \ldots \wedge A_n) \rightarrow B}{(\mathcal{O}A_1 \wedge \ldots \wedge \mathcal{O}A_n) \rightarrow \mathcal{O}B} \, RK \, (n \geq 0)$$

rules are admissible. Our calculi have the subformula property and allow for a straightforward decision procedure by root-first proof search. We proceed as follows: Chapter 2 summarizes the basic notions of axiomatic systems and of neighborhood semantics for the logics **ED**, **MD**, **RD** and **KD**. Chapter 3 introduces sequent calculi for these logics; we show that weakening and contraction are height-preserving admissible; and we give a syntactic proof of the admissibility of cut, which will then be used to show that our sequent calculi are equivalent to the corresponding axiomatic systems. Finally Chap. 4 considers some related works.

2   Deontic Logics

2.1   Axiomatic Systems

We introduce, following [3], the basic notions of axiomatic deontic logics that will be used later on. Given a set of propositional variables $\{p_i : i \in I\}$, the formulas of the (monadic) deontic language – $\mathcal{L}^D$ – are generated by:

$$A ::= p_i \mid \bot \mid A \wedge A \mid A \vee A \mid A \rightarrow A \mid \mathcal{O}A \tag{1}$$

As usual $\neg A$ is a shorthand for $A \rightarrow \bot$, and $A \leftrightarrow B$ for $(A \rightarrow B) \wedge (B \rightarrow A)$. We follow the usual conventions for parentheses. By $\mathcal{O}A$ we mean 'it ought to be the case that A'.

ED is the smallest set of $\mathcal{L}^D$-formulas containing all propositional tautologies and $D := \neg \mathcal{O}\bot$, and closed under modus ponens (MP) and the rule RE of Table 1. **MD** is defined as **ED**, but with rule RE replaced by rule RM. **RD** is is defined as **ED**, but with rule RE replaced by rule RR. **KD** is defined as **ED**, but with rule RE replaced by rule RK. We will use **xD** to talk of a generic logic among them, and we will write **xD** $\vdash A$ whenever $A(\in \mathcal{L}^D)$ is a theorem of the deontic logic **xD**.[2]

The following theorem states the well-known relations between the theorems of such deontic logics, for a proof the reader is referred to [3].

---

[2] Another way of introducing the logics defined above is to take **ED** as basic, and to introduce **MD** as its extensions obtained by adding $\mathcal{O}(A \wedge B) \rightarrow (\mathcal{O}A \wedge \mathcal{O}B)$; **RD** as the extension of **MD** obtained by adding $(\mathcal{O}A \wedge \mathcal{O}B) \rightarrow \mathcal{O}(A \wedge B)$; **KD** as the extension of **RD** obtained by adding $\mathcal{O}\neg\bot$. We have preferred the approach by rules to facilitate the comparison with sequent calculi.

Theorem 1. *For any formula $A \in \mathcal{L}^D$ we have that* **ED** $\vdash A$ *implies* **MD** $\vdash A$; **MD** $\vdash A$ *implies* **RD** $\vdash A$; **RD** $\vdash A$ *implies* **KD** $\vdash A$.

We make few comments before proceeding. **KD**, also known as the *standard deontic logic*, is the minimal normal logic **K** augmented with the axiom $D^* := \neg(\mathcal{O}A \wedge \mathcal{O}\neg A)$, whose correctness has been a big issue in the development of deontic logics. The other logics considered here are non-normal (or classical) logics that have been proposed to repair this and other shortcomings of **KD** [8]. E.g. [3] proposes **MD** as a more plausible system of deontic logic because it allows to have D, but not D^*, as a theorem. In fact the deontic axioms D and D^* have the following relations in the logics we are considering: they are independent in **ED** and **MD**, D^* is derivable from D in **RD**, and they are interderivable in **KD**.

2.2 Semantics

The most widely known semantics for non-normal logics is neighborhood semantics. We sketch its main tenets following [3], where neighborhood models are called *minimal models*.

Definition 1. *A* neighborhood model *is a triple* $\mathcal{M} :=< W, N, P >$, *where W is a non-empty set of possible worlds, $N : W \rightarrow 2^{2^W}$ is a neighborhood function associating with each possible world w a set $N(w)$ of subsets of W, and P gives a truth value to each propositional variable at each world.*

Truth of a formula A at a world w of a neighborhood model $\mathcal{M}$ – is $\models_w^{\mathcal{M}} A$ – is the standard one for the propositional part of $\mathcal{L}^D$ with the addition of

$$\models_w^{\mathcal{M}} \mathcal{O}A \quad iff \quad ||A||^{\mathcal{M}} \in N(w), \tag{2}$$

where $||A||^{\mathcal{M}}$ is the truth set of A – i.e. $||A||^{\mathcal{M}} = \{w : \models_w^{\mathcal{M}} A\}$. We say that a formula A is *globally true* in $\mathcal{M} =< W, N, P >$ iff $||A||^{\mathcal{M}} = W$, and that it is *valid* in a class $\mathcal{C}$ of neighborhood models iff it is globally true in every $\mathcal{M} \in \mathcal{C}$. We have the following adequacy results between the deontic logics defined above and validity in (classes of) neighborhood models, see [3] for the proofs.

Theorem 2. **ED** *is adequate w.r.t. the class $\mathcal{C}$ of all neighborhood models that are* non-blind *– i.e. where for all $X \in 2^W$, if $X \in N(w)$ then $X \neq \emptyset$.* **MD** *is adequate w.r.t. the class of all neighborhood models that are non-blind and* supplemented *– i.e. where for all $X, Y \in 2^W$, if $X \cap Y \in N(w)$ then $X \in N(w)$ and $Y \in N(w)$.* **RD** *is adequate w.r.t. the class of all neighborhood models that are non-blind, supplemented and closed under intersection – i.e. where for all $X, Y \in 2^W$, if $X \in N(w)$ and $Y \in N(w)$ then $X \cap Y \in N(w)$.* **KD** *is adequate w.r.t. the class of all neighborhood models that are non-blind, supplemented, closed under intersection and* contain the unit *– i.e. where for all $w \in W, W \in N(w)$.*

Table 2. The sequent calculus **G3cp**

Initial sequents:	$p_i, \Gamma \Rightarrow \Delta, p_i$

| Propositional rules: | |

$$\frac{A, B, \Gamma \Rightarrow \Delta}{A \wedge B, \Gamma \Rightarrow \Delta} \, L\wedge \qquad \frac{\Gamma \Rightarrow \Delta, A \quad \Gamma \Rightarrow \Delta, B}{\Gamma \Rightarrow \Delta, A \wedge B} \, R\wedge$$

$$\frac{A, \Gamma \Rightarrow \Delta \quad B, \Gamma \Rightarrow \Delta}{A \vee B, \Gamma \Rightarrow \Delta} \, L\vee \qquad \frac{\Gamma \Rightarrow \Delta, A, B}{\Gamma \Rightarrow \Delta, A \vee B} \, R\vee$$

$$\frac{\Gamma \Rightarrow \Delta, A \quad B, \Gamma \Rightarrow \Delta}{A \rightarrow B, \Gamma \Rightarrow \Delta} \, L\rightarrow \qquad \frac{A, \Gamma \Rightarrow \Delta, B}{\Gamma \Rightarrow \Delta, A \rightarrow B} \, R\rightarrow$$

$$\frac{}{\bot, \Gamma \Rightarrow \Delta} \, L\bot$$

3   Sequent Calculi for Deontic Logics

3.1   Sequent Calculi

We introduce deontic sequent calculi that extend the sequent calculus **G3cp** for classical propositional logic – see Table 2 – by adding some modal rules from Table 3. We will call **G3ED** the calculus obtained by adding LR-E and LD; **G3MD** the calculus obtained by adding LR-M and LD; **G3RD** the calculus obtained by adding LR-R and $LD^\star$; **G3KD** the calculus obtained by adding LR-K and $LD^\star$. As we did for deontic logics, we will use **G3xD** to talk of them all. As should already be clear, the calculus **G3xD** is meant to capture the logic **xD**. A derivation of a sequent $\Gamma \Rightarrow \Delta$ – where Γ and Δ are finite multisets of $\mathcal{L}^D$-formulas, and where if Π is a multiset $A_1, \ldots, A_m$ then $\mathcal{O}\Pi$ is $\mathcal{O}A_1, \ldots, \mathcal{O}A_m$ – in **G3xD** is a tree of sequents having $\Gamma \Rightarrow \Delta$ as root, initial sequents as leaves, and all edges obtained by an application of a rule of **G3xD**. The *height* of a derivation is the length of its longest branch. A rule of inference is said to be (*height-preserving*) *admissible* in **G3xD** if, whenever its premisses are derivable, then also its conclusion is derivable (with at most the same derivation height).

We are now going to prove that weakening and contraction are height-preserving admissible in **G3xD**, and that cut is admissible in **G3xD**. As measures for inductive proofs we use the weight of a formula and the height of a derivation, which are defined as usual [11]. We begin by showing that the restriction to atomic initial sequents, which is needed to have the propositional rules invertible, is not limitative in that initial sequents with arbitrary formulas are derivable in **G3xD**.

Lemma 1. *Every instance of $A, \Gamma \Rightarrow \Delta, A$ is derivable in* **G3xD**.

Proof. A standard induction on the weight of A.   □

Table 3. Modal rules

$$\frac{A \Rightarrow B \quad B \Rightarrow A}{\mathcal{O}A, \Gamma \Rightarrow \Delta, \mathcal{O}B} \; LR\text{-}E \qquad \frac{A \Rightarrow B}{\mathcal{O}A, \Gamma \Rightarrow \Delta, \mathcal{O}B} \; LR\text{-}M \qquad \frac{A \Rightarrow}{\mathcal{O}A, \Gamma \Rightarrow \Delta} \; LD$$

$$\frac{\Pi, A \Rightarrow B}{\mathcal{O}\Pi, \mathcal{O}A, \Gamma \Rightarrow \Delta, \mathcal{O}B} \; LR\text{-}R \qquad \frac{\Pi \Rightarrow B}{\mathcal{O}\Pi, \Gamma \Rightarrow \Delta, \mathcal{O}B} \; LR\text{-}K \qquad \frac{\Pi \Rightarrow}{\mathcal{O}\Pi, \Gamma \Rightarrow \Delta} \; LD^{\star}$$

Lemma 2. *The left and right rules of weakening:*

$$\frac{\Gamma \Rightarrow \Delta}{A, \Gamma \Rightarrow \Delta} \; LW \qquad \frac{\Gamma \Rightarrow \Delta}{\Gamma \Rightarrow \Delta, A} \; RW$$

are height-preserving admissible in **G3xD**.

Proof. The proof is a straightforward induction on the height of the derivation δ of $\Gamma \Rightarrow \Delta$. If the last step of δ is by a propositional rule, we proceed as for **G3cp**. If it is by a modal one, we proceed by adding A to the appropriate context of the conclusion of that instance of a modal rule. □

Lemma 3. *All propositional rules are height-preserving invertible in* **G3xD**, *that is the derivability of a conclusion of a propositional rule entails the derivability, with at most same derivation height, of its premiss(es).*

Proof. Same as for **G3cp**, see [11, Thm. 3.1.1]. □

Lemma 4. *The left and right rules of contraction:*

$$\frac{A, A, \Gamma \Rightarrow \Delta}{A, \Gamma \Rightarrow \Delta} \; LC \qquad \frac{\Gamma \Rightarrow \Delta, A, A}{\Gamma \Rightarrow \Delta, A} \; RC$$

are height-preserving admissible in **G3xD**.

Proof. The proof is by simultaneous induction on the height of the derivation δ of the premiss for left and right contraction. The base case is straightforward. For the inductive steps, if the last rule in δ is a propositional one, we use the height-preserving invertibility – Lemma 3 – of that rule. If it is a modal one, either one or no instance of the contraction formula A is principal in it, and the others are introduced in the context Γ for LC (Δ for RC) of its conclusion; we apply that instance of a modal rule with one less occurrence of A in Γ (Δ). □

Theorem 3. *The rule of cut:*

$$\frac{\Gamma \Rightarrow \Delta, D \quad D, \Pi \Rightarrow \Sigma}{\Gamma, \Pi \Rightarrow \Delta, \Sigma} \; Cut$$

is admissible in **G3xD**.

Proof. The proofs, one for each calculus, are by induction on the weight of the cut formula D with a sub-induction on the sum of the heights of the derivations of the two premisses (cut-height for short.). The proofs are, mostly, analogous to that for **G3cp**, for which we refer the reader to [11, Thm. 3.2.3]. We consider explicitly only the new cases that arises by the addition of the modal rules – i.e. those with $\mathcal{O}D$ as cut formula. We keep the numbering of cases in [11], therefore cases **(1)-(3)** are left out.

- **G3ED** **(4)** The cut formula $\mathcal{O}D$ is principal in the left premiss only.
Subcase (a). The right premiss is by rule $LR\text{-}E$, we transform

$$\cfrac{\Gamma \Rightarrow \Delta, \mathcal{O}D \quad \cfrac{A \Rightarrow B \quad B \Rightarrow A}{\mathcal{O}D, \mathcal{O}A, \Pi \Rightarrow, \Sigma, \mathcal{O}B} LR\text{-}E}{\Gamma, \mathcal{O}A, \Pi \Rightarrow \Delta, \Sigma, \mathcal{O}B} Cut \quad \text{into} \quad \cfrac{A \Rightarrow B \quad B \Rightarrow A}{\Gamma, \mathcal{O}A, \Pi \Rightarrow \Delta, \Sigma, \mathcal{O}B} LR\text{-}E$$

where the application of *Cut* has disappeared.
Subcase (b). The right premiss is by rule LD, we transform

$$\cfrac{\Gamma \Rightarrow \Delta, \mathcal{O}D \quad \cfrac{A \Rightarrow}{\mathcal{O}D, \mathcal{O}A, \Pi \Rightarrow, \Sigma} LD}{\Gamma, \mathcal{O}A, \Pi \Rightarrow \Delta, \Sigma} Cut \quad \text{into} \quad \cfrac{A \Rightarrow}{\mathcal{O}A, \Pi \Rightarrow \Delta, \Sigma} LD$$

(5) The cut formula $\mathcal{O}D$ is principal in both premisses.
Subcase (a). Both premisses are by rule $LR\text{-}E$, we have

$$\cfrac{\cfrac{A \Rightarrow D \quad D \Rightarrow A}{\mathcal{O}A, \Gamma \Rightarrow \Delta, \mathcal{O}D} LR\text{-}E \quad \cfrac{D \Rightarrow B \quad B \Rightarrow D}{\mathcal{O}D, \Pi \Rightarrow \Sigma, \mathcal{O}B} LR\text{-}E}{\mathcal{O}A, \Gamma, \Pi \Rightarrow \Delta, \Sigma, \mathcal{O}B} Cut$$

and we transform it into the following derivation that has two admissible cuts of lower cut-height and a cut formula of lesser weight

$$\cfrac{\cfrac{A \Rightarrow D \quad D \Rightarrow B}{A \Rightarrow B} Cut \quad \cfrac{B \Rightarrow D \quad D \Rightarrow A}{B \Rightarrow A} Cut}{\mathcal{O}A, \Gamma, \Pi \Rightarrow \Delta, \Sigma, \mathcal{O}B} LR\text{-}E$$

Subcase (b). The left premiss is by $LR\text{-}E$, and the right one by LD (observe that if the cut formula is principal in the left premiss, the left rule cannot be an instance of LD). We transform

$$\cfrac{\cfrac{A \Rightarrow D \quad D \Rightarrow A}{\mathcal{O}A, \Gamma \Rightarrow \Delta, \mathcal{O}D} LR\text{-}E \quad \cfrac{D \Rightarrow}{\mathcal{O}D, \Pi \Rightarrow \Sigma} LD}{\mathcal{O}A, \Gamma, \Pi \Rightarrow \Delta, \Sigma} Cut \quad \text{into} \quad \cfrac{\cfrac{A \Rightarrow D \quad D \Rightarrow}{A \Rightarrow} Cut}{\mathcal{O}A, \Gamma, \Pi \Rightarrow \Delta, \Sigma} LD$$

- **G3MD** **(4)** The cut formula is principal in the left premiss only.
Subcase (a). Right premiss by $LR\text{-}M$, similar to the case for **G3ED**.
Subcase (b). Right premiss by LD, same as for **G3ED**.

(5)   The cut formula is principal in both premisses.
Subcase (a).   Both premisses are by rule LR-M, we transform

$$\cfrac{\cfrac{A \Rightarrow D}{\mathcal{O}A, \Gamma \Rightarrow \Delta, \mathcal{O}D}\; LR\text{-}M \quad \cfrac{D \Rightarrow B}{\mathcal{O}D, \Pi \Rightarrow \Sigma, \mathcal{O}B}\; LR\text{-}M}{\mathcal{O}A, \Gamma, \Pi \Rightarrow \Delta, \Sigma, \mathcal{O}B}\; Cut \quad \text{into} \quad \cfrac{\cfrac{A \Rightarrow D \quad D \Rightarrow B}{A \Rightarrow B}\; Cut}{\mathcal{O}A, \Gamma, \Pi \Rightarrow \Delta, \Sigma, \mathcal{O}B}\; LR\text{-}M$$

Subcase (b).   The left premiss is by LR-M, and the right one by LD, we transform

$$\cfrac{\cfrac{A \Rightarrow D}{\mathcal{O}A, \Gamma \Rightarrow \Delta, \mathcal{O}D}\; LR\text{-}M \quad \cfrac{D \Rightarrow}{\mathcal{O}D, \Pi \Rightarrow \Sigma}\; LD}{\mathcal{O}A, \Gamma, \Pi \Rightarrow \Delta, \Sigma}\; Cut \quad \text{into} \quad \cfrac{\cfrac{A \Rightarrow D \quad D \Rightarrow}{A \Rightarrow}\; Cut}{\mathcal{O}A, \Gamma, \Pi \Rightarrow \Delta, \Sigma}\; LD$$

- **G3RD** **(4)**   The cut formula is principal in the left premiss only.
Subcase (a). Right premiss by LR-R, Similar to the case for **G3ED**.
Subcase (b). Right premiss by $LD^\star$, we transform

$$\cfrac{\Gamma \Rightarrow \Delta, \mathcal{O}D \quad \cfrac{\Pi' \Rightarrow}{\mathcal{O}D, \mathcal{O}\Pi', \Pi'' \Rightarrow, \Sigma}\; LD^\star}{\Gamma, \mathcal{O}\Pi', \Pi'' \Rightarrow \Delta, \Sigma}\; Cut \quad \text{into} \quad \cfrac{\Pi' \Rightarrow}{\mathcal{O}\Pi', \Pi'' \Rightarrow \Delta, \Sigma}\; LD^\star$$

(5)   The cut formula is principal in both premisses.
Subcase (a).   Both premisses are by rule LR-R, we transform

$$\cfrac{\cfrac{\Gamma' \Rightarrow D}{\mathcal{O}\Gamma', \Gamma'' \Rightarrow \Delta, \mathcal{O}D}\; LR\text{-}R \quad \cfrac{D, \Pi' \Rightarrow B}{\mathcal{O}D, \mathcal{O}\Pi', \Pi'' \Rightarrow \Sigma, \mathcal{O}B}\; LR\text{-}R}{\mathcal{O}\Gamma', \Gamma'', \mathcal{O}\Pi', \Pi'' \Rightarrow \Delta, \Sigma, \mathcal{O}B}\; Cut \quad \text{into} \quad \cfrac{\cfrac{\Gamma' \Rightarrow D \quad D, \Pi' \Rightarrow B}{\Gamma', \Pi' \Rightarrow B}\; Cut}{\mathcal{O}\Gamma', \mathcal{O}\Pi', \Gamma'', \Pi'' \Rightarrow \Delta, \Sigma, \mathcal{O}B}\; LR\text{-}R$$

Subcase (b).   The left premiss is by LR-R, and the right one by $LD^\star$,[3] we transform

$$\cfrac{\cfrac{\Gamma' \Rightarrow D}{\mathcal{O}\Gamma', \Gamma'' \Rightarrow \Delta, \mathcal{O}D}\; LR\text{-}M \quad \cfrac{D, \Pi' \Rightarrow}{\mathcal{O}D, \mathcal{O}\Pi', \Pi'' \Rightarrow \Sigma}\; LD^\star}{\mathcal{O}\Gamma', \Gamma'', \mathcal{O}\Pi', \Pi'' \Rightarrow \Delta, \Sigma}\; Cut \quad \text{into} \quad \cfrac{\cfrac{\Gamma' \Rightarrow D \quad D, \Pi' \Rightarrow}{\Gamma', \Pi' \Rightarrow}\; Cut}{\mathcal{O}\Gamma', \mathcal{O}\Pi', \Gamma', \Pi'' \Rightarrow \Delta, \Sigma}\; LD^\star$$

- **G3KD**   All cases **(4a)-(5b)** are similar to the respective ones for **G3RD**, for a proof see [10].   □

As a corollary of the admissibility of the structural rules, we have that each of the sequent calculi defined above has the subformula property, and, therefore,

---

[3] Even though the formula $D^\star := \mathcal{O}A \rightarrow \neg\mathcal{O}\neg A$ is derivable from $D := \neg\mathcal{O}\bot$ in axiomatic systems for logics at least as strong as **RD**, the following transformation would not have been possible if we had taken LD instead of $LD^\star$ as a rule of **G3RD**. The same holds for the case **(5b)** of **G3KD**. In this respect our approach is not completely modular.

derivability of a sequent in it is decidable: we have to apply an exhaustive root-first proof search procedure. Observe that, given that the modal rules are not invertible, in the root-first decision procedure we may need back-tracking when they are applied.

3.2 Equivalence with the Axiomatic Calculi

It is now time to show that our sequent calculi are equivalent to the deontic logics of Chap. 2. We write **G3xD** $\vdash \Gamma \Rightarrow \Delta$ if the sequent $\Gamma \Rightarrow \Delta$ is derivable in **G3xD**, and we say that a formula A is derivable in **G3xD** whenever **G3xD** $\vdash \Rightarrow A$. We begin by proving the following

Lemma 5. *All the axioms of the axiomatic system* **xD** *are derivable in* **G3xD**.

Proof. A straightforward root-first application of the rules of the appropriate sequent calculus, possibly using Lemma 1. □

Next we prove the equivalence of the axiomatic systems for deontic logics and of our sequent calculi for them in the sense that whenever a sequent $\Gamma \Rightarrow \Delta$ is derivable in **G3xD**, its characteristic formula $\bigwedge \Gamma \to \bigvee \Delta$ is derivable in **xD**, where the empty antecedent stands for $\neg\bot$ and the empty succedent for $\bot$. As a consequence we will get that each of our sequent calculi is sound and complete with respect to the appropriate class of neighborhood models of Chap. 2.

Theorem 4. *Derivability in the sequent system* **G3xD** *and in the axiomatic system* **xD** *are equivalent, i.e.*

$$\textbf{G3xD} \vdash \Gamma \Rightarrow \Delta \qquad \textit{iff} \qquad \textbf{xD} \vdash \bigwedge \Gamma \to \bigvee \Delta$$

Proof. Let $\Gamma \Rightarrow \Delta$ be $A_1, \ldots, A_n \Rightarrow B_1, \ldots, B_m$. To prove the right-to-left implication, we argue by induction of the height of the axiomatic derivation in **xD**. The base case is covered by Lemma 5. For the inductive step, the case of MP follows by the admissibility of Cut and the invertibility of rule $R \to$. For the modal rule Rx of **xD**, we present explicitly only the more convoluted case of rule RE, all the others being simplifications thereof. If the last step is by RE, then $B_1 \vee \ldots \vee B_m = \mathcal{O}C \leftrightarrow \mathcal{O}D$. We know that (in **ED**) we have derived $\mathcal{O}C \leftrightarrow \mathcal{O}D$ from $C \leftrightarrow D$. Remember that $C \leftrightarrow D$ is defined as $(C \to D) \wedge (D \to C)$. Thus we assume, by inductive hypothesis (IH), that **G3ED** $\vdash \Rightarrow C \to D$ and **G3ED** $\vdash \Rightarrow D \to C$, and we proceed as follows

$$
\cfrac{
  \cfrac{
    \cfrac{\overline{\Rightarrow C \to D}\ IH}{C \Rightarrow D}\ \text{Lem. 3} \quad
    \cfrac{\overline{\Rightarrow D \to C}\ IH}{D \Rightarrow C}\ \text{Lem. 3}
  }{
    \cfrac{\mathcal{O}C, A_1, \ldots A_n \Rightarrow \mathcal{O}D}{A_1, \ldots A_n \Rightarrow \mathcal{O}C \to \mathcal{O}D}\ R \to
  }\ LR\text{-}E
  \qquad
  \cfrac{
    \cfrac{\overline{\Rightarrow D \to C}\ IH}{D \Rightarrow C}\ \text{Lem. 3} \quad
    \cfrac{\overline{\Rightarrow C \to D}\ IH}{C \Rightarrow D}\ \text{Lem. 3}
  }{
    \cfrac{\mathcal{O}D, A_1, \ldots A_n \Rightarrow \mathcal{O}C}{A_1, \ldots A_n \Rightarrow \mathcal{O}D \to \mathcal{O}C}\ R \to
  }\ LR\text{-}E
}{
  A_1, \ldots, A_n \Rightarrow (\mathcal{O}C \to \mathcal{O}D) \wedge (\mathcal{O}D \to \mathcal{O}C)
}\ R\wedge
$$

For the converse implication, we assume **G3xD** $\vdash \Gamma \Rightarrow \Delta$, and show, by induction on the height of the derivation in sequent calculus, that $\mathbf{xD} \vdash \bigwedge \Gamma \to \bigvee \Delta$. If the derivation has height 0, we have an initial sequent – so $A_{i(\leq n)} = B_{j(\leq m)} = p_k$ – or an instance on $L\bot$ – thus $A_{i(\leq n)} = \bot$. In both cases the claim holds. If the height is $n + 1$, we consider the last rule applied in the derivation. If it is a propositional one, the proof is straightforward. If it is a modal rule, we argue by cases.

We begin by considering rules LD and $LD^\star$. Suppose we are in **G3ED** or **G3MD** and we have derived $A_1, \ldots, \mathcal{O}A_i, \ldots, A_n \Rightarrow B_1, \ldots B_m$ from $A_i \Rightarrow$. By IH, $\mathbf{xD} \vdash A_i \to \bot$, and we know that $\mathbf{xD} \vdash \bot \to A_i$. Thus by RE (or RM), we get $\mathbf{xD} \vdash \mathcal{O}A_i \to \mathcal{O}\bot$. Now, by a MP with the axiom D, $\mathbf{xD} \vdash \neg\mathcal{O}A_i$. By some easy propositional steps we conclude $\mathbf{xD} \vdash (A_1 \wedge \ldots \wedge \mathcal{O}A_i \wedge \ldots \wedge A_n) \to (B_1 \vee \ldots \vee B_m)$. The case of $LD^\star$ can be treated analogously.

If the last step of a derivation in **G3ED** is by rule $LR\text{-}E$, then $A_{i(\leq n)} = \mathcal{O}A$ and $B_{j(\leq m)} = \mathcal{O}B$, and we have derived $A_1, \ldots, \mathcal{O}A, \ldots, A_n \Rightarrow B_1, \ldots \mathcal{O}B, \ldots B_m$ from $A \Rightarrow B$ and $B \Rightarrow A$. By IH, $\mathbf{ED} \vdash A \leftrightarrow B$, thus $\mathbf{ED} \vdash \mathcal{O}A \leftrightarrow \mathcal{O}B$. We conclude $\mathbf{ED} \vdash (A_1 \wedge \ldots \wedge \mathcal{O}A \wedge \ldots \wedge A_n) \to (B_1 \vee \ldots \vee \mathcal{O}B \vee \ldots \vee B_m)$. The cases of the other LR-rules, in the respective calculi, are left to the reader. $\square$

4   Related Works

Our deontic rules $LR\text{-}x$ are inspired by the rule $LR\text{-}\Box$ – here called $LR\text{-}K$ – introduced in [10]; the rules LD and $LD^\star$ are inspired by a rule in [13]. Observe that we have also introduced a sequent calculus **G3x** for the non-normal logic **x** that is obtained by dropping the deontic axiom $D^{(\star)}$ from **xD**. In the literature there are other sequent calculi for non-normal modal logics [6,7]. The rules we have introduced are similar to the ones of [6,7], where there is no rule that corresponds to the deontic axiom $\neg\mathcal{O}\bot$. One major point of difference is that in our approach contraction is height-preserving admissible, whereas in [6,7] it is in general not even admissible. Given that contraction can be as bad as cut for root-first proof search – we may continue to duplicate some formula forever – we believe this is a substantial improvement. In particular the calculi introduced in [7] have contraction as an implicit rule because sequents are defined as sets and not as multisets; furthermore the rule of weakening is not eliminable because the modal rules are not weakened – i.e. their conclusion doesn't introduce contexts. In [6], where sequents are defined as multisets, contraction is not eliminable from the calculus SC-**ED** because the deontic axiom $D^\star$ is expressed by the rule

$$\frac{\Rightarrow A, B \qquad A, B \Rightarrow}{\mathcal{O}A, \mathcal{O}B, \Gamma \Rightarrow \Delta} \; D\text{-}2 \tag{3}$$

where the premises have exactly two formulas, and therefore it is not possible to permute upward contraction. Observe that the presence of a primitive, and not eliminable, rule of contraction makes the elimination of cut more problematic: in most cases we cannot eliminate the cut directly, but we have to consider the rule known as multicut, see [11, p. 88]. These considerations show how an apparently minor change in the formulation of some rule can have major effects on the structural rules of the calculus.

For a different route towards sequent calculi for non-normal modal logics, see [4] where labelled sequent calculi for several non-normal logics are introduced. The approach in [4], which has the advantage of being completely modular, proceeds indirectly by introducing labelled sequent calculi not for non-normal modal logics, but for provably equivalent multi-modal normal logics, see [12] for labelled sequent calculi for normal modal logics.

We know of no systematic treatment of non-normal deontic logics by means of sequent calculi or natural deduction. [5] introduces hyper-sequent calculi for the interaction of modal logic with standard deontic logic and with a non-normal logic where obligation is weak permission – where the interaction is based on the Ought-implies-Can-principle, and proves that cut is eliminable from both calculi.

Although monadic deontic logics are a useful formal tool in many cases, there are other situations where dyadic ones seem preferable [1,14]. In the future we plan to introduce labelled sequent calculi for dyadic deontic logics following the approach used in [9] for the logic of counterfactuals.

References

1. van Benthem, J., Grossi, D., Fenrong, L.: Priority Structures in Deontic Logic. Theoria 80, 116–152 (2014)
2. Calardo, E.: Non-normal Modal Logics, Quantification, and Deontic Dilemmas. Ph.D. Thesis, Università di Bologna (2013)
3. Chellas, B.F.: Modal Logic: An Introduction. CUP, Cambridge (1980)
4. Gilbert, D., Maffezioli, P.: Modular Sequent Calculi for Classical Modal Logics. Studia Logica (forthcoming)
5. Gratzl, N.: Sequent Calculi for Multi-modal Logic with Interaction. In: Grossi, D., Roy, O., Huang, H. (eds.) LORI. LNCS, vol. 8196, pp. 124–134. Springer, Heidelberg (2013)
6. Indrzejczak, A.: Admissiblity of Cut in Congruent Modal Logics. Logic and Logical Philosophy 21, 189–203 (2011)
7. Lavendhomme, R., Lucas, L.: Sequent Calculi and Decision Procedures for Weak Modal Systems. Studia Logica 65, 121–145 (2000)
8. McNamara, P.: Deontic Logic. In: Gabbay, D.M., Woods, J. (eds.) Handbook of the History of Logic, vol. 7, pp. 197–288. Elsevier, Amsterdam (2006)
9. Negri, S.: Extending the Scope of Labelled Sequent Calculi: the Case of Classical Counterfactuals (Abstract) (2013),
   http://www.lix.polytechnique.fr/colloquium2013/abstracts/sarneg.pdf
10. Negri, S., Hakli, R.: Does the Deduction Theorem Fail for Modal Logic? Synthese 184, 849–867 (2012)
11. Negri, S., von Plato, J.: Structural Proof Theory. CUP, Cambridge (2001)
12. Negri, S., von Plato, J.: Proof Analysis. CUP, Cambridge (2011)
13. Valentini, S.: The Sequent Calculus for the Modal Logic D. Bollettino dell'Unione Matematica Italiana 7, 455–460 (1993)
14. van der Torre, L.: Reasoning about Obligations. Tinbergen Institute Research Series, Amsterdam (1997)

"Sing and Dance!"
Input/Output Logics without Weakening

Xavier Parent and Leendert van der Torre

University of Luxembourg
x.parent@mail.com, leonvandertorre@uni.lu

Abstract. Makinson and van der Torre [13] introduce a number of input/output (I/O) logics to reason about conditional norms. The key idea is to make obligations relative to a given set of conditional norms. The meaning of the normative concepts is, then, given in terms of a set of procedures yielding outputs for inputs. Using the same methodology, Stolpe [19,20] has developed some more I/O logics to include systems without the rule of weakening of the output (or principle of inheritance). We extend Stolpe's account in two directions. First, we show how to make it support reasoning by cases—a common form of reasoning. Second, we show how to inject a new (as we call it, "aggregative") form of cumulative transitivity, which we think is more suitable for normative reasoning. The main outcomes of the paper are soundness and completeness theorems for the proposed systems with respect to their intended semantics.

1 Introduction

Makinson and van Torre [13] introduce a number of input/output (I/O) logics to reason about conditional norms. The key idea is to make obligations relative to a given set of conditional norms. The meaning of the normative concepts is, then, given in terms of a set of procedures yielding outputs for inputs. A number of I/O operations are studied in the aforementioned paper [13]. It is shown that they correspond to a series of proof systems of increasing strength. I/O logic promotes a paradigm shift from modal logic to what has recently been called "norm-based semantics" by Hansen [9, p. 288]. The core idea is to explain the truths of deontic logic, not by some set of possible worlds among which some are ideal or at least better than others, but with reference to an explicit set of given norms or existing moral standards. The founders of I/O logic mostly criticized the modal logic paradigm for—to use Quine's famous expression—having been "conceived in the sin": the sin of assuming that norms bear truth-values. Thus, their main motivation was philosophical. Still, one reason why modal logic has been so popular in deontic logic is that it is a general framework, which provides us with plenty of freedom to pick and choose the axiom schemata we think are right. Whatever philosophical reservations one may have about the use of modal logic in deontic logic, one would like to know if, or to what extent, norm-based

F. Cariani et al. (Eds.): DEON 2014, LNAI 8554, pp. 149–165, 2014.
© Springer International Publishing Switzerland 2014

semantics in general, and I/O logic in particular, can offer the same kind of flexibility.

In this paper, we focus on the so-called rule of Weakening of the Output (WO), which all the I/O operations defined by Makinson and van Torre [13] satisfy. The rule may be given the following form, where the conditional obligation for x given a is written as (a, x), and $\vdash$ stands for the deducibility relation in propositional logic:

$$\text{WO } \frac{(a, x) \qquad x \vdash y}{(a, y)}$$

This is also known as the "principle of inheritance". It has been called into question, mostly in connection with the deontic paradoxes [4,5,10,8] and the question of how to accommodate conflicts between obligations—see, e.g., [6,7]. This raises the question whether the framework may be generalized to include systems without output weakening; if yes, how.

A first step towards answering the above question was made by Stolpe [19,20]. He considers two of the four standard I/O operations defined by Makinson and van der Torre [13], namely the so-called simple-minded I/O operation out_1, and the so-called reusable I/O operation out_3. Both develop output by detachment. While out_1 spells out the basic mechanism used to achieve this, out_3 extends it to cover iteration of successive detachments. For both operations, a suitable semantics is given, for which the rule WO fails. Each semantics comes with a sound and complete axiomatic characterization. The present paper extends Stolpe's account in two ways.

First, we show how to refine Stolpe's account to make the latter one support reasoning by cases—this is a common form of reasoning. We look at the I/O operation out_2, called "basic" by Makinson and van der Torre [13]. Its distinctive feature is that it validates the rule OR:

$$\text{OR } \frac{(a, x) \qquad (b, x)}{(a \vee b, x)}$$

The present paper shows how to incorporate such a rule. We provide a suitable semantics for the I/O operation, a proof system for it, and a completeness result linking the two.

Second, we show how to integrate other forms of cumulative transitivity. There is no doubt that some form of transitivity is required for an adequate account of norms and normative systems. One reason is that transitivity serves as a means of binding together different parts of a code. For instance, a legal system is invariably organized into different modules which are interrelated. Rules from one module stipulate legal consequences that are used as premises for some rules part of another module. Penal law may, e.g., state that grand larseny ought to be added to a person's criminal record, whereas administrative law may stipulate that an unblemished record is a prerequisite for public office. Some form of transitivity is required to go from grand larseny to the bar to holding public office.

Stolpe uses the rule of (as he calls it) "mediated cumulative transitivity" (MCT):

$$\text{MCT} \quad \frac{(a, x') \qquad x' \vdash x \qquad (a \wedge x, y)}{(a, y)}$$

As we will see in Section 4, given the other rules of his system, MCT turns out to be equivalent to the rule of cumulative transitivity (CT), as initially used by Makinson and van der Torre [13]:

$$\text{CT} \quad \frac{(a, x) \qquad (a \wedge x, y)}{(a, y)}$$

We look at the following alternative (call it "aggregative") variant, first introduced in a companion paper [16]:

$$\text{ACT} \quad \frac{(a, x) \qquad (a \wedge x, y)}{(a, x \wedge y)}$$

The counterexamples usually given to CT in the literature [15,11,12] no longer work, when ACT is used in place of CT. This is because they all rely on the intuition that the obligation of y ceases to hold when the obligation of (a, x) is violated. The following example, due to Broome [1, § 7.4], may be used to illustrate this point:

You ought to exercise hard everyday	$(\top, x)$
If you exercise hard everyday, you ought to eat heartily	(x, y)
?* You ought to eat heartily	?* $(\top, y)$

Intuitively, the obligation to eat heartily no longer holds, if you take no exercise. In this example, the correct conclusion is $(\top, x \wedge y)$, and not $(\top, y)$. Thus, ACT appears to be more suitable for normative reasoning, because it keeps track of what has been previously detached.

The layout of this paper is as follows. Section 2 lays the groundwork, tackling out_1 in essentially the same way as Stolpe does. Section 3 extends the account so that reasoning by cases is supported. Section 4 shows how to inject the aggregative form of cumulative transitivity mentioned above. The main achievement of the paper is the establishment of soundness and completeness theorems for the proposed systems with respect to their intended semantics. We do not give all the details of the soundness and completeness proofs, but we outline the main steps.[1] Section 5 discusses some properties satisfied by the I/O operations defined in this paper.

---

[1] The detailed proofs will be included in the journal version of the present paper.

2   Developing the Output by Detachment (out_1)

We start with the simple-minded I/O operation out_1. The I/O operation to be defined here is noted $\mathcal{O}_1$. It is essentially a variation on the I/O operation PN_1 put forth by Stolpe [19,20]. The main reason for including such an operation in our study is that the completeness result for it will be needed for subsequent developments.

First, some definitions are needed. A normative code is a set N of conditional obligations. A conditional obligation is a pair $(a,\ x)$, where a and x are two formulae of classical propositional logic. We use this notation instead of $\bigcirc(x \mid a)$, because the latter has distinct interpretations in the literature. In the notation $(a,\ x)$, the first element a is called the body of the rule, and is thought of as an input, representing some condition or situation. The second element x is called the head of the rule, and is thought of as an output, representing what the norm tells us to be obligatory in that situation. We use the standard notation $(\top, x)$ for the unconditional obligation of x, where $\top$ is a zero-place connective standing for 'tautology'. In I/O logic, the main construct has the form

$$x \in out(N, a)$$

Intuitively: given input a (state of affairs), x (obligation) is in the output under norms N. An equivalent notation is: $(a, x) \in out(N)$. The I/O operations to be defined in this paper will be denoted by the symbol $\mathcal{O}$ in order to avoid any confusion with out and $\bigcirc$.

Some further notation. $\mathcal{L}$ is the set of all formulae of classical propositional logic. Given an input $A \subseteq \mathcal{L}$, and a set N of norms, $N(A)$ denotes the image of N under A, i.e., $N(A) = \{x : (a, x) \in N$ for some $a \in A\}$. $Cn(A)$ denotes the set $\{x : A \vdash x\}$, where $\vdash$ is the deducibility relation used in classical propositional logic. The notation $x \dashv\vdash y$ is short for $x \vdash y$ and $y \vdash x$. We use PL as an abbreviation for (classical) propositional logic. Given $M \subseteq N$, we denote by $h(M)$ the set of all the heads of elements of M, viz $h(M) = \{x : (a, x) \in M\}$.

Definition 1 (Semantics). $x \in \mathcal{O}_1(N, A)$ *if and only if there is some finite* $M \subseteq N$ *such that*

- $M(Cn(A)) \neq \emptyset$, *and*
- $x \dashv\vdash \wedge M(Cn(A))$

Intuitively: x is equivalent to the conjunction of heads of rules in some $M \subseteq N$ that are all triggered by input A.

The main difference between $\mathcal{O}_1$ and PN_1 arises when A does not trigger any norm, viz. $M(Cn(A)) = \emptyset$ for all $M \subseteq N$. In this limiting case, PN_1 outputs the set of all tautologies, while $\mathcal{O}_1$ outputs nothing. Von Wright [21, pp. 152-4] argues, rightly in our view, that the obligation of $\top$ does not express a genuine prescription.

$\mathcal{O}_1$ is monotonic with respect to the input set. The latter claim requires a careful and detailed proof, because there is a pitfall to avoid.

Theorem 1 (Factual Monotony). *We have $\mathcal{O}_1(N, A) \subseteq \mathcal{O}_1(N, B)$ whenever $Cn(A) \subseteq Cn(B)$.*

Proof. Assume $x \in \mathcal{O}_1(N, A)$ and $Cn(A) \subseteq Cn(B)$. From the former, there is some finite $M_1 \subseteq N$ such that $M_1(Cn(A)) \neq \emptyset$, and

1. $x \dashv\vdash \wedge M_1(Cn(A))$

There is no guarantee that input set B does not trigger more pairs in M_1 than A does. To circumvent this problem, the argument takes a detour through the set

$$M_1^- = \{(c, y) \in M_1 : c \in Cn(A)\}$$

Thus, M_1^- is M_1 "stripped of" all the pairs that are not triggered by A. We have $M_1(Cn(A)) = M_1^-(Cn(A))$. We also have $M_1^-(Cn(A)) = M_1^-(Cn(B))$, viz.

$$\{y : (c, y) \in M_1^-, c \in Cn(A)\} = \{y : (c, y) \in M_1^-, c \in Cn(B)\}$$

The $\subseteq$-direction follows from the second opening assumption, $Cn(A) \subseteq Cn(B)$. The $\supseteq$-direction follows from the definition of M_1^-. The argument may, then, be continued thus:

2. $x \dashv\vdash \wedge M_1^-(Cn(A))$
3. $x \dashv\vdash \wedge M_1^-(Cn(B))$

Thus, $x \in \mathcal{O}_1(N, B)$ as required. □

It immediately follows that $\mathcal{O}_1(N, A) \subseteq \mathcal{O}_1(N, B)$ whenever $A \subseteq B$.

We set $\mathcal{O}_1(N) = \{(A, x) : x \in \mathcal{O}_1(N, A)\}$. The notion of derivation is defined as in standard I/O logic except that $(\top, \top)$ is not allowed to appear in a derivation unless it is explicitly given in the set N of assumptions.

Definition 2 (Proof System). $(a, x) \in \mathcal{D}_1(N)$ *if and only if there is a derivation of (a, x) from N using the rules $\{SI, EQ, AND\}$:*

$$SI \frac{(a, x) \qquad b \vdash a}{(b, x)} \qquad\qquad EQ \frac{(a, x) \qquad x \dashv\vdash y}{(a, y)}$$

$$AND \frac{(a, x) \qquad (a, y)}{(a, x \wedge y)}$$

Where A is a set of formulae, $(A, x) \in \mathcal{D}_1(N)$ means that $(a, x) \in \mathcal{D}_1(N)$, for some conjunction a of formulae, all taken from a finite subset of A. $\mathcal{D}_1(N, A)$ is $\{x : (A, x) \in \mathcal{D}_1(N)\}$.

Theorem 2. *$\mathcal{O}_1$ validates the rules of $\mathcal{D}_1$ (for individual formulae a).*

Proof. The argument is straightforward, and left to the reader. (For SI, the same trick as in the proof of Theorem 1 must be used.) □

Theorem 3 (Soundness). $\mathcal{D}_1(N, A) \subseteq \mathcal{O}_1(N, A)$

Proof. The proof is by induction on the length of the derivation, using Theorems 1 and 2. □

Theorem 4 (Completeness). $\mathcal{O}_1(N, A) \subseteq \mathcal{D}_1(N, A)$

Proof. Assume $x \in \mathcal{O}_1(N, A)$. So there exists some finite $M \subseteq N$ such that $M(Cn(A)) = \{x_1, ..., x_n\} \neq \emptyset$ and $x \Vdash \wedge_{i=1}^n x_i$. For each x_i, there is some $a_i \in Cn(A)$ such that $(a_i, x_i) \in M$. For each a_i, there is also a conjunction b_i of elements in A such that $b_i \vdash a_i$. A derivation of (A, x) from M, and hence from N, is shown below.

$$\text{EQ} \frac{\text{AND} \frac{\text{SI} \dfrac{(a_1, x_1)}{(\wedge_{i=1}^n b_i, x_1)} \quad \cdots\cdots \quad \text{SI} \dfrac{(a_n, x_n)}{(\wedge_{i=1}^n b_i, x_n)}}{(\wedge_{i=1}^n b_i, \wedge_{i=1}^n x_i)}}{(\wedge_{i=1}^n b_i, x)}$$

This is a derivation of (A, x), as $\wedge_{i=1}^n b_i$ is a conjunction of elements in A. □

3   Reasoning by Cases

In this section, the account described in the previous section is extended to the basic operation out_2, which supports reasoning by cases. The I/O operation is denoted $\mathcal{O}_2$, and the corresponding proof system is called $\mathcal{D}_2$. We call a set of formulae complete if it is either equal to $\mathcal{L}$ or maximal consistent (the set is consistent, and none of its proper extensions is consistent).

Definition 3. $\mathcal{O}_2(N, A) = \cap\{\mathcal{O}_1(N, V) : A \subseteq V, V \, complete\}$.

Theorem 5. $\mathcal{O}_1(N, A) \subseteq \mathcal{O}_2(N, A)$.

Proof. Let $x \in \mathcal{O}_1(N, A)$. Let V be a complete set such that $A \subseteq V$. By Theorem 1, $x \in \mathcal{O}_1(N, V)$. By Definition 3, $x \in \mathcal{O}_2(N, A)$ as required. □

Theorem 6 (Factual Monotony). $\mathcal{O}_2(N, A) \subseteq \mathcal{O}_2(N, B)$ *if* $Cn(A) \subseteq Cn(B)$

Proof. Assume $x \in \mathcal{O}_2(N, A)$ and $Cn(A) \subseteq Cn(B)$. Let V be a complete set such that $B \subseteq V$. We have $Cn(B) \subseteq Cn(V) = V$. From this and the second opening assumption, $Cn(A) \subseteq V$. So, $A \subseteq V$. From this and the first opening assumption, $x \in \mathcal{O}_1(N, V)$. Thus, $x \in \mathcal{O}_2(N, B)$. □

Definition 4. $(a, x) \in \mathcal{D}_2(N)$ *if and only if there is a derivation of* (a, x) *from N using the rules of $\mathcal{D}_1$ supplemented with*

$$\text{OR} \frac{(a, x) \qquad (b, x)}{(a \vee b, x)}$$

The next theorem appeals to the fact that $\mathcal{O}_1$ validates AND and EQ for an input set of arbitrary cardinality rather than just a singleton set. The argument is virtually the same in both cases. Details are omitted.

Theorem 7. $\mathcal{O}_2$ *validates the rules of $\mathcal{D}_2$ (for individual formulae a).*

Proof. For SI. Assume $x \in \mathcal{O}_2(N, a)$ with $b \vdash a$. Let V be a complete set such that $b \in V$. From $b \vdash a$, we get $a \in V$. By Definition 3, we infer $x \in \mathcal{O}_1(N, V)$. This shows that $x \in \mathcal{O}_2(N, b)$.

For AND. Assume $x \in \mathcal{O}_2(N, a)$ and $y \in \mathcal{O}_2(N, a)$. Let V be a complete set such that $a \in V$. By Definition 3, $x \in \mathcal{O}_1(N, V)$ and $y \in \mathcal{O}_1(N, V)$. Since $\mathcal{O}_1$ validates AND, $x \wedge y \in \mathcal{O}_1(N, V)$. This shows that $x \wedge y \in \mathcal{O}_2(N, a)$.

For OR. Assume $x \in \mathcal{O}_2(N, a)$ and $x \in \mathcal{O}_2(N, b)$. Let V be a complete set containing $a \vee b$. Since V is complete, either $a \in V$ or $b \in V$. Assume that the first applies. In that case, $x \in \mathcal{O}_1(N, V)$, by the first opening assumption and Definition 3. Assume the second applies. In that case $x \in \mathcal{O}_1(N, V)$, by the second opening assumption and Definition 3. Either way, $x \in \mathcal{O}_1(N, V)$, and thus $x \in \mathcal{O}_2(N, a \vee b)$ as required.

For EQ, assume $x \in \mathcal{O}_2(N, a)$ and $x \dashv\vdash y$. Let V be a complete set containing a. By Definition 3, $x \in \mathcal{O}_1(N, V)$. Since $\mathcal{O}_1$ validates EQ, $y \in \mathcal{O}_1(N, V)$, and so $y \in \mathcal{O}_2(N, a)$ as required. $\qquad\square$

Theorem 8 (Soundness). $\mathcal{D}_2(N, A) \subseteq \mathcal{O}_2(N, A)$.

Proof. Same argument as before, but using Theorems 6 and 7. $\qquad\square$

Theorem 9 (Completeness). $\mathcal{O}_2(N, A) \subseteq \mathcal{D}_2(N, A)$.

Proof. We give an outline of the proof for a singleton input set $\{a\}$. The proof may easily be generalized to an input set of arbitrary cardinality. For ease of exposition, throughout the proof we write (SI,AND) to indicate an application of SI followed by that of AND. We break the argument into two cases.

Case 1: a is inconsistent. In this case, there is exactly one complete set V containing a; it is $\mathcal{L}$. So $\mathcal{O}_2(N, a) = \mathcal{O}_1(N, \mathcal{L})$. Let $x \in \mathcal{O}_1(N, \mathcal{L})$. This means that $x \dashv\vdash \wedge_{i=1}^{n} x_i$, for $x_1, ..., x_n \in h(N)$. Let $a_1, ..., a_n$ be the bodies of the rules in question. We have $a \vdash \wedge_{i=1}^{n} a_i$. A derivation of (a, x) from N may, then, be obtained as shown below.

$$\cfrac{\cfrac{\cfrac{(a_1, x_1) \quad \cdots \quad (a_n, x_n)}{(\wedge_{i=1}^{n} a_i, \wedge_{i=1}^{n} x_i)} \text{ (SI,AND)} \qquad \cfrac{\wedge_{i=1}^{n} x_i \dashv\vdash x}{} \text{ EQ}}{(\wedge_{i=1}^{n} a_i, x)} \text{ SI} \qquad a \vdash \wedge_{i=1}^{n} a_i}{(a, x)}$$

Case 2: a is consistent. Assume (for reductio) that $x \in \mathcal{O}_2(N, a)$ and that $x \notin \mathcal{D}_2(N, a)$. From the former, $x \dashv\vdash \wedge_{i=1}^{n} x_i$, for $x_1, ..., x_n \in h(N)$. In order to derive the contradiction that $x \notin \mathcal{O}_2(N, a)$, we start by showing that $\{a\}$ can be extended to some "maximal" $V \supseteq \{a\}$ such that $x \notin \mathcal{D}_2(N, V)$. By maximal, we mean that for all $V' \supset V$, $x \in \mathcal{D}_2(N, V')$. Thus, V is amongst the "biggest" input sets V containing a and not making x derivable.

V is built from a sequence of sets $V_0, V_1, V_2, \ldots$ as follows. Consider an enumeration $x_1, x_2, x_3, \ldots$ of all the formulae. We define:

$$V_0 = \{a\}$$

$$V_n = \begin{cases} V_{n-1} \cup \{x_n\}, & \text{if } x \notin \mathcal{D}_2(N, V_{n-1} \cup \{x_n\}) \\ V_{n-1}, & \text{otherwise} \end{cases}$$

$$V = \cup\{V_n : n \geq 0\}$$

It is a straightforward matter to show the following:

Fact 1. $x \notin \mathcal{D}_2(N, V_n)$, for all $n \geq 0$.

Fact 2. $V_n \subseteq V$, for all $n \geq 0$.

Fact 3. For every finite subset $V' \subseteq V$, $V' \subseteq V_n$, for some $n \geq 0$.

By Fact 2, V includes $\{a\}$ $(=V_0)$. The argument may be continued thus:

Claim 1. $x \notin \mathcal{D}_2(N, V)$.

Proof of the claim. Assume, to reach a contradiction, that $x \in \mathcal{D}_2(N, V)$. By compactness for $\mathcal{D}_2$, $x \in \mathcal{D}_2(N, V')$ for some finite $V' \subseteq V$. By Fact 3, $V' \subseteq V_n$ for some $n \geq 0$. By monotony in the right argument, $x \in \mathcal{D}_2(N, V_n)$. This contradicts Fact 1.

Claim 2. For all $V' \supset V$, $x \in \mathcal{D}_2(N, V')$.

Proof of the claim. Let $V' \supset V$. So, there is some y such that $y \in V'$ but $y \notin V$. Any such y is such that $y = x_n$, for some $n \geq 1$. By Fact 2, $V_n \subseteq V$. So, $y \notin V_n$. By construction, $V_{n-1} = V_n$, and $x \in \mathcal{D}_2(N, V_{n-1} \cup \{y\}) = \mathcal{D}_2(V, V_n \cup \{y\})$. But $V_n \cup \{y\} \subseteq V \cup \{y\} \subseteq V'$. By monotony in the right argument for $\mathcal{D}_2$, we get that $x \in \mathcal{D}_2(N, V')$, as required.

Claim 3. V is consistent.

Proof of the claim. Assume not. Since $x \Vdash \wedge_{i=1}^n x_i$, for $x_1, \ldots, x_n \in h(N)$, a derivation of (V, x) from N may be obtained by reiterating the argument under case 1, contradicting Claim 1.

Claim 4. V is $\neg$-complete; that is, for all y, either $y \in V$ or $\neg y \in V$.

Proof of the claim. Assume $y \notin V$ and $\neg y \notin V$ for some y. By Claim 2, it follows that $x \in \mathcal{D}_2(N, V \cup \{y\})$ and $x \in \mathcal{D}_2(N, V \cup \{\neg y\})$. Thus, $(b \wedge y, x)$ and $(c \wedge \neg y, x)$ are both derivable from N, where b and c are conjunctions of elements of V. The following is, then, derivable:

$$\frac{\dfrac{(b \wedge y, x) \qquad (c \wedge \neg y, x)}{((b \wedge y) \vee (c \wedge \neg y), x)} \text{ OR}}{(b \wedge c, x)} \text{ SI}$$

Thus, $x \in \mathcal{D}_2(N, V)$, in contradiction with Claim 1.

Claim 5. V *is maximal consistent; that is, if* $V \cup \{y\}$ *is consistent, then* $y \in V$.

Proof of the claim. Assume $y \notin V$. By Claim 4, $\neg y \in V$. It, then, follows that $V \cup \{y\}$ is inconsistent, as required.

We are almost finished. By Theorem 3 and Theorem 4, we have $\mathcal{O}_1(N, V) = \mathcal{D}_1(N, V) \subseteq \mathcal{D}_2(N, V)$. So $x \notin \mathcal{O}_1(N, V)$. Hence, $x \notin \mathcal{O}_2(N, A)$. $\square$

4   Aggregative Cumulative Transitivity

This section shows how to redefine Makinson and van der Torre's reusable output operation out_3 so that it validates neither WO nor CT but ACT:

$$\text{ACT} \; \frac{(a, x) \quad (a \wedge x, y)}{(a, x \wedge y)} \qquad\qquad \text{CT} \; \frac{(a, x) \quad (a \wedge x, y)}{(a, y)}$$

ACT and WO together imply CT.

Stolpe [19,20] named "PN_3" his own variant of out_3. The distinctive rule of PN_3 is the rule MCT mentioned in the introduction:

$$\text{MCT} \; \frac{(a, x') \quad x' \vdash x \quad (a \wedge x, y)}{(a, y)}$$

We said that, given the other rules in Stolpe's system, MCT is equivalent to CT. This is easily checked. The other rules are: SI, AND and EQ. On the one hand, given reflexivity for $\vdash$, MCT entails CT. For assume (a, x) and $(a \wedge x, y)$. Since $x \vdash x$, a direct application of MCT yields (a, y). On the other hand, given SI, CT entails MCT:

$$\text{CT} \; \cfrac{(a, x') \qquad \cfrac{(a \wedge x, y) \quad \cfrac{x' \vdash x}{a \wedge x' \vdash a \wedge x} \text{ SI}}{(a \wedge x', y)}}{(a, y)}$$

Note that, given SI, AND and EQ (we will keep them all), ACT is equivalent to:

$$\text{AMCT} \; \frac{(a, x') \quad x' \vdash x \quad (a \wedge x, y)}{(a, x' \wedge y)}$$

In this respect, weakening has still a "ghostly" role to play for iteration of successive detachments.

For the sake of conciseness, throughout this section $\mathcal{B}_A^M$ will denote the set of all the Bs such that $A \subseteq B = Cn(B) \supseteq M(B)$. Intuitively, $\mathcal{B}_A^M$ gathers all the Bs that contain A and are closed under both Cn and M.

Definition 5 (Semantics). $x \in \mathcal{O}_3(N, A)$ *if and only if there is some finite* $M \subseteq N$ *such that,*

- $M(Cn(A)) \neq \emptyset$, *and*
- *for all* B, *if* $B \in \mathcal{B}_A^M$, *then* $x \dashv\vdash \wedge M(B)$.

We do not single out any particular B as "proper". But we highlight two very useful such Bs, which we call the smallest and the largest: $\cap \mathcal{B}_A^M$; $\mathcal{L}$.

A subset M of N that makes $x \in \mathcal{O}_3(N, A)$ true is called an "A-witness for x". Unlike with $\mathcal{O}_1$, we have the guarantee that such a M does not contain any rule that is superfluous, viz. not required to get output x:

Theorem 10. *If* M *is an* A-witness for x, *then* $x \dashv\vdash \wedge h(M)$.

Proof. Let M be an A-witness for x. By Definition 5, $M(Cn(A)) \neq \emptyset$, and $x \dashv\vdash \wedge M(B)$ for all $B \in \mathcal{B}_A^M$. Consider $B = \mathcal{L}$. We have $x \dashv\vdash \wedge M(\mathcal{L})$. But $M(\mathcal{L}) = h(M)$, and thus $x \dashv\vdash \wedge h(M)$. $\qquad\square$

Theorem 11 (Factual Monotony). *We have* $\mathcal{O}_3(N, A_1) \subseteq \mathcal{O}_3(N, A_2)$ *whenever* $Cn(A_1) \subseteq Cn(A_2)$.

Proof. Assume $x \in \mathcal{O}_3(N, A_1)$ and $Cn(A_1) \subseteq Cn(A_2)$. From the first, we get: there is some finite $M_1 \subseteq N$ such that $M_1(Cn(A_1)) \neq \emptyset$ and, for all $B \in \mathcal{B}_{A_1}^{M_1}$,

$$M_1(B) = \{x_1, ..., x_n\} \text{ and } x \dashv\vdash \wedge_{i=1}^n x_i \tag{1}$$

Note that, by Theorem 10, $x \dashv\vdash \wedge h(M_1)$, and so the trick used for the proof of Theorem 1 is no longer needed.

From $Cn(A_1) \subseteq Cn(A_2)$, we get $M_1(Cn(A_1)) \subseteq M_1(Cn(A_2))$. Therefore, $M_1(Cn(A_2)) \neq \emptyset$. Now, consider some $B_1 \in \mathcal{B}_{A_2}^{M_1}$. We have $A_2 \subseteq B_1$. Therefore, $Cn(A_2) \subseteq Cn(B_1) = B_1$. From $A_1 \subseteq Cn(A_1) \subseteq Cn(A_2)$, we then get $A_1 \subseteq B_1$, and hence $B_1 \in \mathcal{B}_{A_1}^{M_1}$. By (1), $x \dashv\vdash \wedge M_1(B_1) \dashv\vdash \wedge h(M_1)$. So, $x \in \mathcal{O}_3(N, A_2)$ as required. $\qquad\square$

We define $\mathcal{O}_3(N) = \{(A, x) : x \in \mathcal{O}_3(N, A)\}$. Example 1 shows that $\mathcal{O}_3$ does not validate the rule of deontic detachment, and hence does not validate CT.

Example 1 (Deontic Detachment). Consider the set of norms $N = \{(\top, a), (a, x)\}$. We have $a \in \mathcal{O}_3(N, \top)$, since $M = \{(\top, a)\}$ is a $\top$-witness for a. We also have $x \in \mathcal{O}_3(N, a)$, since $M = \{(a, x)\}$ is an a-witness for x. But we do not have $x \in \mathcal{O}_3(N, \top)$. This may be verified in two steps. First, you identify all the non-empty subsets M of N that are triggered by the input, $M(Cn(\emptyset)) \neq \emptyset$. Next, you go through the list of all these subsets, and check that, for none of them, the smallest relevant B outputs heads whose conjunction is equivalent to x:

M	B	$M(B)$
$\{(\top, a)\}$	$Cn(a)$	$\{a\}$
$\{(\top, a)(a, x)\}$	$Cn(a, x)$	$\{a, x\}$

We illustrate the account with two examples from the literature.

Example 2 ("Change your mind!"). Hansson [11] gives the following example, with credit to Pörn:

> "Consider the hoary example of the man who ought to go to a meeting on August 5 and who ought to send, on August 2, a note explaining his absence, if and only if he is in fact going to be absent." [11, p.425-6]

The example is structurally identical to the Chisholm example [2]. The norms involved may be rendered as $N = \{(\top, m), (m, \neg s)(\neg m, s)\}$, where m and s are for attending the meeting and sending a note, respectively. Given input $\top$, $m \wedge \neg s$ is outputted, but not $\neg s$. This is as it should be. The obligation of $\neg s$ will not be triggered unless the agent is going to fulfil his primary obligation of m. In the violation context $\neg m$, $m \wedge \neg s$ is still outputted. If not, then the following intuitive deontic reasoning pattern would not be supported:

> "August 2 arrives, and though he is able to attend the meeting, he has no intention of doing so. He argues: 'I ought to change my mind, forbear note-writing, and attend the meeting.... My present fulfillment of this obligation will help make up for my sinfully staying at home on the fifth!'." [11, p. 426]

Example 3 ("Sing and dance!"). Conjunction elimination refers to the move from "it ought to be the case that $x \wedge y$" to "it ought to be the case that x". Consider $N = \{(\top, x \wedge y)\}$. Given input $\top$, $x \wedge y$ is outputted, but not x. Goble [5, p.183-184] and Hansen [8, §6.2], among others, have argued against conjunction elimination. There are cases where the two states of affairs (mentioned in the obligation) are only conjunctively required. If the obligation of x was outputted, then (when assessing how well or badly the agent did) a strange consequence would follow, in the event that the agent made x, but not y, true. One would have to acknowledge that (to quote Goble) "he's not a complete scoundrel" [5, p.183], since at least one obligation (albeit a derived one) was fulfilled. Intuitively, one would like to be able to say that *no* obligations have been fulfilled, and that *nothing* right has happened. This may be illustrated with the sing-and-dance example, due to Goble. If the agent makes only one conjunct true, he makes things worse than if he does nothing. Suppose there is a party of song and dance performers given in honour of someone called Gene. Everyone ought to perform a song and dance routine, because Gene loves them both, and cannot tolerate either without the other. One guest, call him Fred, chooses not to sing but only to dance. Gene is appalled. The party is ruined, because of Gene's tantrum.

Definition 6 (Proof System). $(a, x) \in \mathcal{D}_3(N)$ *if and only if there is a derivation of (a, x) from N using the rules $\{SI, EQ, ACT\}$.*

$$ACT \ \frac{(a, x) \qquad (a \wedge x, y)}{(a, x \wedge y)}$$

AND is derivable from SI and ACT. We define $(A, x) \in \mathcal{D}_3(N)$ and $\mathcal{D}_3(N, A)$ as we did for $\mathcal{D}_1$.

Theorem 12. $\mathcal{O}_3$ *validates the rules of* $\mathcal{D}_3$ *(for individual formulae a).*

Proof. The argument for SI is virtually the same as in the proof of Theorem 11. The argument for EQ is straightforward, and is omitted. We show ACT. Assume that $x \in \mathcal{O}_3(N, a)$, $y \in \mathcal{O}_3(N, a \wedge x)$ and $x \wedge y \notin \mathcal{O}_3(N, a)$. From the first two, it follows that there are finite $M_1, M_2 \subseteq N$ such that $M_1(Cn(a)) \neq \emptyset$, $M_2(Cn(a, x)) \neq \emptyset$, and

$$x \dashv\vdash \wedge M_1(B) \text{ for all } B \in \mathcal{B}_a^{M_1} \tag{2}$$

$$y \dashv\vdash \wedge M_2(B) \text{ for all } B \in \mathcal{B}_{a \wedge x}^{M_2} \tag{3}$$

By Theorem 10,

$$x \dashv\vdash \wedge h(M_1) \tag{4}$$

$$y \dashv\vdash \wedge h(M_2) \tag{5}$$

Therefore,

$$x \wedge y \dashv\vdash \wedge h(M_1) \wedge (\wedge h(M_2)) \tag{6}$$

$$\dashv\vdash \wedge h(M_3) \tag{7}$$

where $M_3 = M_1 \cup M_2$. From the third opening assumption, since $M_3(Cn(a)) \neq \emptyset$, it follows that there is some $B_1 \in \mathcal{B}_a^{M_3}$ such that

$$\text{not-}(x \wedge y \dashv\vdash \wedge M_3(B_1)) \tag{8}$$

We have $M_1(B_1) \subseteq M_3(B_1)$, and so $B_1 \in \mathcal{B}_a^{M_1}$. Therefore $x \in B_1$, and hence $a \wedge x \in B_1$. So $B_1 \in \mathcal{B}_{a \wedge x}^{M_2}$ too, since $M_2(B_1) \subseteq M_3(B_1)$. Now,

$$M_3(B_1) = M_1(B_1) \cup M_2(B_1)$$

where $\wedge M_1(B_1) \dashv\vdash x$ and $\wedge M_2(B_1) \dashv\vdash y$. Thus, $\wedge M_3(B_1) \dashv\vdash x \wedge y$, a contradiction. □

Theorem 13 (Soundness). $\mathcal{D}_3(N, A) \subseteq \mathcal{O}_3(N, A)$

Proof. Same argument as for Theorem 3 using Theorems 11 and 12. □

Theorem 14 (Completeness). $\mathcal{O}_3(N, A) \subseteq \mathcal{D}_3(N, A)$

Proof. We give an outline of the proof for the particular case where A is a singleton set $\{a\}$. Suppose that $x \in \mathcal{O}_3(N, a)$. To show: $x \in \mathcal{D}_3(N, a)$. From the former, there is some finite $M \subseteq N$ such that $M(Cn(a)) \neq \emptyset$ and, for all $B \in \mathcal{B}_a^M$, $x \dashv\vdash \wedge M(B)$.

Put $B_1 = Cn(\{a\} \cup \mathcal{D}_3(M, a))$. We have $a \in B_1 = Cn(B_1)$. We also have $M(B_1) \neq \emptyset$, because $Cn(a) \subseteq B_1$. A phasing result from [13] allows, then, to establish that $M(B_1) \subseteq B_1$, so that $B_1 \in \mathcal{B}_a^M$. The opening assumption, then, yields, $x \dashv\vdash \wedge M(B_1)$.

Based on this, one gets a derivation of (a, x) from N as follows. First, note that $M(B_1) \neq \emptyset$. By Definition 1, one gets $x \in \mathcal{O}_1(N, \{a\} \cup \mathcal{D}_3(M, a))$. By Theorem 4, $x \in \mathcal{D}_1(N, \{a\} \cup \mathcal{D}_3(M, a))$, and thus $x \in \mathcal{D}_3(N, \{a\} \cup \mathcal{D}_3(M, a))$. This means that $x \in \mathcal{D}_3(N, \{a, a_1, ..., a_n\})$, where, for each a_i, $a_i \in \mathcal{D}_3(M, a)$. By AND, $\wedge_{i=1}^{n} a_i \in \mathcal{D}_3(M, a)$. Since $M \subseteq N$, $\wedge_{i=1}^{n} a_i \in \mathcal{D}_3(N, a)$. A derivation of (a, x) from N is shown below.

$$\text{EQ} \frac{\text{ACT} \dfrac{(a, \wedge_{i=1}^{n} a_i) \qquad (a \wedge (\wedge_{i=1}^{n} a_i), x)}{(a, \wedge_{i=1}^{n} a_i \wedge x)} \qquad \dfrac{x \vdash \wedge_{i=1}^{n} a_i}{\wedge_{i=1}^{n} a_i \wedge x \dashv\vdash x}}{(a, x)}$$

The argument for $x \vdash \wedge_{i=1}^{n} a_i$ appeals to two lemmas:

- $x \dashv\vdash \wedge h(M)$, Theorem 10
- $h(M) \vdash a_i$, for all $1 \leq i \leq n$ – the proof of this is by induction on the length of the derivation of (a, a_i)

The argument may be generalized to an input set A of arbitrary cardinality. $\square$

5   Properties

In a companion paper [16], we identify some desirable properties, which are all satisfied by $\mathcal{O}_3$. These are listed in Table 1. We refer the reader to the aforementioned paper for the motivation and a discussion of these properties. These also hold for $\mathcal{O}_1$ and $\mathcal{O}_2$, when replacing out_3 by out_1 or out_2, respectively. On the left hand side of the table, exact factual detachment (EFD) and violation detection (VD) characterise what is special about *deontic* logic, while substitution (SUB), replacements of logical equivalents (RLE), implication (IMP) and paraconsistency (PC) say something about *logic*. We use the notation $x[\sigma]$ to denote a substitution instance of x. Thus, $x[\sigma]$ is obtained from x by replacing uniformly, in x, all occurrences of a propositional letter by the same propositional formula. $A[\sigma]$ and $N[\sigma]$ extend the notion of substitution instance to sets of formulae, and sets of norms in the straightforward way. We write $N \approx M$ whenever M is obtained from N, by replacing each $(b, y) \in N$ with some (c, z) such that b is equivalent with c, and y is equivalent with z. Implication makes use of the so-called materialisation $m(N)$ of a normative system N, which means that each norm (a, x) is interpreted as a material conditional $a \rightarrow x$, i.e. as the propositional sentence $\neg a \vee x$. We distinguish between violations $V(N, A) = \{x \in \mathcal{O}_3(N, A) \mid \neg x \in Cn(A)\}$ and non-violations (or cues for action) $\overline{V}(N, A) = \mathcal{O}_3(N, A) \setminus V(N, A)$.

On the right hand side of the table, norm monotony (NM) and norm induction (NI) are called "norm change properties", because the normative system N is no longer held constant. Together, exact factual detachment, norm monotony and norm induction are equivalent to saying that $\mathcal{O}_3(N)$ is a closure operator. Finally, the reusability properties relate the system to standard I/O logic: inclusion in reusable output (IO), redundancy (R) and strong redundancy (SR). Their formulation appeals to some key notions of so-called constrained input/output

Table 1. Properties [16]

EFD $(x, y) \in N \Rightarrow y \in \mathcal{O}_3(N, x)$		NM $\mathcal{O}_3(N) \subseteq \mathcal{O}_3(N \cup M)$
VD $(A, y) \in \mathcal{O}_3(N) \Rightarrow (A \cup \{\neg y\}, y) \in \mathcal{O}_3(N)$		NI $M \subseteq \mathcal{O}_3(N) \Rightarrow$
SUB $x \in \mathcal{O}_3(N, A) \Rightarrow x[\sigma] \in \mathcal{O}_3(N[\sigma], A[\sigma])$		$\mathcal{O}_3(N) = \mathcal{O}_3(N \cup M)$
RLE $N \approx M \Rightarrow \mathcal{O}_3(N) \subseteq \mathcal{O}_3(M)$		IO $\mathcal{O}_3(N, A) \subseteq out_3(N, A)$
IMP $\mathcal{O}_3(N, A) \subseteq Cn(m(N) \cup A)$		R $outf_3(N, A) = outf_3(\mathcal{O}_3(N), A)$
PC $x \in \overline{V}(N, A) \Rightarrow \exists M \subseteq N : x \in \mathcal{O}_3(M, A)$		SR $outf_3(N \cup M, A) =$
and $\mathcal{O}_3(M, A) \cup A$ consistent		$outf_3(\mathcal{O}_3(N) \cup M, A)$

logic, developed by Makinson and van der Torre [14] in order to reason about norm violation:

$$conf(N, A) = \{N' \subseteq N \mid out(N', A) \cup A \text{ consistent}\}$$
$$maxf(N, A) = \{N' \in conf(N, A) \mid N' \subseteq \text{-maximal}\}$$
$$outf(N, A) = \{out(N, A) \mid N' \in maxf(N, A)\}$$

It is worth recalling the reason why consistency checks were introduced in I/O logic. This was done in relation to contrary-to-duty reasoning. In unconstrained input/output logic, a violation leads to outputting the whole propositional language. This deontic explosion is not a property of the logics we introduce in this paper, as a direct consequence of the lack of the weakening rule. We believe that the unconstrained logics introduced in this paper can capture some aspects of contrary-to-duty-reasoning.

There is another property that acts as a bridge between the logics defined in this paper and the traditional input/output logics. It was not listed in [16] because it may not necessarily be considered a desirable property. This is the property: $out_1(N, A) = Cn(\mathcal{O}_1(N, A))$ and $out_3(N, A) = Cn(\mathcal{O}_3(N, A))$. Somewhat surprisingly, we do not have in general $out_2(N, A) = Cn(\mathcal{O}_2(N, A))$. For a counter-example, take $N = \{(a, x), (b, x \wedge y)\}$ and $A = \{a \vee b\}$. We leave it for future research to define a logic $\mathcal{O}'_2$ satisfying not only the properties in Table 1, but also the requirement $out_2(N, A) = Cn(\mathcal{O}'_2(N, A))$.

6   The Way Forward

This paper has extended Stolpe's results on I/O logics without weakening in two directions. First, we have shown how to account for reasoning by cases. Second, we have shown how to inject a new ("aggregative") form of cumulative transitivity, which we think is more suitable for normative reasoning. Soundness and completeness theorems for the proposed systems have been reported.

More work is to be carried out. First, it would be interesting to know if the two semantics proposed here may be merged to yield a new basic reusable operation out_4, with ACT, but not WO, amongst its primitive rules. Second, we have found that ACT has two drawbacks. The first one is that ACT derives the so-called pragmatic oddity [17]. The second one is that, in a violation context, ACT creates 'irrelevant' obligations, and thus the account faces an over-generation problem: more obligations are generated than it seems right. Let us call this the irrelevant obligation problem. The derivation to the left illustrates the pragmatic oddity with the dog-and-sign scenario—the letters d and s are for "there is a dog" and "there is a warning sign," respectively. The derivation to the right illustrates the irrelevant obligation problem.

$$\text{ACT} \cfrac{\text{SI} \cfrac{(\top, \neg d)}{(d, \neg d)} \quad \text{SI} \cfrac{(d, s)}{(d \wedge \neg d, s)}}{(d, \neg d \wedge s)} \qquad\qquad \text{ACT} \cfrac{\text{SI} \cfrac{(\top, \neg x)}{(x, \neg x)} \quad \text{SI} \cfrac{(b, y)}{(x \wedge \neg x, y)}}{(x, \neg x \wedge y)}$$

Pragmatic oddity   Irrelevant obligation

There are similarities between the two problems. However, we feel that the two should be distinguished. While it is clear that the derivation to the right should always be blocked, it is less clear whether the one to the left should always be blocked too. Indeed, one can think of examples in which the pragmatic oddity does make sense. For instance, if you do not pay the tax you own, you usually have to pay both a fine and your tax. Furthermore, as we will see in a moment, a solution to the irrelevant obligation problem may not be a solution to the pragmatic oddity.

This irrelevant obligation problem was pointed out by Stolpe [20, p. 134], in relation to MCT/CT. His diagnosis is that plain transitivity is more suitable for normative reasoning than cumulative transitivity. Plain transitivity is the rule "From (a, x) and (x, y), infer (a, y)". We will not follow up on his suggestion: the counter-examples alluded to in the introduction discredit both forms of transitivity. We are presently studying other ways around. A number of solutions naturally come into mind. These are listed below.

One first obvious possibility is to restrict the application of ACT, allowing it to be applied only if, e.g., the output is consistent with the input. This would solve both problems. This solution is proof-theoretical in nature. It would remain to see how to build it in the semantics.

A second possibility is to adopt a more procedural approach, by incorporating 'backtesting' into the account:

Backtesting
$$(A, x) \in \mathcal{O}_3'(N) \text{ iff } \exists A' \subseteq Cn(A) \text{ with } (A', x) \in \mathcal{O}_3(N) \text{ and } A' \cup \{x\} \not\vdash \bot$$

Intuitively, the definition says: for x to be obligatory in context A, it must have been the case that x was obligatory before the violation occurred, viz in context $A' \subseteq Cn(A)$ with A' consistent with x. Thus, obligations do not 'drown' in a violation context. We leave it to the reader to verify that backtesting filters out pragmatic oddities.

A third option is to change the base logic from classical logic to some suitable sub-classical logic. In order to resolve the irrelevant obligation problem, any logic that rejects the principle *ex falso quodlibet*, $\{x, \neg x\} \vdash y$, will do. A number of paraconsistent logics are available (for an overview, see [18]). Devised by Dosen[3], the so-called system **N** is amongst the simplest ones. It may fruitfully be used to illustrate the latter point. System **N** comes with a Kripke-type possible worlds semantics similar to that used for intuitionistic logic. The main difference is that the evaluation rules for $\rightarrow$ and $\neg$ use separate accessibility relations. The system is strictly included in Johansson's well-known system for minimal negation. One key difference is that, unlike the latter, the former does not keep *ex falso* in the following modified form: $\{x, \neg x\} \vdash \neg y$. The fact that such a system would do the job can easily be checked. Put $N = \{(\top, \neg x), (b, y)\}$ and $A = \{x\}$. We have

	M	B	$M(B)$
1.	$\{(\top, \neg x)\}$	$Cn(x, \neg x)$	$\{\neg x\}$
2.	$\{(\top, \neg x)(b, y)\}$	$Cn(x, \neg x)$	$\{\neg x\}$

The bottom line is this. System **N** keeps the principle *verum ex quodlibet*. This is the law $\Gamma \vdash \top$, where Γ is a set of formulae. So, on line 2, $\top \in B = Cn(x, \neg x)$, and thus $\neg x \in M(B)$. But $y \notin M(B)$ because, in the absence of *ex falso*, $b \notin B = Cn(x, \neg x)$. We use system **N** for illustrative purposes only. It could be that a more sophisticated paraconsistent logic is needed. Futhermore, to handle the pragmatic oddity, we need to do more than just let *ex falso* go away.

Acknowledgments.   We thank two anonymous reviewers for valuable comments.

References

1. Broome, J.: Rationality Through Reasoning. Wiley-Blackwell, West Sussex (2013)
2. Chisholm, R.: Contrary-to-duty imperatives and deontic logic. Analysis 24, 33–36 (1963)
3. Dosen, K.: Negation in the light of modal logic. In: Gabbay, D., Wansing, H. (eds.) What is Negation?, pp. 77–86. Springer, Heidelberg (1999)
4. Forrester, J.: Gentle murder, or the adverbial Samaritan. Journal of Philosophy 81, 193–197 (1984)
5. Goble, L.: A logic of good, should, and would: Part I. Journal of Philosophical Logic 19, 169–199 (1990)
6. Goble, L.: A proposal for dealing with deontic dilemmas. In: Lomuscio, A., Nute, D. (eds.) DEON 2004. LNCS (LNAI), vol. 3065, pp. 74–113. Springer, Heidelberg (2004)
7. Goble, L.: Prima facie norms, normative conflicts and dilemmas. In: Gabbay, D., Horty, J., Parent, X., van der Meyden, R., van der Torre, L. (eds.) Handbook of Deontic Logic and Normative Systems, pp. 241–352. College Publications, London (2013)
8. Hansen, J.: Imperative logic and its problems. In: Gabbay, D., Horty, J., Parent, X., van der Meyden, R., van der Torre, L. (eds.) Handbook of Deontic Logic and Normative Systems, pp. 499–544. College Publications, London (2013)

9. Hansen, J.: Reasoning about permission and obligation. In: Hansson, S.O. (ed.) David Makinson on Classical Methods for Non-Classical Problems, pp. 287–333. Springer (2014)
10. Hansson, S.O.: Preference-based deontic logic (PDL). Journal of Philosophical Logic 19, 75–93 (1990)
11. Hansson, S.O.: Situationist deontic logic. Journal of Philosophical Logic 26(4), 423–448 (1997)
12. Makinson, D.: On a fundamental problem in deontic logic. In: Namara, P.M., Prakken, H. (eds.) Norms, Logics and Information Systems. Frontiers in Artificial Intelligence and Applications, pp. 29–54. IOS Press, Amsterdam (1999)
13. Makinson, D., van der Torre, L.: Input/output logics. Journal of Philosophical Logic 29(4), 383–408 (2000)
14. Makinson, D., van der Torre, L.: Constraints for input/output logics. Journal of Philosophical Logic 30(2), 155–185 (2001)
15. McLaughlin, R.N.: Further problems of derived obligation. Mind 64(255), 400–402 (1955)
16. Parent, X., van der Torre, L.: Aggregative deontic detachment for normative reasoning (short paper). In: Eiter, T., Baral, C., Giacomo, G.D. (eds.) Proceedings of the 14th International Conference on Principles of Knowledge Representation and Reasoning, KR 2014. AAAI Press (2014)
17. Prakken, H., Sergot, M.: Contrary-to-duty obligations. Studia Logica 57, 91–115 (1996)
18. Priest, G.: Paraconsistent logic. In: Gabbay, D., Guenthner, F. (eds.) Handbook of Philosophical Logic, vol. 6, pp. 287–393. Springer (2002)
19. Stolpe, A.: Normative consequence: The problem of keeping it whilst giving it up. In: van der Meyden, R., van der Torre, L. (eds.) DEON 2008. LNCS (LNAI), vol. 5076, pp. 174–188. Springer, Heidelberg (2008)
20. Stolpe, A.: Norms and Norm-System Dynamics. Ph.D. thesis, Department of Philosophy, University of Bergen, Norway (2008)
21. von Wright, G.: Norm and Action: A Logical Enquiry. Routledge & Kegan Paul PLC (1963)

The Logical Structure of Scanlon's Contractualism

Martin Rechenauer[1] and Olivier Roy[2]

[1] Universität Konstanz
[2] Universität Bayreuth

Abstract. In this paper we use fixed-point modal logic to study the logical properties of justified norms in Scanlonian contractualism. We show a natural connection between Scanlon's test for justifiability and the computation of the smallest fixed point; we rebut a common charge of vacuity based on the recursive character of the proposal; we show that the resulting justification operator is not a normal modality, and we explore the epistemic component often left implicit in the justification procedure. Given that such procedural justifications have so far not been investigated using modern logical tools, this paper contributes both to deontic logic and to the literature on contractualism in ethics.

Procedural justifications play a large role in contemporary ethics. Their basic idea is that norms are justified if they have been approved by the application of an appropriate *procedure*. Kantian ethics is a classical example. A norm or precept is justified if it passes the test of the categorical imperative. Modern approaches include Rawlss theory of justice, discourse ethics and *Scanlonian contractualism*, on which we focus in this paper.

Up to now, procedural ethics has stayed beyond the scope of deontic logic. This is an important gap. It is not known which principles govern the deontic modals stemming from procedural justification, how complex these justification procedures are, and whether they fall prey to known deontic paradoxes. On the other hand, by leaving out procedurally justified norms, deontic logic misses what is arguably one of the most important views on norms in contemporary ethics.

This paper takes some steps towards filling this gap. Using apparatus from modern fixed-point logics, we study the logical properties of justification of norms in Scanlonian contractualism. We rebut a common charge of vacuity based on the recursive character of the proposal, we show that the resulting justification operator is not a normal modality, and we explore the epistemic component often left implicit in the justification procedure.

1 Scanlon's Contractualism

The basic tenet of Scanlon's contractualism is that *a norm is justified if it can't be reasonably rejected by anybody who is motivated to find such norms that others,*

F. Cariani et al. (Eds.): DEON 2014, LNAI 8554, pp. 166–176, 2014.
© Springer International Publishing Switzerland 2014

similarly motivated, could not reasonably reject. Most interpreters focus on the first part of this claim, the criterion of reasonable rejectability. In this paper we rather focus on the motivational condition and the recursive character of the proposal. Logically they are the most interesting conditions (see Section 3). But there is also a philosophical reason to look at them more carefully. Just as in all procedural justification models, there is a condition of impartiality in Scanlon's conception. It is embodied by the motivational, recursive condition. The motivation condition also delineates the membership of those to whom we owe compliance with the norms established and thus might help avoid moral skepticism. Its peculiar role seems not to have been adequately understood within the literature.

Here is a canonical statement of Scanlon's contractualism:

> When I ask myself what reason the fact that an action would be wrong provides me with not to do it, my answer is that such an action would be one that I could not justify to others on grounds I could expect them to accept. This leads me to describe the subject matter of judgements of right and wrong by saying that they are judgements about what would be permitted by principles that could not reasonably be rejected, by people who were moved to find principles for the general regulation of behavior that others, similarly motivated, could not reasonably reject. [5, p.4]

In this passage, we have several elements collected together that are germane to Scanlon's version of contractualism. We will explain them briefly in order to point out what we are and are not going to discuss, and also where we depart from the letter though not the spirit of Scanlonian contractualism.

To begin with wrongness vs. justification in general. In the quoted passage, Scanlon is presenting his account as one that specifically deals with the explication of moral wrongness. But his version of contractualism is often viewed as a general test for fundamental normative principles regulating behavior broadly conceived. Scanlon himself says nothing that speaks against such a generalization; others have already moved in this direction, a prominent example being Brian Barry [1] who instead of moral wrongness talks about injustice. This is the view we take here. The approach is general enough to apply to many kinds of practical justification. A principle is justified if it passes the test provided by Scanlon's formula. One could call this a test concerning rightness or wrongness, but it might not only be confined to acts properly, but also to states of affairs and judgments of their normative quality.

The objects of justification are principles. We understand principles very liberally here. Any rule for normative regulations of behavior, but also assessments of states of affairs (such as distributional patterns within a society) will do. The scope of justification is to be taken rather broadly.

An interesting aspect is Scanlon's notion of *reasonableness*. There is a long-standing discussion of the confrontation between *rational* and *reasonable*. This is already to be found in the work of John Rawls, and Scanlon has added a twist of his own. We do not enter this discussion. Here *reasonable rejection* is a primitive. But just as a guide to intuition, one way to make this distinction is to say that *rational* might be explicated in a rather minimal way, as just referring to basic coherence assumptions like transitivity of value orderings or preferences, and probabilistic coherence of partial beliefs. Reasonableness, on the other hand, has to do with a more general idea of responsiveness to good reasons [5, p.33]

Rejection: One might well ask why the contractualist formula is given in terms of rejection rather than acceptance. Scanlon himself is not too explicit about this. In a sense, rejection harmonizes better with our understanding of Scanlon's ideas about consensus. The fact that all agents would agree on a principle is not the basis of validity, in contrast to the main interpretation of, e.g., discourse ethics. Nevertheless, consensus is important but only as a by-product of the justification procedure. In this paper we only make a minimal assumption about the properties of rejection. We take it to be an intentional action, i.e. done for a motivational reason, and we claim that it should not be closed under logical consequence.

The motivation condition is important but has not been discussed much. We take it as one contribution of this paper to highlight the key role played by this condition in the logic of justification. It introduces recursion into the justification procedure. Justified principles are to be unrejectable by people who are such as to be motivated by the search for principles that are unrejectable by people like themselves. The motivation condition also helps delineate the group of people to whom justification is owed. We will also see that it has an implicit, interesting doxastic component.

2   Modal Fixed-Point Logic

2.1   Syntax

Modal fixed-point logics extend basic modal logic with operators to talk about fixed-points. The *modal μ-calculus* is one of such logics, containing operators μ and ν to talk about smallest and largest fixed-points, respectively (c.f. [8,2] for a general presentation). For reasons that will become clear later, the language $\mathcal{L}$ we work with also contains three additional modalities, one for belief, one for reasonable rejection, and one for counterfactual conditional.

$$\varphi := p \mid \neg\varphi \mid \varphi \wedge \varphi \mid B_i\varphi \mid RR_i\varphi \mid \varphi \supset \varphi \mid \mu x.\varphi(x) \mid \nu x.\varphi(x)$$

The modal μ-calculus comes with a set of propositional variables $(x, y, z...)$. These are distinct from atomic propositions (p, q, r). Propositional variables are to be used in the evaluation of fixed-point operators. They should always occur bounded in formulas of $\mathcal{L}$.

The modal $\mu - calculus$ makes an important syntactic restriction on propositional variables. They should be positive. This means that they must be within the scope of an even number of negations. The formula $\mu x.(x \vee p)$, for instance, is well-formed, but not $\mu x.(\neg x \vee p)$. This restriction makes all formulas monotonic operators in the intended classes of models. Such operators have smallest and greatest fixed-points. This makes truth for formulas of the form $\mu x.\varphi(x)$ and

$\nu x.\varphi(x)$ always well-defined. For simplicity we also impose another restriction on the syntax: formulas within the scope of an RR_i operator do not contain propositional variables.

Formulas of the form $B_i\varphi$ should be read "agent i believes that φ". Formulas of the form $RR_i\varphi$ read "agent i reasonably rejects φ." To simplify matters we will read $\neg RR_i\varphi$ as agent i reasonably accepts φ, leaving out the possibility of unreasonable rejection. Formulas of the form $\varphi \supset \psi$ should be read "if φ were the case then ψ would be the case."

This language does not contain any formulas stating that a principle is justified. Such formulas will be defined as fixed-points of some formulas containing belief and rejection operators. The computation of these fixed points will represent the justification procedure. Norms are the output of procedures.

2.2 Semantics

This language is interpreted using *game-theoretical semantics*, a standard tool in fixed-point logic. The details are in the appendix. The models we employ are a combination of Kripke and neighborhood models. A model $\mathcal{M}$ is a tuple $\langle W, I, \{R_i, N_i\}_{i \in I}, V \rangle$, where W is a set of states, I a finite set of agents, and V a valuation function on the set of atomic propositions. The belief modality B_i receives its standard Kripke semantics, using the binary R_i, which we assume to be serial. For counterfactuals we make an important simplifying assumption. The standard Lewis-Stalnaker semantics using a similarity relation do not square well with the syntactic restriction of the μ-calculus. Roughly, it makes the antecedent of the counterfactual occurring both positively and negatively in the evaluation clause. In our proposal below the crucial recursion happens in such an antecedent. So in this paper we interpret counterfactual statements as necessary implications. In Lewis-Stalnaker semantics this boils down to saying that all states are maximally similar to each other. This is an important simplification, to be investigated further in future work.

For the rejection operator RR_i we use the neighborhood function N_i, which assigns to each state a set of set of states. Intuitively, reasonable rejection shouldn't be a normal modality. If i reasonably rejects breaking one's promises, and also reasonably rejects apologizing for having broken a promise—because there is no reason to do so unless a promise has indeed been broken— this doesn't mean that she will reasonably reject the conjunction of these, quite the contrary. The other way around seems also plausible. If i reasonably rejects helping the poor in conjunction with stealing from the rich, she might nonetheless not reasonably reject, or even reasonably not reject helping the poor.

2.3 Some Mathematical Properties

The belief operator is a KT modality. We make no assumption on the properties of reasonable rejection. So its minimal logic is E. The strict conditional is a normal binary modality. A natural property to assume is that reasonable rejection is introspective, although we do not use it explicitly here:

$$RR_i\varphi \to B_i(RR_i\varphi)$$

$$\neg RR_i\varphi \to B_i(\neg RR_i\varphi)$$

The μ-calculus with a single normal modality is finitely axiomatizable, decidable and has the finite model property. Are these properties retained in the present language? We leave this for future work.

3   Scanlon's Condition in Fixed-Point Logic

Scanlon's proposal has two main components: *reasonable rejection* and the *motivation condition*. Reasonable rejection is a primitive operator in our language. As argued, this operator should not be normal. This is important. We will see later on that the properties of justified norms defined in this language depend on the properties of the rejection operator.

3.1   The Motivation Condition

The motivation condition is the most interesting one for our analysis. Logically, it plays two roles. First, it bounds the quantification that determines which agents are going to be considered in the test for justifiability. These are those who are motivated to find norms that pass that very test, those who are "similarly motivated." And this brings us to the second key feature of the motivation condition: it introduces recursion.

We render the motivation condition implicitly, using a broadly Humean view of intentional agency, e.g. in [6]. We take reasonable rejection to be an intentional action, one for which the agent has a motivational reason.[1] Humeanism about motivational reasons takes many forms, but its core is the idea that a motivational reason for action is a combination of two families of attitudes, representational and motivational. Beliefs are paradigmatic representational attitudes. We use the corresponding modality to capture them. Desires are typical motivational attitudes, but there might be others.[2].

On this view an intentional action is one done for, and caused by a motivational reason. Turning back to reasonable rejection, this means that an agent will be disposed to reasonably reject a principle when she considers it possible that someone similarly motivated would reasonably reject it. The agent will

---

[1] We emphasize "motivational". The notion of reasonableness suggests that this reason might also be normative. In Scanlonian terminology, that it is a good reason with respect to the set of issues at hand. For the present analysis we do not need to take a stance on the issue of whether this motivational reason is also normative.

[2] Humeans considering motivational reasons usually use desires as the main or primitive motivational attitudes (see e.g. in [6]). Scanlon is critical of that view [5]. We do not take a stance on this issue and stick to the general denomination "motivational attitude."

be responsive to this belief when she is appropriately motivated. According to many Humean views of motivational reasons the converse is also true. The agent being responsive to the belief just mentioned implies that she has the required motivational attitude.

Responsiveness to motivational reasons is a causal notion here. The belief that some similarly motivated agent would reasonably reject a principle φ, given that the agent is herself motivated to find such a principle, should causally influence rejection. According to a standard analysis of causation, A is a cause of B whenever A, if it hadn't occurred, B wouldn't have either, c.f. [3]. In the present context this gives:

$$B_i x \supset \neg RR_i \varphi \tag{M}$$

In words: if the agent believed that no one similarly motivated would reasonably reject φ, she would not reject it herself. This formula, strictly speaking, is not a well-formed formula of our language $\mathcal{L}$. It has a free variable x. The crux of our proposal is to embed this condition into a well-formed formula which will introduce recursion on this variable. We now turn to this.

3.2   A Recursive Account of Justification

In Scanlon's proposal the notion of a justified norm is used in the very procedure that is used to determine whether a norm is justified. So this is a recursive definition. A norm φ is justified, written $J\varphi$, when *no* appropriately motivated agent rejects φ. Our proposal is to render this as follows:

$$J\varphi \equiv_{df} \mu x. \bigwedge_{i \in I} ((B_i x \supset \neg RR_i \varphi) \to \neg RR_i \varphi) \tag{S}$$

The formula bounds to those who are motivated to find such norms the range of agents whose reasonable rejection is taken into account while checking whether a norm φ is justified. It says that a norm is justified in exactly those cases where no such agent rejects φ. In other words, exactly in the cases where the set of appropriately motivated agents is included in the set of those who do not reasonably reject φ. The usual combination of Boolean conjunction ($\bigwedge_{i \in I}$) and material implication ($\to$) captures this bounded quantification.

The recursive clause is used within the motivation condition (M), here in the antecedent of the implication. Recall that (M) establishes a causal dependency between one's belief that no similarly motivated agent reasonably rejects φ and one's own reasonable rejection. We explained in the previous section why we see this as a plausible rendering of the motivation condition. For now it is the recursion that is important. Here the game-theoretical semantics for the μ-calculus will help. In a nutshell, "μ means finite unfolding." [8] In the evaluation game for a formula $J\varphi$, the variable x gets unfolded when reached. The game continues on $\bigwedge_{i \in I}((B_i x \supset \neg RR_i \varphi) \to \neg RR_i \varphi)$. But x itself is in the scope of a doxastic operator. So being appropriately motivated means being responsive to the belief that no appropriately motivated agent rejects φ. But these agents are precisely those who are themselves responsive to that belief in non-rejection by appropriately motivated peers. So the unfolding continues. The content of this second-order belief, i.e. belief about how others would respond to a certain belief of theirs, itself uses the motivation condition, and so involves a third-order belief about how others similarly motivated would respond to that very belief, and so on. This is a genuine recursion. We will show in the next section that it is grounded, i.e. not vicious.

(M) uses the smallest fixed-point operator μ. Why? It is well-known that for monotone functions smallest and largest fixed-points can be respectively computed by "bottom-up" or "top-down" procedures. For the smallest fixed-point of a monotone function f on an algebra of propositions, one starts by computing the value of a given function at $\bot$, re-applies that function $f(\bot)$, and so on. By Tarski's fixed point theorem [7] this procedure always reaches a fixed-point, which will be the smallest.

Scanlon's test for justifiability is captured by this abstract algorithm plugged in (S). When is a principle φ justified? One natural point to start answering that question is to ask whether anyone could reasonably reject φ in the first place. That is, check *all* possible reasonable rejections. This can be checked by fixing the antecedent of (S) to $\top$. Plugging in $\bot$ as the value of x in that formula does precisely that, because we assume R_i to be reflexive. One then asks which of these reasonable rejections comes from agents who would be responsive to the belief that no one reasonably rejects φ. This is step 2. The procedure continues (step 3) by asking who among those would be responsive to the belief that no one identified in step 2 rejects φ, and so on until one reaches a point where no further iteration is needed, i.e. where the set of relevant reasonable rejections is the same as those rejections coming from agents who would be responsive to the belief that no one in that very set reasonably rejects φ. In Scanlon's terminology, where one has identified the set of reasonable rejections coming from agents similarly motivated.

This procedure makes justification context dependent. Justification would be context independent if $J\varphi$ would be true either at all or at none of the states w, for any φ. This need not be the case (see example in the next section). Indeed, whether φ is a justified principle in w depends on who are adequately motivated agents, and whether they reasonably reject φ. The motivation condition in turn depends on whether the counterfactual statement holds, i.e. whether agents with the relevant beliefs would reject φ.

3.3   Circularity, Normality, Epistemics

The very syntactic structure of (M) delivers an important result for Scanlon's contractualism. Its justification procedure is recursive, but not viciously circular. It has a fixed point, and so the set of "similarly motivated agents" is well-defined.[3] Recall the standard syntactic constraints that propositional variables should be positive. In (M) x is in the antecedent of a material implication, but also in the antecedent of a conditional, which we in turn interpreted as

---

[3] Note that this does not mean that it is never empty.

necessary implication. So x remains positive in (M), this formula meets the syntactic restriction, and so it has a fixed point. This answers a common worry about Scanlon's proposal, voiced for instance in [4].

The rejection operator is not a normal modality. In particular it fails the following upward closure property:

$$J\varphi \rightarrow J(\varphi \vee \psi) \qquad\qquad \text{(Up)}$$

For a counter-model to (Up), take a model $\mathcal{M}$ with one agent i, a single R_i-reflective state w. Set $N_i(w) = \{W\}$. Verifier wins $\mathcal{G}(J\bot, \mathcal{M}, w)$. After a trivial choice by Falsifier her winning strategy is to choose $\neg RR_i\bot$. But she loses $\mathcal{G}(J\top, \mathcal{M}, w)$. Observe that w satisfies $RR_i\top$. So Verifier is forced to choose $\mathcal{G}^{-1}(B_ix \supset \neg RR_i\top, \mathcal{M}, w)$. In that dual game she is forced to re-choose w and the choice next is for Falsifier, whose winning strategy is to go for the dual game of $\mathcal{G}^{-1}(B_ix, \mathcal{M}, w)$, i.e. $\mathcal{G}(B_ix, \mathcal{M}, w)$. Because w is R_i reflexive the game continues on $\mathcal{G}(x, \mathcal{M}, w)$, and so x is unfolded and we are back where we started. So Falsifier's winning strategy is to force Verifier into an infinite path where x gets unfolded infinitely often, and so ensures him a win.

This is an important result for the logic of the justification operator. The type of upward monotony expressed in (Up) that is at the root of Rosss paradox. It is generally seen as an undesired property for deontic modalities. So it is good news that Scanlonian justifications are non-monotonic.

Our logical reconstruction of Scanlon's test for justifiability unveils interesting epistemic phenomena related to the justification procedure. We only mention one here. That a principle φ is justified does not mean that this is a common or even a shared belief. Justification is not introspective. This is so even when reasonable rejection is individually introspective, i.e. when each agent has always correct beliefs regarding what she reasonably rejects.[4] The model in Figure 1 provides a counter-example. There the function $N_i(w)$, for each i and w, is constrained in such a way that the RR_i formulas gets the truth value as indicated.

(S) is true at w_1. But agent 1 considers it possible that it might be false, i.e. at w_2. That (M) is true at w_1 follows directly from the fact that no agent rejects p in that situation. The winning strategy for Falsifier in the game played on (S) is the following. First he chooses agent 2's conjunct. Verifier is forced to continue with the dual game on $B_2x \supset \neg RR_2p$, otherwise she loses because 2 reasonably rejects p. Now she can choose to continue the dual game at w_1 or at w_2. If she chooses the first then Falsifier wins at the next move by choosing to continue

---

[4] Individual introspection of reasonable rejection is a debatable assumption. There might be an objective component in the notion of a good reason that seems inherent to reasonableness. Then even when rejection itself is introspective, reasonable rejection might not be. An agent might reject but be uncertain whether it is reasonable to do so. This would be a failure of positive introspection. An agent might also reject but wrongly believe that this is a reasonable rejection. This would constitute a failure of negative introspection. We use introspection of rejection here only to strengthen the result. Even if reasonable rejection were introspective, justification need not be.

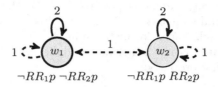

Fig. 1. A counter-model to the introspection of (S)

on $\neg RR_2p$. So Verifier must stay at w_2. The choice is now for Falsifier, between the non-dual game on B_2x or stay in the dual game but on $\neg RR_2p$. He loses the latter (recall that this is the dual game). But if he chooses the former then he stays at w_2 at the next move, x gets unfolded, and we are back where we started. We are engaged on an infinite path, where Falsifier wins because x gets unfolded infinitely often.

4   Conclusion

This paper takes a first step in the logical investigation of procedural views of norms and justification. We looked at the case of Scanlon's contractualism. A relatively simple formalization in modal fixed-point logic yields important insights for this theory. It shows that the definition is recursive but not viciously circular and that the test for justification closely follows the computation of a smallest fixed point. Our rending of Scanlon's formula also highlights some epistemic aspects up to now left implicit in the proposal. It finally brings this contractualist view into the deontic logic area, and does so in an interesting way. Justifications turn out to be non-monotonic. We take these to be valuable insights both for ethics and deontic logic.

Further directions for research abound. An immediate question is whether common knowledge that appropriately motivated agents reject (or not) a norm simplifies the computation of the fixed point. Computability results for this procedure would also be insightful, both logically and philosophically speaking. Modal fixed-point logic over a single K modality is decidable. But it is not known whether this also holds for the richer language we use here. Other procedural views of justification should also be investigated. Discourse ethics, with its emphasis on reaching agreement in an ideal speech situation, is an obvious place to apply the logical tools we have used here.

Acknowledgements. The authors would like to thank the participants of the first meeting of the Bavarian Deontic Group for helpful comments and suggestions. Part of this work has been done while both authors were affiliated with the Munich Center for Mathematical Philosophy. Financial support of the Alexander von Humboldt and the LMU Munich is gratefully acknowledged.

References

1. Barry, B.M.: Justice as impartiality. Oxford UP (1995)
2. Fontaine, G.M.M.: Modal fixpoint logic: some model theoretic questions. Ph.D. thesis, Institute for Logic, Language and Computation (2010)
3. Menzies, P.: Counterfactual theories of causation. In: Zalta, E.N. (ed.) The Stanford Encyclopedia of Philosophy, Spring 2014 edn. (2014)
4. Pogge, T.W.: What we can reasonably reject. Nous 35(s1), 118–147 (2001)
5. Scanlon, T.: What we owe to each other. Harvard University Press (1998)
6. Smith, M.: The moral problem (1994)
7. Tarski, A., et al.: A lattice-theoretical fixpoint theorem and its applications. Pacific Journal of Mathematics 5(2), 285–309 (1955)
8. Venema, Y.: Lectures on the modal μ-calculus (2008) (manuscript)

Appendix - Game Theoretical Semantics

A formula φ is evaluated through a game $\mathcal{G}(\varphi, \mathcal{M}, w)$, with $\mathcal{M}, w$ a pointed model as described above. The game is defined inductively on the structure of φ. We use $\mathcal{G}^{-1}(\varphi, \mathcal{M}, w)$ for the game in which Verifier and Falsifier switch roles.

If φ is a disjunction $\psi \vee \chi$ then Verifier chooses one of the disjuncts, and the players play accordingly either $\mathcal{G}(\psi, \mathcal{M}, w)$ or $\mathcal{G}(\chi, \mathcal{M}, w)$. Negation $\neg\psi$ is a role switch. The players play $\mathcal{G}^{-1}(\psi, \mathcal{M}, w)$.

If φ is a belief modality $B_i\psi$ then Falsifier must choose a state w' such that wR_iw' and they play $\mathcal{G}(\psi, \mathcal{M}, w')$.

If φ is a rejection modality $RR_i\psi$ then Verifier must first choose a set $X \in N_i(w)$. Then the game splits into two parallel sub-games $\mathcal{G}_1$ and $\mathcal{G}_2$. In the first one Falsifier must choose a state $w' \in X$ and they play $\mathcal{G}_1(\psi, \mathcal{M}, w')$. In the second one Falsifier chooses a state $w'' \notin X$ and they play $\mathcal{G}_2^{-1}(\psi, \mathcal{M}, w'')$. Verifier needs to win both sub-games.

If φ is a conditional modality $\psi \supset \chi$ then Falsifier chooses a state w' and Verifier chooses between playing $\mathcal{G}^{-1}((\psi, \mathcal{M}, w')$ or $\mathcal{G}((\chi, \mathcal{M}, w')$.

For formulas of the form $\mu x.\psi(x)$, the game continues on $\psi(x)$, but after the next occurrence of x the game will continue on $\varphi(x)$, at whatever pair $\mathcal{M}, w'$ that has been reached meanwhile. Observe that this last condition makes evaluation games potentially infinite.

If a play of that game is finite then it is a win for Verifier if the last move is winning for her. Verifier wins an infinite play starting with $\mathcal{G}(\mu x.\varphi(x), \mathcal{M}, w)$ if x is only unfolded finitely often, and she wins an infinite play starting with $\mathcal{G}(\nu x.\varphi(x), \mathcal{M}, w)$ if x is unfolded infinitely often. A play ends if φ is a propositional constant or if there is nothing to choose for one of the players at her/his choice point. If a player cannot choose at a point then the other wins. For propositional constants Verifier wins if and only if $w \in V(p)$. A winning strategy for Verifier at $\mathcal{M}, w$ in the game induced by φ is a strategy that guaranties a win, whatever Falsifier chooses.

Before proving the adequacy theorem we recall the standard Kripke semantics for the standard modal part of our language. Introducing the set-theoretic semantics for fixed point operators would require too much new machinery here. We refer the interested reader to [8,2]. $\mathcal{M}, w \models \varphi$ is defined as follows:

- $\mathcal{M}, w \models p$ iff $p \in V(w)$.
- $\mathcal{M}, w \models \psi \wedge \chi$ iff both $\mathcal{M}, w \models \psi$ and $\mathcal{M}, w \models \chi$.
- $\mathcal{M}, w \models \neg\psi$ iff $\mathcal{M}, w \not\models \psi$.
- $\mathcal{M}, w \models B_i\psi$ iff $\mathcal{M}, w' \models \psi$ for all w' s.t. wR_iw'
- $\mathcal{M}, w \models RR_i\psi$ iff $||\psi|| \in N_i(w)$
- $\mathcal{M}, w \models \psi \supset \psi$ iff for all w', either $\mathcal{M}, w' \not\models \varphi$ or $\mathcal{M}, w' \models \psi$.

Observation 1. *The following is equivalent:*

- $\mathcal{M}, w \models \varphi$.
- *Verifier has a winning strategy in* $\mathcal{G}(\varphi, \mathcal{M}, w)$.

Proof. By induction on φ. The basic case is trivial. The induction hypothesis is that the observation is true for ψ, χ and any state w'. The clauses for the Boolean connectives are standard, as well as for the modalities B_i. The case of the fixed point operator can be found in [2]. We present only the cases for the non-normal modalities RR_i and the binary modality $\supset$.

Suppose φ is of the form $RR_i\psi$. If $||\psi|| \in N_i(w)$ then Verifier's winning strategy is to choose precisely that set. By the inductive hypothesis she directly has a winning strategy in both sub-games. Now suppose $||\psi|| \notin N_i(w)$. If Verifier cannot choose she loses. Otherwise suppose that she chooses set X. If $X \cap ||\neg\psi|| \neq \emptyset$ then Falsifier chooses a state in that intersection in the first sub-game, for which he has a winning strategy by our inductive hypothesis. $X \cap ||\neg\psi|| = \emptyset$ then by assumption $||\psi||$ is not empty and so Falsifier can choose a state in there and wins the second sub-game.

Suppose that $\mathcal{M}, w \models \varphi \supset \psi$ and that falsifier chooses w'. By induction hypothesis Verifier wins by choosing to play $G^{-1}(\varphi, \mathcal{M}, w')$ whenever $\mathcal{M}, w' \not\models \varphi$ and otherwise we know by assumption and our inductive hypothesis that she wins in $G(\psi, \mathcal{M}, w')$. Now suppose Verifier has a winning strategy in $G(\varphi \supset \psi, \mathcal{M}, w)$. Take an arbitrary w'. Verifier's winning strategy must specify what to do when Falsifier chooses w'. It is either to choose to continue the game on $G^{-1}(\varphi, \mathcal{M}, w')$ or $G(\psi, \mathcal{M}, w')$. Depending on which one it is we know by inductive hypothesis that either $\mathcal{M}, w' \not\models \varphi$ or $\mathcal{M}, w' \models \psi$.

Self-governance by Transfiguration: From Learning to Prescriptions

Régis Riveret[1], Alexander Artikis[2], Dídac Busquets[1], and Jeremy Pitt[1]

[1] Imperial College London, United Kingdom
{r.riveret,didac.busquets,j.pitt}@imperial.ac.uk
[2] NCSR Demokritos, Greece
a.artikis@iit.demokritos.gr

Abstract. Norms are commonly understood as guides for the conduct of autonomous agents, thereby easing their individual decision-making and coordination. However, their study exhibits a polarity between (i) norms as behavioural patterns emerging from repeated agents' (inter)actions and (ii) norms as explicit prescriptions. In this paper, we attempt to build a bridge between these two conceptual poles of norms: it takes the form of a mental function for prescriptive transfiguration allowing reinforced learning agents to express their learning experiences into prescriptions. The population of transfigurative agents are then equipped with a consensus system to build and enforce prescriptive systems to self-govern on-line. Simple simulations suggest the pertinence of the approach and shows its weaknesses, in particular prescriptions stalling learning, and timeliness in norm construction.

1 Introduction

Norms are commonly understood as guides for the conduct of autonomous agents, thereby easing their individual decision-making and coordination. However, their study exhibits a polarity in their conception. On the one hand, jurists concentrate on norms as prescriptions promulgated by institutional powers and enforced by explicit sanctions. On the other hand, social researchers study norms as tacit behavioural patterns emerging from expectations and enforced by entwined sanctions. This polarisation is reflected by the treatment of norms by computer scientists. Prescriptions and legal reasoning are investigated in formal logics (typically deontic logics and argumentation, see e.g. [13]) to represent and reason upon explicit norms, leading eventually to architecture for cognitive agents (see e.g. [7]) while social norms are accounted as patterns emerging from repeated interactions amongst agents (typically learning agents see e.g. [14]).

Scholars have thus investigated the influence of social norms and prescriptions on each other, but the conceptual gap remains hardly explored by computer scientists, in particular with regard to applied systems, c.f. [12]. To address it, we propose a simple mental apparatus to perform a prescriptive transfiguration allowing reinforced learning agents to express their learning experiences into prescriptions.

F. Cariani et al. (Eds.): DEON 2014, LNAI 8554, pp. 177–191, 2014.
© Springer International Publishing Switzerland 2014

As to the practical relevance, our proposal regards self-organising systems and in particular self-governing systems, specifically the on-line construction of prescriptive systems for and by reinforced learning agents. Indeed, while reinforcement learning is a prevalent mean for the adaptation of autonomous agents with incomplete information on their environment [16], norms are an attractive manner to guide the conduct of these agents. Furthermore, explicit norms are commonly advocated to facilitate their updates, and consequently system maintenance, improve system transparency and ease system governance. Unfortunately, as remarked by [8], the manual construction of prescriptive systems is often time-consuming and error prone, the construction at design time (i.e. off-line construction) is computationally complex, and both are unsuited for dynamic systems with unpredictable changes. Therefore we opt for on-line construction. Since systems of multiple autonomous agents have their essence into decentralised control and computation, this on-line construction shall occur in a distributed manner in the sense there is no entity with complete information taking the role of a central legislative body. We will focus on explicit primary norms and in particular regulative norms, i.e. those guiding the ideal behaviour of agents, leaving other primary constitutive norms (used to constitute institutions) and secondary norms (managing primary norms) for future work.

The practical challenge in this paper regards thus the self-governance of learning agents, or more specifically the domain-independent construction at run-time of explicit regulative norms from scratch, for and by learning agents, without any agent having a complete information on the system. Our solution, inspired by direct democracy, is a consensus system coupled with the mental function of prescriptive transfiguration so that every agent shall propose and vote for prescriptions meant to govern themselves. The overall system results thereby into a direct self-governance taking advantage of every agents' learning experiences.

Noting there is no obvious or immediate utility for a reinforced learning agent to share his own experiences to influence the construction of a prescriptive system (paradox of voting, also called Downs paradox), our proposal of direct self-governance is imposed to the agents (i.e. hard-coded). Nevertheless, as every agent is learning with respect to the qualities of behaviours, the construction of norms occurs in the same spirit. Every possible proposal and vote is associated with a probability reflecting a scalar potential and we assume that every agent is endowed with a mental apparatus described in this paper to compute these potentials. This apparatus is light so that it is compatible with the presumption of agents with bounded cognition.

The simulations of reinforced learning agents equipped with such legislative apparatus suggest the pertinence of such approach but also its weaknesses, in particular prescriptions stalling learning and timeliness in norm construction.

The remainder of this paper is organised as follows. In the next Section 2, we base our system of learning agents on the model of stochastic games and we define the problem of self-governance we are interested in. Our proposal for direct self-governance is given in Section 3. It is illustrate in Section 4 with some simulation results and related with other work in Section 5, before concluding.

2   Stochastic Games

We base our framework on the common model of stochastic games. A stochastic game can be considered as an extension of a Markov decision process with multiple agents with possibly conflicting goals, and where the joint actions of agents determine state transitions and payoffs. A stochastic game consists of a tuple $< \mathcal{G}, \mathcal{S}, \mathcal{A}, T, R >$ where:

- $\mathcal{G}$ is a set of N agents indexed by i;
- $\mathcal{S} = \{s_1, \ldots s_n\}$ is a finite non-empty set of global states;
- $\mathcal{A} = \prod_i \mathcal{A}_i$ is a set of joint actions. $\mathcal{A}_i$ is a set of individual actions available to agent i.
- $T : \mathcal{S} \times \mathcal{A} \times \mathcal{S} \rightarrow [0, 1]$ is a function of transition, $T(s_r, A, s_q) = p(s_{t+1} = s_r | A, s_t = s_q)$ is the probability of resulting in a state s_r at time $t + 1$ when attempting the joint action A in a state s_q at time t.
- $R : \mathcal{S} \times \mathcal{A} \times \mathcal{S} \rightarrow \mathbb{R}^N$ is a payoff function, $R_i(s_r, A, s_q) = r_i(s_{t+1} = s_r, s_t = s_q)$ is the payoff of agent i upon transition from a state s_q at time t to state s_r at time $t + 1$ under joint action A.

Though this setting implies that the possible states, transition and payoff functions are known by the investigator when specifying a game, it offers nevertheless a setting where we assume they are unknown by the agents.

The control of behaviours of agent i is described by a policy denoted π_i. It is a mapping from agent i's state history to individual behaviours. The objective of any agent i is to maximize *the infinite horizon discounted return*:

$$R_{i,t} = r_i(s_t, s_{t+1}) + \gamma . r_i(s_{t+1}, s_{t+2}) + \gamma^2 . r_i(s_{t+2}, s_{t+3}) + \ldots$$

where γ is a discount rate.

Since the probabilities and payoffs are unknown by the agents, and sanctions play an important role in normative multi-agent systems, we consider individual reinforced learning agents [16] meant to pursue the best policies. At each time step, every agent senses its environment, and, given the observed state, every agent simultaneously selects the best behaviour on the basis of past experiences (exploitation) and also by trying new options (exploration). No agent is informed about the actions performed and payoffs received by the other agents.

A behaviour j of an agent i, denoted by a pair state-action $(s, a_{i,j})$, is associated with a real number $Q(s, a_{i,j})$ representing the quality of this behaviour over time. The quality $Q(s, a_{i,j})$ is the discounted moving average of the payoffs associated to the individual action $a_{i,j}$ in state s. Let $a_{i,t} = a_{i,j}$ be an action j selected by agent i at time t in a state s_t, the quality $Q(s_t, a_{i,t})$ is updated as follows:

$$Q(s_t, a_{i,t}) \leftarrow Q(s_t, a_{i,t}) + \alpha_i . [\delta + \gamma_i . Q(s_{t+1}, a_{i,t+1})]$$

with $\delta = r_i(s_t, s_{t+1}) - Q(s_t, a_{i,t})$, where α_i is a learning rate, and γ_i a discount factor trading off the importance of recent versus later payoffs. For each agent, the selection of a behaviour at time t, is simulated by a Gibbs-Boltzmann probability distribution over all the behaviours available for the agent i:

$$\pi_i^t(s_t, a_{i,t}) = \frac{e^{Q(s_t, a_{i,t})/\tau_i}}{\sum_{a_{i,j}} e^{Q(s_t, a_{i,j})/\tau_i}}$$

where τ_i is a positive real number balancing the exploitation and the exploration of behaviours.

A stochastic game can have diverse objectives. A very popular is to find a behavioural profile (a set of policies π_i) for which no agent can benefit from unilaterally changing its behaviour, i.e. a Nash equilibrium. Stochastic games can have several Nash equilibria thus those maximising social measures such as welfare or fairness shall be preferred.

The challenge addressed in this paper is twofold: firstly we investigate the problem of prescriptive transfiguration, that is the transfiguration of agents' policies (i.e. learning experiences and thereby behavioural patterns) into prescriptions (a prescription is a conditioned obligation or prohibition with an associated sanction), secondly we consider the problem of self-governance, that is the on-line construction of a set of prescriptions for and by agents to govern themselves. As a first approach, the objective of self-governing agents is to maximise a social return possibly defined as the sum of agents' infinite horizon discounted social returns:

$$\sum_i R_{i,t}^* = \sum_i r_i^*(s_t, s_{t+1}) + \gamma.r_i^*(s_{t+1}, s_{t+2}) + \gamma^2.r_i^*(s_{t+2}, s_{t+3}) + \ldots$$

where r_i^* is a payoff accounting for social measures, for example those catering for the notion of justice. In a simple case, the overall social return may only deal with the global wealth of the system, accordingly r_i^* shall be a material payoff disregarding the sanctions of violated prescriptions.

Since the essence of systems of multiple autonomous agents is to limit centralised control, we look at the problem in which there is no agent having complete information about the game to design the prescriptive system. So, we base our mechanism on the idea that every agent shall participate on the construction of prescriptions. Prescriptions shall be constructed for and by agents. Accordingly, we use a voting system: the set of messages are the possible motions (the explicit norm to be voted) and the votes; the results of social decisions are the enter of force (and thus reinforcement) of these explicit norms. Remark that though the implemented system implied a central entity implementing the voting system for accepting agent's motions, votes and for deliberations, other distributed consensus mechanisms could be employed.

Example 1. We will illustrate and evaluate the prescriptive transfiguration and the proposed self-governance of learning agents with an example inspired by accident law (we do not aim at legal precision, c.f. [10]). Consider a population of agents acting in two possible global states: one is safe and the other is dangerous. In any state, every agent can act with care or with negligence. Whatever the state, if all the agents act with care then the next state will be safe. If an agent acts with negligence then there is a risk of an accident and the next state is dangerous. The probability of an accident is higher when the negligent act is performed in a dangerous state. Hence it suffices that only one agent acts with negligence and that an accident occurs to bring the population in a dangerous state. The Markov decision problem graph is drawn in Figure 1 for a system populated by a single agent. The unique Nash equilibrium takes place when all

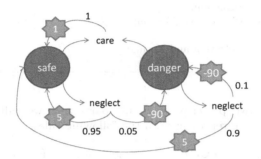

Fig. 1. The Markov decision process graph for a system populated by a single agent. Each transition from an action to a state is represented by an arrow labelled with its probability and associated payoff.

agents act with care. Reinforced learning agents may or may not learn to act with care, in any case we will investigate a system of self-governance where agents construct prescriptions to guide themselves.

3   Direct Self-governance

To address the problem of transfiguration and self-governance as presented in the previous section, we endow agents agents with a mental apparatus to transform learning experiences into prescriptions and this apparatus is coupled with a consensus mechanism so that agents make a social choice on those prescriptions meant to govern themselves. The pseudo-code animating the population in its environment is given in Algorithm 1.

   In the remainder of this section, we describe the step regarding transfiguration, and the steps concerning self-governance, i.e. submissions, motion selection and voting.

3.1   Individual Prescriptive Transfiguration

Prescriptive transfiguration is based on a mapping from a learning policy to prescriptions. In practice any behaviour B in a state s resulting in an action

---

Algorithm 1. Animation of self-governed learning agents for an episode

---

Initialise the system;
for each step of an episode **do**
  for each agent **do**
    Choose an action amongst alternatives;
  end for
  Compute the environment;
  for each agent **do**
    Observe the individual payoffs;
    Update quality of behaviours;
    Individual prescriptive transfiguration;
  end for
  Submissions, Motion selection and Voting;
end for

---

a is associated with two prescriptive counterparts that we call possible self-prescriptions and that we represent with the following rules:

$$r_{\mathsf{Obl}\,(B)} : s \Rightarrow \mathsf{Obl}\,a \qquad\qquad r_{\mathsf{Forb}\,(B)} : s \Rightarrow \mathsf{Forb}\,a$$

where $r_{\mathsf{Obl}\,(B)}$ ($r_{\mathsf{Forb}\,(B)}$) is an identifier of the self-obligation (self-prohibition), s represents the conditions and $\mathsf{Obl}\,a$ ($\mathsf{Forb}\,a$) is the consequent. The identifier may be dropped when its omission does not raise any ambiguity.

Example 2. For every agent, there are four possible self-prescriptions:

$$\text{safe} \Rightarrow \mathsf{Obl}\,\text{care} \qquad\qquad \text{danger} \Rightarrow \mathsf{Obl}\,\text{care}$$
$$\text{safe} \Rightarrow \mathsf{Forb}\,\text{care} \qquad\qquad \text{danger} \Rightarrow \mathsf{Forb}\,\text{care}$$

$$\text{safe} \Rightarrow \mathsf{Obl}\,\text{neglect} \qquad\qquad \text{danger} \Rightarrow \mathsf{Obl}\,\text{neglect}$$
$$\text{safe} \Rightarrow \mathsf{Forb}\,\text{neglect} \qquad\qquad \text{danger} \Rightarrow \mathsf{Forb}\,\text{neglect}$$

Notice that we assume no equivalence between the obligation to act with care and the prohibition to act with negligence. The adopted logic is thus light on this aspect. Nevertheless a kind of quantitative equivalence shall appear when we will introduce potentials to prescriptions (see below).

Possible self-prescriptions are not active: every agent shall propose the most relevant amongst all them as a motion to the whole population before voting for its enforcement. The construction of the prescriptive system occurs in three activities:

1. Individual prescriptive transfiguration: every agent shall individually transfigure learning experiences into (self-)prescriptions,
2. Submissions and Motions: every agent shall submit a prescription and the most common proposal becomes the motion,
3. Voting: every agent votes for the motion with respect to its self-prescriptive background (the set of agent's self-prescriptions).

In each activity, every agent has to make a choice about (self-)prescriptions (self-prescribe, make a proposal, vote for the motion). Since we have learning agents, every agent will make its choice by taking into consideration the quality or potential of the (self-)prescriptions with a flavour of reinforcement learning, as we will see in the remainder of this section.

Once and every time an agent observes the current state and considers the set of alternative behaviours, this agent shall individually transfigure learning experiences into submissible prescriptions. This phase is decomposed in two steps: the agent decides or not to self-prescribe, then eventually, a submissible self-prescription is drawn.

Self-prescribe or Not. This step is meant to avoid an agent to transfigure learning policies when alternative behaviours have similar qualities. Indeed, there is no advantage to oblige or prohibit a behaviour with respect to the others when they all result in similar payoffs. There are many manners to avoid the prescription of behaviours with similar qualities. We chose to do so by using an entropic threshold. Every agent i computes the entropy S_i of the distribution of the alternative behaviours in a state. If S_i is less than a threshold τ_i^S then the agent will draw a self-prescription. We propose no calculus here to set up this threshold τ^S, but we can give some basic considerations. If it set to high, then the agent i may not gain enough experiences before considering prescriptions and thus non-optimal prescriptions may be selected in the next phase. At the opposite, if the threshold is set to low, then the agent may have so much experiences that prescriptions shall appear useless.

Example 3. Suppose the agent named Tom is in a safe state. Tom has two behavioural alternatives: either behave with care or behave with negligence. Assume that the careful behaviour has a quality 4 and the negligent behaviour has quality 2, thus their respective probability is:

$$p(\text{care}|\text{safe}) = \frac{e^4}{e^4 + e^2} \sim 0.88 \qquad p(\text{neglect}|\text{safe}) \sim 0.12$$

The entropy is $S_{Tom} \sim -0.88 \cdot \ln(0.88) - 0.12 \cdot \ln(0.12) \,(\sim 0.37)$. Consider a threshold $\tau_{Tom}^S = 0.5$, then Tom will consider alternative prescriptions to elevate one to the rank of submissible prescription (see below). If the entropy was higher than this threshold, then Tom would consider no prescription for the safe state.

Selection of Submissible Prescriptions. If an agent decides to transfigure learning experiences into self-prescriptions then it will draw a self-prescription that becomes a *submissible prescription*. To do so, every possible self-prescription is associated with a scalar measure that we call the submissible potential. The higher the quality of a behaviour with respect to the quality of other behaviours, the higher its potential to be considered as an obligation. At the opposite, the lower the quality of a behaviour with respect to the quality of other behaviours, the higher its potential to be considered as a prohibition. Let's capture formally these ideas. Let $\widehat{Q}_i$ denote the average quality of alternative behaviours in a state according to agent i. For a self-obligation $r_{\mathsf{Obl}(B)}$, its submissible potential according to agent i, denoted $\delta_i(r_{\mathsf{Obl}(B)})$, is the difference between the quality for behaviour B and the average quality of alternative behaviours. For a self-prohibition, we have the opposite:

$$\delta_i(r_{\mathsf{Obl}(B)}) = Q_i(B) - \widehat{Q}_i \qquad\qquad \delta_i(r_{\mathsf{Forb}(B)}) = \widehat{Q}_i - Q_i(B)$$

Consequently $\delta_i(r_{\mathsf{Obl}(B)}) = -\delta_i(r_{\mathsf{Forb}(B)})$.

At every step, every agent will consider a set of self-prescriptions compatible with the prescriptions in force regulating the states. A self-obligation is compatible with the prescriptions in force if:

- there is no prohibited alternative,
- there is no obliged alternative.

A self-prohibition is compatible if:

- there is another alternative not being prohibited,
- there is no obliged alternative.

As a matter of compactness of the prescriptive system, the above items assume we won't oblige an action and explicitly prohibit one of its alternative. On this basis, every agent i shall draw a self-prescription r amongst n compatible self-prescriptions $\{r_1, \ldots, r_n\}$ with a probability $p_i^\delta(r)$ using a Boltzmann-Gibbs distribution over the submissible potentials:

$$p_i^\delta(r) = \frac{e^{\delta_i(r)/\tau_i^\delta}}{\sum_{i=1}^n e^{\delta_i(r_i)/\tau_i^\delta}}$$

where τ_i^δ is a parameter balancing the exploitation and exploration for submissions. If this parameter tends to 0, then the agent shall pick up the prescription with the highest submissible potential. In this case, the potential of the selected prescription shall be positive, $0 \leq \delta_i(r)$. The choice of this distribution is meant to pave the way for learning prescriptive agents, in particular for frameworks where the repeal of prescriptions is possible.

Example 4. Table 1 illustrates Tom's measure of submissible potentials and the associated probabilities. We suppose in the remainder that Tom has selected two submissible prescriptions: the obligation to act with care when the state is safe, and the prohibition to act with negligence when the state is dangerous.

3.2   Submissions and Motions

Once some agents have transfigured some learning experiences into a set of submissible prescriptions, these agent shall submit each a prescription. The most common submission becomes a motion, and agents vote for its enforcement.

Table 1. Illustration of submissible qualities δ_i and associated probabilities p_i^δ to consider the prescription as submissible. The qualities of corresponding behaviours (Q_{Tom}) are arbitrary given (its average is 3) and the parameter τ_{Tom}^δ balancing the exploitation and exploration for submissions is set at 0.1.

Prescription	Q_{Tom}	δ_{Tom}	p_{Tom}^δ
safe $\Rightarrow$ Obl care	4	1	0.5
safe $\Rightarrow$ Forb care	4	- 1	0
safe $\Rightarrow$ Obl neglect	2	-1	0
safe $\Rightarrow$ Forb neglect	2	1	0.5

A submitted prescription is a submission. Every agent will draw a submission from the set of submissible prescriptions using again a Boltzmann-Gibbs distribution. Let $\{r_1, \ldots, r_n\}$ be the set of submissible prescriptions of agent i (drawn in the previous step), the agent i will draw a submission r from this set with a probability $p_i^D(r)$ from a Gibbs-Boltzmann distribution over the potentials $\delta_i(r)$ with a temperature τ_i^D balancing the exploitation and exploration of submissions amongst submissible self-prescriptions. Amongst all the submissions within a population of agents, the most common submission becomes a motion, and in the next phase every agent will vote or not for this motion.

Example 5. Amongst the obligation to act with care when the state is safe, and the prohibition to act with negligence when the state is dangerous, we assume that Tom draws the obligation to act with care. We further assume that the most common proposals by the population is the prohibition to act with negligence when the state is safe. Consequently, this proposal becomes a motion.

At this stage, the prescription of a motion is not associated to any sanction. There is a well-accepted principle in retributive justice according which the level of the sanction should be scaled relative to the severity of the offending behaviour. In our framework, a simple mean to evaluate the severity of an offending behaviour is to consider the potential δ_i of the proposals meant to guide this behaviour. Thus, the higher the potential of a proposal, the higher the severity of a violation, the higher the sanction.

So, we associate any motion m with a potential $\widehat{\delta}(m)$ which is the average of the potentials of the proposals unifying with m. This average potential is meant to feature the value of a scalar sanction. Accordingly, we choose in this paper to define the sanction as $\widehat{\delta}(m).\mu$ where μ is a positive real number (typically set superior to 1).

Example 6. Suppose that 3 agents proposed the prohibition to act with negligence when the state is safe (the motion), and they proposed it with the potential 2, 3 and 4. The average potential is 3 and thus the quality of the motion m is $\widehat{\delta}(m) = 3$. Assuming $\mu = 10$ the associated scalar sanction associated to this motion is 30.

3.3   Voting

Once there is a motion about a prescription with its sanction, every agent is invited to vote for it. The cognitive process resulting in a vote against or in favour is not trivial to model. In a utilitarian setting, we could argue that an agent shall vote for a globally useful motion and vote against a useless motion. We assumed that the 'global potential' of a motion m is measured by its average potential $|\widehat{\delta}(m)|$ (featuring its associated sanction - see previous section). Since agents have to vote about the motion and the associated sanction $|\widehat{\delta}(m)|.\mu$, then we suppose that agents are communicated $\widehat{\delta}(m)$. We further assume that an agent shall vote in favour or against a motion by comparing the average potential of this motion $\widehat{\delta}(m)$, with the potential of this motion according to this agent $\delta_i(m)$. The lower the difference between the potential $\delta_i(m)$ of the motion and the average potential $\widehat{\delta}(m)$, the higher the probability for agent i to vote in favour of the motion m. In punishment terms, an agent shall vote in favour of a motion if the associated sanction corresponds to a sanction "as it should be" according to this agent. Furthermore, an agent shall vote in favour of the motion m only if its corresponding individual potential $\delta_i(m)$ matches the positive or minus sign of the average potential $\widehat{\delta}(m)$, in other words an agent will not vote in favour of a motion with a positive average potential if this agent holds that this motion has a negative potential. Accordingly, we capture these considerations with a scalar measure called the *individual potential of the motion*. The agent i's individual potential of the motion m is denoted $\Delta_i(m)$:

$$\Delta_i(m) = \frac{|\delta_i(m) - \widehat{\delta}(m)|}{\tau_i^\Delta.|\widehat{\delta}(m)| + \epsilon} \cdot \frac{2}{1 + sgn(\delta_i(m) \cdot \widehat{\delta}(m)) + \epsilon}$$

where ϵ tends towards 0 and τ_i^Δ is a strictly positive real number. An agent i will vote in favour of a motion m with a probability $p_i^\Delta(m)$ using a folded sigmoid function:

$$p_i^\Delta(m) = \frac{1}{1 + \Delta_i(m)}$$

The higher τ_i^Δ, the higher the probability for agent i to vote in favour of the motion m. If τ_i^Δ is large then agents shall vote for any motion (the most common proposal) at the risk of being ruled by a minority.

Example 7. Recall the most common submitted prescription by the population is the prohibition to act with negligence when the state is safe. Hence every agent is invited to vote about this motion. We computed that the average agents' quality over this motion is 3, $\widehat{\delta}(m) = 3$. Let $\tau_{Tom}^\Delta = 0.1$, the individual potential of Tom for this motion is thus: $\Delta_{Tom}(m) \sim |1 - 3|/0.1 \cdot 3$. Tom will vote in favour of this motion with a probability $p_{Tom}^\Delta(m) \sim 0.01$.

The consensus can take many different forms, it can be distributed or centralised for example, but for our purposes we arbitrary considered a majority rule. Accordingly a prescription and its enforcement voted by the majority enters in force. The abrogation of a prescriptions shall be possible but we reserve its presentation in another work. Any prescription in force is enforced by applying its associated sanction to any non-compliant agent (modifying thus the payoffs of the underlying stochastic game).

4   Simulation Results

To evaluate and get more insights into the proposed prescriptive transfiguration and associated self-governance, we animated the stochastic game of Example 1 with a homogeneous population of reinforced learning agents with no initial pre-scriptions. The environment, the agents, their interactions and the prescriptions were implemented as a development of the platform based on a probabilistic rule-based argumentation and machine learning [11], so that the system speci-fications were directly executed. The results are averaged over 100 runs of 250 time steps of a population of 50 agents.

The probability of careful behaviours in the safe and dangerous state with or without self-governance is shown in the figures 2 and 3. When self-governance is deactivated, agents learn to behave with care in both states, but the convergence is slower in the safe state as the careful and negligent behaviours in this state have closer expected utilities. When self-governance is activated, the enforce-ment of careful behaviours guided the agents towards careful behaviours with a higher speed of convergence in both states. The possible prescriptions and

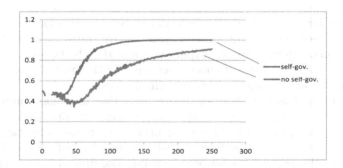

Fig. 2. Average probability of careful behaviours in the safe state with self-governance (red) and without self-governance (blue) vs. time

their empirical probability of enforcement with respect to the parameter τ_i^Δ (see Section 3.3) are shown in Table 2. Remark that the probabilities with respect to a state may not add up to one as few simulations did not end up with prescribed states. The reason holds in the choice of a low value (see e.g. $\tau_i^\Delta = 0.1$) so agents appeared quite picky in their vote. At the opposite, when this parameter was set large, e.g. $\tau_i^\Delta = 1$, all the simulations ended with prescribed states. The benefits of the system are thus illustrated by these simulations: an increase of global

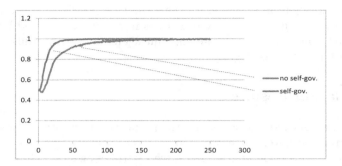

Fig. 3. Average probability of careful behaviours in the dangerous state with self-governance (red) and without self-governance (blue) vs. time

Table 2. Prescriptions with their empirical probability of enforcement

τ_i^Δ	0.1	0.5	1			0.1	0.5	1
safe $\Rightarrow$ Obl care	0.54	0.42	0.46		danger $\Rightarrow$ Obl care	0.46	0.50	0.50
safe $\Rightarrow$ Forb care	0	0	0		danger $\Rightarrow$ Forb care	0	0	0
safe $\Rightarrow$ Obl neglect	0	0.02	0.02		danger $\Rightarrow$ Obl neglect	0	0	0
safe $\Rightarrow$ Forb neglect	0.40	0.50	0.52		danger $\Rightarrow$ Forb neglect	0.40	0.50	0.50

wealth (since careful agents shall accumulate more wealth when behaving with care) while addressing the problem of (i) prescriptive transfiguration, that is the transfiguration of agents' learning experiences and thereby behavioural patterns into prescriptions, and (ii) self-governance, that is the on-line construction of prescriptions for and by agents to govern themselves.

Weaknesses exist as well. For instance, remark that an obligation to act with negligence was voted in one simulation: its enforcement occurred at the time step 41 when the probability of careful behaviour in the safe state was low enough to let a minority of negligent agents to pass this obligation. This shows a weakness of the present framework regarding the difficulty to appropriately prescribe behaviours with close qualities at voting time. There is indeed a risk of a consensus for policies enforcing undesirable behaviours when the quality of these behaviours is close to desirable behaviours. This occurs when the expected utilities of alternatives are close or when the dynamics is such that undesired behaviours appear with relatively high quality for a period of time during which a vote occurred. This later unfortunate condition emphasis the importance of timeliness in norm construction. In conditions where accidents are sparse but very harmful, if a vote occurs too early then there is risk that agents vote for policies enforcing undesired behaviours. At the opposite, a late vote may imply new explicit policies enforcing a well-established social norm; in this case policies shall be nevertheless useful to newcomers. The good timeliness shall necessary occur between the 'too early' and the 'could have been earlier', but the probabilistic setting implies that the vote of optimal policies cannot be ensured. This is particularly annoying when one reckons the difficulties to get rid of policies impeding opportunistic exploration of better behaviours.

Another weakness regards the dissonance arousing from reinforcement learning agents and norm-governed agents. On the one hand, learning agents are supposed to pursue a maximisation of individual wealth by balancing the exploitation of promising strategies and the exploration of other options. On the other hand, norms tend to imped opportunistic exploration. Norms stall learning, and thereby agents may get trapped into suboptimal prescriptive systems.

5   Related Work

Social norms are often studied in two extremes: in game theoretical settings of strategic agents and in simulation of thoughtless agents like evolutionary game theoretical investigations. In both types of approaches, the convergence to an equilibrium is interpreted as the emergence of a social norm: norms are not explicitly represented and agents do not have a mental representation of them.

On the contrary, formal logics (typically deontic logics and argumentation, see e.g. [13]) are commonly investigated to represent and reason upon explicit norms, leading eventually to architecture for cognitive agents (see e.g. [7,2]). These architectures are usually based on a BDI template and without learning abilities, while our agents are logic-based and reinforced learners but they have no explicit desires or intentional features (their implicit desire is to maximize the accumulation of payoffs). BDI frameworks usually assume that prescriptions are built-in whereas our agents have to ability to learn best behaviours and thereby generate new prescriptions (though prescriptions could be also built-in). The limitation of BDI architecture with regard to norm recognition has been addressed by Conte at al. in [12] where BDI agents recognise norms by observing other agents, c.f. [3]. Our agents transfigure individual experiences into prescriptions without the need to observe other agents, and the utility of these individual experiences are the results of the (inter)actions with other agents.

Multi-agent learning is an active field of research where agents are meant to coordinate by learning joint actions, typically using individual reinforcement learning or its extensions to collective tasks. Partalas et al. proposed in [9] to combine reinforcement learning with voting. Their agents learn predefined strategies (joint actions) while our agents learn individual actions. When their agents are in a strategic state they vote for a common strategy: there is no transfiguration and no construction of prescriptions.

With regard to norm-synthesis, the problem was pioneered by the work of Shoham and Tennenholtz [15]. Fitoussi and Tennenholtz [6], for example, described the synthesis of 'minimal' and 'simple' prohibitions. The rationale for minimality is that a minimal norm provides the agents more freedom in choosing their behaviour (that is, it prohibits fewer actions) while ensuring that they conform to the system specification. The rationale for simplicity is that a simple norm relies less on the agents capabilities rather than a non-simple one. Agotnes and Wooldridge [1] included the implementation costs of norms and multiple design goals with different priorities. Christelis and Rovatsos [4] proposed a first-order planning approach to better cope with the size of the state-space. The approaches mentioned above are typically applied off-line. However, off-line design is not appropriate for coping with open systems, that are inherently dynamic and the state space may change over time. To address this issue, Morales et al [8] proposed a mechanism called IRON for the on-line synthesis of norms. IRON employs designated agents, often called 'institutional agents' [5], representing a norm-governed system/institution, and observing the interactions of the members of the system in order to synthesise conflict-free norms without lapsing into over-regulation. Our work is fundamentally different: we target multi-agent systems without designated agents receiving updates about the system interactions and the authority to enforce norms.

6   Conclusion and Future Directions

We tackled the challenge regarding prescriptive transfiguration and self-governance. We proposed a simple cognitive apparatus with which learning agents can transfigure learning policies (and thus behavioural patterns) into prescriptions. This apparatus was coupled to a consensus system so that agents can submit prescriptions for a vote and vote eventually for their enforcement.

The simulations of a self-governed population of learning agents suggested the benefits of our approach with regard to the convergence to desirable behaviours. However, simulations with large stochastic games have to confirm these benefits. Timeliness in run-time construction with learning agent appeared of the most importance. A vote may indeed occur when there is a risk agents consider inadequate prescriptions, or when useless prescriptions shall enforce behaviours already adopted by agents. Nevertheless, these useless prescriptions shall ease the decision-making and coordination of newcomers.

In practice, our proposal illustrates an alternative of off-line construction of prescriptive systems: a domain-independent construction at run-time of explicit primary regulative prescriptions from scratch, for and by learning agents, without any agent having a complete information on the system.

Future directions can be multiple. They include learning of joint actions and the transfiguration of these collective into complex prescriptive systems, distributed consensus systems (possibly in network) to avoid a central body collecting the votes. An important point regards learning of norms modifications so that agents can escape from unfortunate prescriptive systems. But how could agents change prescriptions without having the possibility to explore and without jeopardizing the coherence and the temporal stability of the overall system? A solution holds in agents simulating the system to explore "without moving" but it implies computational resources a priori incompatible with bounded agents. Maverick agents on whose payoffs santions have a less significant effect may be an interesting line of research.

Eventually, we hope the reader found inspiration in a manner to bridge the gap between social norms and prescriptions, and its use for run-time constructions of prescriptive systems in a population of learning agents and thereby for self-organisation and in particular self-governance.

Acknowledgements. The authors would like to thank the anonymous reviewers. Part of this work is supported by the Marie Curie Intra-European Fellowship PIEF-GA-2012-331472.

References

1. Ågotnes, T., Wooldridge, M.: Optimal social laws. In: AAMAS, pp. 667–674 (2010)
2. Broersen, J., Dastani, M., Hulstijn, J., Huang, Z., van der Torre, L.: The boid architecture: Conflicts between beliefs, obligations, intentions and desires. In: Proceedings of the Fifth International Conference on Autonomous Agents, AGENTS 2001, pp. 9–16. ACM, New York (2001)
3. Castelfranchi, C., Giardini, F., Lorini, E., Tummolini, L.: The prescriptive destiny of predictive attitudes: from expectations to norms via conventions. In: Proceedings of CogSci 2003, 25th Annual Meeting of the Cognitive Science Society (2003)
4. Christelis, G., Rovatsos, M., Petrick, R.P.A.: Exploiting domain knowledge to improve norm synthesis. In: AAMAS, pp. 831–838 (2010)
5. Esteva, M., Rodriguez-Aguilar, J., Arcos, J., Sierra, C., Garcia, P.: Institutionalising open multi-agent systems. In: Durfee, E. (ed.) Proceedings of the International Conference on Multi-agent Systems (ICMAS), pp. 381–382. IEEE Press (2000)
6. Fitoussi, D., Tennenholtz, M.: Choosing social laws for multi-agent systems: minimality and simplicity. Artificial Intelligence 119(1-2), 61–101 (2000)
7. Governatori, G., Rotolo, A.: Bio logical agents: Norms, beliefs, intentions in defeasible logic. Autonomous Agents and Multi-Agent Systems 17(1), 36–69 (2008)
8. Morales, J., López-Sánchez, M., Rodríguez-Aguilar, J.A.: Michael Wooldridge, and Wamberto Vasconcelos, 'Automated synthesis of normative systems'. In: AAMAS, pp. 483–490 (2013)
9. Partalas, I., Feneris, I., Vlahavas, I.P.: Multi-agent reinforcement learning using strategies and voting. In: ICTAI (2), pp. 318–324. IEEE Computer Society
10. Mitchell Polinsky, A., Shavell, S.: Handbook of Law and Economics, 1st edn., vol. 1. Elsevier (2007)
11. Riveret, R., Rotolo, A., Sartor, G.: 'Probabilistic rule-based argumentation for norm-governed learning agents'. Artif. Intell 20(4), 383–420 (2012)
12. Andrighetto, G., Conte, R., Campennl, M.: Minding Norms: Mechanisms and dynamics of social order in agent societies. Oxford Scholarship (2013)
13. Sartor, G.: Legal Reasoning: A Cognitive Approach to Law. Springer (2005)
14. Sen, S., Airiau, S.: Emergence of norms through social learning. In: Proceedings of the 20th International Joint Conference on Artifical Intelligence, IJCAI 2007, pp. 1507–1512. Morgan Kaufmann Publishers Inc., San Francisco (2007)
15. Shoham, Y., Tennenholtz, M.: 'On social laws for artificial agent societies: off-line design'. Artificial Intelligence 73(1-2), 231–252 (1995)
16. Sutton, R.S., Barto, A.G.: Reinforcement learning: An introduction, vol. 116. Cambridge Univ. Press (1998)

Chisholm's Paradox and Conditional Oughts

Catharine Saint Croix and Richmond H. Thomason

Philosophy Department, University of Michigan, Ann Arbor, MI, USA

Abstract. Since it was presented in 1963, Chisholm's paradox has attracted constant attention in the deontic logic literature, but without the emergence of any definitive solution. We claim this is due to its having no single solution. The paradox actually presents many challenges to the formalization of deontic statements, including (1) context sensitivity of unconditional oughts, (2) formalizing conditional oughts, and (3) distinguishing generic from nongeneric oughts. Using the practical interpretation of 'ought' as a guideline, we propose a linguistically motivated logical solution to each of these problems, and explain the relation of the solution to the problem of contrary-to-duty obligations.

1 Chisholm's Paradox

[1] formulates the problem of contrary-to-duty obligations with the following example.[1]

(1a) It ought to be that Jones go to the assistance of his neighbors.
(1b) It ought to be that if he does go he tells them he is coming.
(1c) If he does not go, then he ought not to tell them he is coming.
(1d) He does not go.

Although the language in (1a) and (1b) is somewhat stilted and unnatural, there is nothing uncommon about the situation it describes. Frequently a secondary obligation results from the violation of a primary obligation. This is only considered to be paradoxical because such examples are difficult to formalize.

We begin with an unconditional version of the Chisholm quartet. Having shed some complication, we argue, it becomes clear that 'oughts' are context-sensitive. Bearing that in mind, we shift to conditional obligation in Section 5. There, we offer considerations in favor of factual, rather than deontic, detachment. In Section 5.3, we show that a narrow-scope deontic conditional is unsuitable for the formalization of reparational obligations unless the conditional is contextualized. This contextualization avoids the problematic inference. We then discuss how generic constructions may be used to recover deontic detachment for wide-scope conditional 'ought's where necessary. Finally, we return to Chisholm's paradox in Section 6, where we demonstrate our proposed solution.

---

[1] This is Chisholm's exact wording, except that we have substituted 'Jones' for 'a certain man' in (1a).

F. Cariani et al. (Eds.): DEON 2014, LNAI 8554, pp. 192–207, 2014.

2   An Unconditional Version

Steps (1b) and (1c) are conditional oughts, conditionals whose main clause involves an 'ought'. Like Chisholm himself, many authors ([2], [3], [4], [5], for instance) have felt that this example reveals the inadequacy of naive formalizations of conditional obligation, and that an adequate logic of conditional obligation will solve the problem.

This can't be entirely right, because of examples like (2), which are similar to (1) but do not involve the conditional.

(2a) Jones ought to assist his neighbors.
(2b) But he will not go.
(2c) So he ought to tell his neighbors he is not going to assist them.
(2d) Therefore, he ought to assist his neighbors and to tell them he is not going to assist them.

(2d) follows from (2a) and (2c) in standard deontic logics. But, although (2a–c) appear to be mutually consistent and provide a plausible description of the Chisholm scenario, (2d) is clearly false.

We don't deny that the conditional version of Chisholm's paradox illustrates logical difficulties having to do with conditional obligation. But the unconditional version reveals a more fundamental problem that needs to be cleared up before turning to the conditional case.

3   A Methodology

Work in deontic logic tends to concentrate on examples with moral overtones, like promise-keeping. But 'ought' has many uses. If we assume, with [6], that these uses differ only in the sort of possibilities that are in play, these differences will not affect the underlying logic. Practical or prudential uses of 'ought' provide intuitions that in general are crisper than moral uses, and moreover are readily restricted to simple domains or scenarios.

We propose to use a game that we'll call *Heads Up* as a laboratory for testing deontic intuitions. A number of playing cards are set down side by side. A player, Jones, gets to choose a card. The player's payoffs are dependent on whether he chooses a face card. Simple versions of this game will involve just one choice, while more complicated versions will involve successive choices.

4   Unconditional Oughts

We begin with a simple three-card version of *Heads Up*. In this game, there will always be at least one face card and at least one non-face card on the table. If Jones chooses a face card, he gets $50. Otherwise, he gets nothing.

Suppose Jones is presented with $\langle \mathsf{Jack}, 3, 9 \rangle$ as a layout. Clearly, he ought to choose the leftmost card, ought not to choose the middle card, and ought not to choose the rightmost card. Similarly, if presented with $\langle \mathsf{Queen}, \mathsf{King}, 4 \rangle$, Jones ought not to choose the rightmost card. And it's not the case that he ought to choose the leftmost card, since choosing the middle card would result in the same payoff. So, he ought to choose *either* left or middle.

Following Lewis' semantics for deontic operators ([7,8]), $\bigcirc(\phi)$ is true just in case there is an outcome u satisfying ϕ such that any outcome at least as good as u also satisfies ϕ.

Under the $\langle \mathsf{Queen}, \mathsf{King}, 4 \rangle$ layout, then, let $\phi = $ 'choose left or middle'. In this case, both outcomes satisfying ϕ, left and middle, reward \$50 and the only other possible outcome of Jones' choice rewards \$0. So, $\bigcirc(\phi)$ is true. $\bigcirc$(choose left), however turns out to be false. This is because there is another outcome (middle), that doesn't satisfy 'choose left' but has a reward as good as that of 'choose left'.

We can formalize this by augmenting a boolean propositional logic with $\bigcirc$ as a modal operator. Let a Lewis frame be a structure $\mathcal{F} = \langle W, \preceq, f \rangle$, where W is a non-empty set (of outcomes), $\preceq$ is a preorder over W, and f is a function taking members of W to subsets of W. A model over $\mathcal{F}$ is a structure $\langle W, \preceq, f, V \rangle$, where V is a function taking propositional atoms to subsets of W.

Definition 1. Satisfaction in a model for $\bigcirc(\phi)$.
   Given a model $\mathfrak{M}$ and a world $w \in W$, $\mathfrak{M}, w \models \bigcirc(\phi)$ **iff** there is some $u \in f(w)$ such that $\mathfrak{M}, v \models \phi$ for all $v \in f(w)$ such that $v \preceq u$.

By allowing f to pick out deontic alternatives from w, we can restrict the outcomes evaluated to those that are relevant given the situation. When the layout is $\langle \mathsf{Queen}, \mathsf{King}, 4 \rangle$, we don't need to check irrelevant possibilities in which Jones chooses the left and it isn't a Queen.

4.1   Knowledge of Circumstances and Uncertainty

So far, we have said nothing about what Jones knows: *Heads Up* doesn't specify whether the cards are dealt face down. If they are dealt face up, Jones will know all of the relevant information, but otherwise we can't ignore his epistemic state. Suppose now that the cards are dealt face down and that the layout is $\langle \mathsf{Queen}, 5, 2 \rangle$. Perhaps surprisingly, it is still natural to say that Jones ought to choose the leftmost card, even though he doesn't *know* this.

If the dealer were to say, "Jones, there's no card you ought to choose" in this case, the natural interpretation would be that there is no unique face card on the table, rather than that there may be a unique face card although Jones doesn't know this. Practical oughts act, in fact, as if they want to ignore the agent's epistemic situation, even though agents must find the best choice in light of their knowledge.

We can, however, find cases where practical oughts are relativized to the agents' knowledge. If, for instance, Jones detects a bias for queens on the left in the face-down version of the game he might say to himself, as he chooses, "I ought to choose the leftmost card." But this epistemically conditioned usage is evanescent. If the cards are turned over after Jones plays and the layout is $\langle 5, 2, \mathsf{Queen} \rangle$, Jones could well say "Damn! I ought to have chosen the card on the right!" He can't, however, say "Damn! I ought to have chosen the card on the left, but it wasn't a queen!." The "miners' paradox" scenario presented in [9] illustrates the point: although these knowledge-relative uses can be exhibited, doing so takes a certain amount of work.

4.2   Solving the Unconditional Paradox

Confine attention now to objective oughts that ignore the agent's epistemic state. In the *Heads Up* domain this means, among other things, that if $w' \in f(w)$ then w and w' do not differ in their factual circumstances, in the layout of cards, although they can differ in the choices that the agent makes. Now, consider another version of *Heads Up*, with just two cards in the layout and where the player will make two successive choices. The payoffs are: \$500 if a face card is chosen in both rounds, \$0 if a face card is chosen in the first round but not the second, \$250 if a face card is chosen only in the second round, and, finally, \$350 if no face card is chosen in either round. The relevant choices are now

L_1:   Choose the left card on the first turn
L_2:   Choose the left card on the second turn
R_1:   Choose the right card on the first turn
R_2:   Choose the right card on the second turn

Suppose the layout is $\langle$Queen, 2$\rangle$. Since choosing left allows Jones to reach the optimal outcome, \$500, $O(L_1)$ is true in our model: (a) $\models O(L_1)$. If he does choose left, (b) $\models O(L_2)$: Jones should choose left in the second round, since the outcomes available after L_1 are \$500 for L_2 and \$0 for R_2.

But now suppose that Jones performs poorly in the first round and chooses the card on the right. Then the outcomes available to Jones are different. In light of his poor first choice, Jones' payoffs are \$250 for L_2 or \$350 for R_2. With these options, then, (c) $\models O(R_2)$. Thus, we have:

(3a) $O(L_1)$ is true.
(3b) But $\neg L_1$ is true.
(3c) So, $O(R_2)$ is true

At this point, it has become clear that the satisfaction relation is context dependent. To represent examples like this, we must make this explicit. We modify the function f of a Lewis frame to take a context set X of worlds into account: $f(w, X) \subseteq X$. We then define *contextualized satisfaction* by relativizing satisfaction to this set X, representing the set of alternatives presumed to be open to choice.

Definition 2. Contextualized satisfaction in a model for $O(\phi)$.
Given a Lewis frame $\mathcal{F} = \langle W, \preceq, f \rangle$, a model $\mathfrak{M}$ on $\mathcal{F}$, a subset X of W, and a world $w \in W$, $\mathfrak{M}, X, w \models O(\phi)$ **iff** there is some $u \in f(w, X)$ such that $\mathfrak{M}, X, v \models \phi$ for all $v \in f(w, X)$ such that $v \preceq u$.

Using contextualized satisfaction, (3a–c) can now be stated as follows.

(4a) $\mathfrak{M}, X, w \models O(L_1)$
(4b) But $\mathfrak{M}, X, w \models \neg L_1$
(4c) So, $\mathfrak{M}, X \cap [\![\neg L_1]\!], w \models O(R_2)$

Here $[\![\neg L_1]\!]$ is the set of worlds satisfying $\neg L_1$.

Thus, the assertion in (4b) serves to add its content to the context set used to interpret the next step, (4c). This is similar to the effect of assertion on presupposition noted in Stalnaker's [10] and similar writings.

The scenario in (4) closely resembles (2), the unconditional version of Chisholm's paradox. This is easier to see from the following English version of (4).

(5a) Jones ought to choose left on round one.

(5b) But he will not.

(5c) So he ought to choose right on round two.

The sensitivity of 'ought' and other modals to a context set was noted by Kratzer in [6] and has become a standard part of linguistic theories of modals; see [11,12]. Thus, this contextual solution to the unconditional Chisholm paradox is well motivated in terms of linguistic theories of modals and assertion.

5   Conditional Oughts

We now turn to the conditional version of Chisholm's paradox. Here, the crucial issue is how to formalize statements of conditional obligation.

At the outset, there are two approaches to this issue: (i) take conditional oughts to be primitive 2-place modalities $O(\psi/\phi)$, and (ii) take them to be compositional combinations of 'ought' and 'if'. Although many linguists and logicians take the first approach, a compositional theory is generally to be preferred. With this in mind, we will not consider Approach (i) here, but will explore Approach (ii), with the thought that a linguistically adequate compositional theory will render Approach (i) unnecessary. Our hypothesis, then, is that conditional oughts involve the same conditional that figures generally in other conditional constructions, with or without modals.

We begin with a neutral conditional ' $\bullet\bullet\!\!\rightarrow$ ', making no assumptions about its logic for the moment. Assuming compositionality, there are two options for formalizing 'If ϕ then ought ψ': (i) wide scope O, $O(\phi \bullet\bullet\!\!\rightarrow \psi)$ and (ii) narrow scope O, $\phi \bullet\bullet\!\!\rightarrow O(\psi)$.

Chisholm's somewhat tortured phrasing in (1) suggests that both are involved, with (1b) taking wide scope and (1c) narrow. With this in mind, we might formalize (1) as follows:

(6a) $O(\mathsf{Help})$

(6b) $O(\mathsf{Help} \bullet\bullet\!\!\rightarrow \mathsf{Tell})$

(6c) $\neg\mathsf{Help} \bullet\bullet\!\!\rightarrow O\neg\mathsf{Tell}$

(6d) $\neg\mathsf{Help}$

This formalization uses both the wide scope formalization of conditional 'ought' (6b) and the narrow scope (6c). Ultimately, we will question the formalization of (6b), but this contrast provides a useful way to frame the important issues, which have to do with detachment. (See [13].)

5.1   Detachment

In Example (6) we want to conclude that Jones ought to tell his neighbors he isn't coming to help, $O\neg\mathsf{Tell}$. Now, with few exceptions, logics of the conditional allow us to infer ψ from $\phi \bullet\!\!\bullet\!\!\rightarrow \psi$ and ϕ. And with fewer exceptions, deontic logics with a conditional $\bullet\!\!\bullet\!\!\rightarrow$ validate the inference from $\bigcirc(\phi \bullet\!\!\bullet\!\!\rightarrow \psi)$ and $\bigcirc\phi$ to $\bigcirc\psi$. There are, then, two ways to conclude an 'ought' statement.

Definition 3. *Factual deontic detachment* (FDD)
  Infer $\bigcirc\psi$ from $\phi \bullet\!\!\bullet\!\!\rightarrow \bigcirc\psi$ and ϕ.

Definition 4. *Deontic deontic detachment* (DDD)
  Infer $\bigcirc\psi$ from $\bigcirc(\phi \bullet\!\!\bullet\!\!\rightarrow \psi)$ and $\bigcirc\phi$.

Factual deontic detachment provides the inference of $\bigcirc\neg\mathsf{Tell}$ from (6c) and (6d), and this inference is welcome in the example. The difficulty is that deontic deontic detachment allows us to infer $\bigcirc\mathsf{Tell}$ from (6b) and (6a), and this conclusion is not so welcome.

5.2   Factual Detachment in *Heads Up*

Let's refine our intuitions by returning to a simple version of *Heads Up*. In this version, we still have just two cards in the layout, exactly one of which will be a face card, and there is only a single round.

Intuitively, the following deontic conditionals are true in this scenario:

(7a) If the left card is a queen, Jones should choose it.
(7b) If the right card is a queen, Jones should choose it.
(7c) If the left card is a two, Jones should not choose it.
(7d) If the right card is a two, Jones should not choose it.

The intuitions in favor of FDD are very powerful in this example. As soon as we learn, for instance, that left card is a queen, we think that Jones should choose it. This supports formalizations of (7a–d) as narrow-scope deontic conditionals $\phi \bullet\!\!\bullet\!\!\rightarrow \bigcirc\psi$, as indeed the language suggests.

5.3   Factual Detachment as *modus ponens*

Superficially, FDD may appear to be very simple if it merely amounts to using *modus ponens* with a narrow scope deontic conditional. But if we look more carefully at this matter in model-theoretic terms, the matter is more complex. For definiteness, we will work from now on with Stalnaker's semantics for the conditional ([10]).[2] This invokes a "selection function" s from propositions or sets of worlds to sets of worlds. This function satisfies the following conditions.

(8a) For some $u \in W$, $s(Y, w) \subseteq \{u\}$
(8b) $s(Y, w) \subseteq Y$.
(8c) $s(Y, w) = \{w\}$ if $w \in Y$.
(8d) If $s(Y, w) = \emptyset$ then $Y = \emptyset$.
(8e) If $s(Y, w) \subseteq Y'$ and $s(Y', w) \subseteq Y$ then $Y = Y'$.

---

[2] We mention at this point [4], which also proposes $\phi \bullet\!\!\bullet\!\!\rightarrow \bigcirc\psi$ as a formalization of conditional oughts but without mentioning contextual effects.

To interpret the conditional $>$, we add the function s to our frames and add the following satisfaction clause. (Because in Section 4 we decided to relativize satisfaction with contexts, this definition contains a parameter 'X'.)

Definition 5. Naive Satisfaction for $>$
  Given a model $\mathfrak{M}$, a context set X, and a world $w \in W$,
  $\mathfrak{M}, X, w \models \phi > \psi$ **iff** if $s(\phi, w) = \{u\}$ then $\mathfrak{M}, X, u \models \psi$.

Here, $s(\phi, w)$ is $s([\![\phi]\!], w)$.

It is easy to verify that this definition validates *modus ponens*, so that it supports FDD with narrow scope formalizations of conditional obligation. This depends crucially on Stalnaker's "centering" condition (8c).

But Definition 5 creates problems in formalizing reparational obligation. Suppose, to return to (1), that we formalize (1c), 'If he does not go, then he ought not to tell them he is coming,' as follows.

(9) $\neg$Help $> \bigcirc\neg$Tell

According to the naive satisfaction condition in Definition 5, (9) is true in a world w where, say, Jones has promised to help his neighbors, iff in the closest world u where Jones does not help his neighbors he ought to tell them he will not come to help them. But in this world u, the factual circumstances, including Jones' promise, remain the same, only Jones' choice is changed. (Otherwise, u would not be the closest world.) So in this world u, just as in w, Jones ought to help his neighbors because, in the best alternatives, he does. Of course, in those alternatives, $\neg$Tell is false. So, u isn't an $\bigcirc\neg$Tell world. Thus, (9) turns out to be false and the rather pointless (10) is true.

(10) $\neg$Help $> \bigcirc$Help

This conditional, amounting to 'If Jones does not help his neighbors, then he (still) ought to help them' may make some sense as an admonition, but it is impractical and certainly doesn't correspond to our intuitions concerning secondary obligations.

To solve this problem, we replace the naive satisfaction clause for $>$ with a more sophisticated version that contextualizes the conditional as well as $\bigcirc$. First, we make the selection function s sensitive to context, so that s now inputs a set of worlds (the antecedent proposition), another set of worlds X (the context), and a world, and, as before, returns a unit set of worlds or the empty set. Our new satisfaction condition is this.[3]

Definition 6. Satisfaction for $>$
  Given a model $\mathfrak{M}$, a context set X, and a world $w \in W$,
  $\mathfrak{M}, X, w \models \phi > \psi$ **iff** if $s(\phi, X, w) = \{u\}$ then $\mathfrak{M}, X \cap [\![\phi]\!], u \models \psi$.

This cumulative satisfaction clause adds the antecedent proposition to the context in which the consequent is evaluated. Only worlds satisfying the antecedent are to be taken into account in evaluating a consequent $\bigcirc\psi$. And this solves the problem of formalizing secondary obligations. (9), for instance, is true,

---

[3] [14] argues that this condition provides an improvement on Stalnaker's semantics for the conditional.

because in the closest world u where John doesn't help his neighbors, $\bigcirc\neg\mathsf{Tell}$ is true in the best options in which the background facts are assumed, as well as the proposition that Jones will not help his neighbors. In other words, under the first definition, (9) failed because $\mathfrak{M}, X, v \not\models \bigcirc\neg\mathsf{Tell}$ for the best alternatives, v in X. Under the second we use $X \cap [\![\neg\mathsf{Help}]\!]$ instead of X, and (9) is true.

On the other hand, *modus ponens* is no longer valid on this interpretation of $>$ (see [14] for details), so that FDD is threatened.

It may seem at this point that it is difficult or even impossible to retain (i) a compositional account of the conditional oughts involved in FDD, (ii) a compositional account of the conditional oughts involved in stating secondary obligations, and (iii) a logical endorsement of FDD.

However, we do not believe that things are as bad as this. FDD is *pragmatically valid*, in a sense that was first introduced in [10]. Although the inference

(11a) ϕ

(11b) $\phi > \bigcirc\psi$

(11c) $\bigcirc\psi$

is invalid when all three terms are evaluated with respect to the same context X, (11c) will be true if the first step adds its proposition to the context in which the subsequent steps are validated. In other words, $\bigcirc\psi$ follows from *the assertion* of ϕ and $\phi > \bigcirc\psi$. We feel that pragmatic validity provides an adequate account of the very strong intuitions that favor FDD, and also serves the purposes of FDD in practical reasoning, allowing detachment when the minor premise has been learned and added to the background context.

5.4   Deontic Deontic Detachment

Treating conditional oughts compositionality has the apparent advantage of providing a natural formalization for DDD, by providing a wide scope logical form $\bigcirc(\phi > \psi)$. If the underlying conditional validates *modus ponens* and $\bigcirc$ is a modal operator, then DDD with wide scope is just the modal principle K.

But in fact, this idea is not well supported on linguistic grounds. There is very little linguistic evidence for *any* cases in which modals take wide scope over conditionals. If such cases occur at all, they rarely occur naturally.

As [13] point out, deontic statements should figure in bodies of rules and maxims, and DDD does seem to play a useful role in reasoning in these domains. If employes ought to be paid for every day they work, and employees ought to work on weekdays that aren't holidays, then it might be useful to conclude that employees ought to be paid on weekdays that aren't holidays. We turn now to the formalization of DDD, and to how we might be able to capture inferences such as this, and begin by turning to another difficulty with FDD.

5.5   Worries about Strengthening the Antecedent

Our treatment of conditional oughts as narrow scope conditionals allows for left-nonmonotonicity effects—failure of Strengthening the Antecedent. For instance, suppose that Jones is looking at $\langle \mathsf{Queen}, 4, \mathsf{King} \rangle$. Then, 'If the right card were a

6, I ought to choose the left card' is clearly true. And equally clearly, 'If the right card were a 6 and the middle card were a jack and the left card were an eight, I ought to choose the left card' is false. This is intended, but it creates problems for treating these conditional oughts as standing rules that can be applied in new circumstances.

When we lack Strengthening the Antecedent, it is harder to apply FDD than one might think. Suppose we are in a world u satisfying p. Jones knows this, and has reason to believe $p > \bigcirc q$ and $(p \wedge r) > \bigcirc q$. For instance, Jones may know he has received a bill for cleaning a carpet, and think that he ought to pay the bill if he's received it, but also that he ought *not* to pay it if he's received it and the carpet has not, in fact, been cleaned. Intuitively, the following deontic conditionals are true in this scenario:

(12a) $p > \bigcirc q$ ['I ought to pay the bill if I received it.']

(12b) $(p \wedge \neg r) > \neg \bigcirc q$ ['I ought not to pay the bill if I received it and the carpet wasn't cleaned.']

Using FDD, Jones can reach the following conclusions.

(13a) $p, p > \bigcirc q \therefore \bigcirc q$

(13b) $p, \bigcirc q, (p \wedge r) > \bigcirc \neg q \therefore \neg r$

(13c) $p, r, (p \wedge r) > \bigcirc \neg q \therefore \neg(p > \bigcirc q)$

In our example, Jones can't use FDD to conclude $\bigcirc(q)$ without also committing to $\neg r$. If, on the other hand, he finds that the rug hasn't been cleaned, he must, in view of (13c), give up the conditional obligation $p > \bigcirc q$. In this common situation, it seems that we can't use typical, presumably action-guiding conditional oughts without checking many background facts (and, in particular, all the possible defeaters). It seems that we can't at the same time take conditional obligations to be *standing* oughts—true in the worlds in which we deliberate—and have FDD.

5.6   Standing Oughts as Generics

If at all possible, we should avoid having to resolve a conflict between FDD and standing maxims by choosing one at the expense of the other. In practical reasoning, we need to combine conditional oughts with beliefs about our current circumstances to obtain unconditional, immediate oughts—so we need FDD. But we also need standing oughts, and these too must be applicable to circumstances. This we take to be the real challenge of accommodating DDD.

We propose appealing to *generic constructions*, and in particular, generic conditionals, as a solution to this problem. Generics are quite generally available in the world's languages. In English, non-progressive, non-perfective present tense sentences are quite likely to be generics: 'She jogs home from work', 'It rains in Seattle', 'He likes red wine better than white wine'. The critical feature to observe is that generic claims tolerate exceptions: there's no contradiction in saying 'Even though I jog home from work, I think I'll take the bus today'.

Linguists postulate a GEN operator to provide a logical form for generics, so that GEN(ϕ) is the generic of ϕ. (See [15].)

While there isn't an overt tense marker signaling a generic in the case of 'ought' or 'should', generic uses of modal statements would not be surprising. Suppose you are discussing the monthly bills with a significant other, and say,

(14a) "You ought to pay this bill."
(14b) "Sure, I ought to, but I can't pay it till I have the money."
(14c) ?"No, I ought not to pay it; I don't have the money to pay it."

(14b) is the natural response in this case, rather than (14c). Without a generic interpretation of (14a), this would create a problem for the plausible idea that 'ought' implies 'can'. If (14a) is interpreted as a *generic* ought, however, (14b)'s characteristic 'but' would signal that the clause following it marks an exception to a general rule. Conditional oughts can have the same flavor, as the following variation on (14a–b) illustrates.

(15a) "If you got a bill from the carpet cleaner, you ought to pay it."
(15b) "Yes, if I got the bill I ought to pay it, but I don't have the money."

Generic constructions like (15a), involving a conditional and a deontic modal, do not appear to have been studied much by linguists. However, it's natural to treat them just like other generics, assigning them the same form $\text{GEN}(\phi)$, but where GEN is now operating on a conditional ought.

We committed ourselves in Section 5.3 to $\phi > \bigcirc\psi$ as a formalization of statements of conditional obligation. This means that (15a) will have a fully compositional formalization along the following lines.

(16) $\text{GEN}(\text{BillReceived} > \bigcirc\text{PayBill})$

Unfortunately, this suggestion offers less help with satisfaction conditions than one would like. The semantic interpretation of GEN is chronically problematic. Although linguists have proposed satisfaction conditions,[4] there is no general agreement about what they should be. Even the postulation of GEN is to some extent controversial.[5] Fortunately, however, we do not need to commit ourselves to any specific account of GEN to see that formalizations like (16) can serve the purposes that have made wide scope deontic conditionals and DDD seem attractive to some deontic logicians.

First, like all generics, statements like (16) are standing generalizations, ready to be used in any deontic deliberation.[6] Second, generics support defeasible instantiation. For instance, with no reason to the contrary, from (15a) one can infer that I ought to pay the bill if I received it.

Thus, we can replicate DDD with a two-stage inference: first, from (15a) we defeasibly infer $\text{BillReceived} > \bigcirc\text{PayBill}$. This and BillReceived pragmatically imply

---

[4] See, for instance, [16].

[5] Also, it's natural to think that a singular generic like (14a) is in fact derived from a more general formulation using the generic plural, such as 'People ought to pay their bills'. The semantic interpretation of generic plurals is even more controversial.

[6] In [17] Adam Lerner and Sarah-Jane Leslie explicitly identify ethical maxims with deontic generic constructions. This paper goes into some detail about how such constructions enter into reasoning about what one ought to do.

○PayBill. Thus, a combination of defeasible inference and pragmatic implication delivers the desired consequence.

According to this theory, (12a–b) would now appear as

(12a′) GEN(BillReceived > ○PayBill) ['I ought (normally) to pay the bill if I received it.']

(12b′) GEN((¬CarpetCleaned ∧ BillReceived) > ○¬PayBill) ['I ought not (normally) to pay the bill if I received it but the carpet wasn't cleaned.']

These formalizations do not cause the difficulties we saw in (13a–d), because BillReceived > ○PayBill is only a defeasible consequence of GEN(BillReceived > ○PayBill). This consequence is defeated by BillReceived ∧ ¬CarpetCleaned, in the presence of (12b′).

6   Revisiting Chisholm

With the formal apparatus developed in the preceding sections, we can return to (1a–d), formalizing the example as follows:

(17a) ○Help

(17b) ○(Help > Tell)

(17c) ¬Help > ○¬Tell

(17d) ¬Help

Here we have used a wide-scope ○ to formalize (17b). Although we have said that ○ rarely takes wide scope over a conditional, we can take Chisholm's awkward 'it ought to be that if' to force such a reading, as Chisholm probably intended.

The apparent problem, then, is that

(18a) ○Help, ○(Help > Tell) imply ○Tell, and

(18b) ¬Help, ¬Help > ○¬Tell imply ○¬Tell

which together *pragmatically* imply a deontic contradiction, ○ ⊥.

While (17a–d) do imply this deontic contradiction, we don't need to accept the conclusion. The context sensitivity of ○ provides a natural account of how we can, in a sense, accept (17a–d) without thereby accepting ○ ⊥. In fact, the solution to paradox in its original form mirrors the contextual solution we gave in Section 4.2 to the simpler, unconditional form of the paradox.

Below, in Section 6.2, we provide a more detailed analysis of of the contextual solution, this time with a narrow-scope formalization of the second premise.

6.1   *Heads Up*, One More Time

Recall the two-move version of *Heads Up* introduced in section 4.2. Here, we can find what might be called *impractical oughts* arising from a speaker's choice of context. Suppose the cards are dealt face up and the layout Jones faces is ⟨King, 6⟩. We'll call the four outcomes w_1 ($500), w_2 ($0), w_3 ($250), and w_4

($350). We imagine an observer watching over Jones' shoulder as he makes his choices in the world w_3. A king is on the left and a six is on the right, and Jones has chosen right. This first choice has left him in a suboptimal position; he can no longer get $500, but by choosing right again he can get $350. Choosing left will get him $250. The observer might well say:

(19) $\bigcirc R_2$ ['Jones ought to choose right for the second pick']

The observer has made a good point: Jones will get less if he chooses left.

The observer could also take a more remote and less practical perspective. If at the outset Jones had gone for the best outcome, he would have chosen left. At the second round, then, the correct choice would also be left. With this perspective (and a bit of wishful thinking) the observer could also say:

(20) $\bigcirc L_2$ ['Jones ought to choose left for the second pick']

This is a bit unnatural; this impractical perspective is better expressed as:

(21) Jones ought to be choosing left right now.[7]

If there are semantic differences between (20) and (21), they're subtle. At any rate, Jones can't be choosing left unless he chooses left, and if he chooses left he is choosing left. It may be that the reason that (21) is more felicitous than (20) in the impractical sense is pragmatic, and is similar in many respects to the difference between indicative and subjunctive mood.

Now, there is an apparent contradiction between (19) on the one hand, and (20/21) on the other, both said at the same time in w_3. But the contradiction is only apparent. There is a contextual element in play: the background of possibilities considered to be open alternatives.

The practical 'ought' in (19) should be interpreted with respect to the practical set of possibilities that are open at w_3. This will be a set consisting of two worlds: w_3 and a world w_4 which is like w_3 except for the fact that Jones chooses right on the second turn instead of left in w_4. To evaluate this practical ought, we set the context, X, to $[\![R_1]\!]$, or $\{w_3, w_4\}$. Then, $\mathfrak{M}, X, w_3 \models \bigcirc R_2$, since $\{w_4\}$ is preferable to $\{w_3\}$.

But the impractical 'ought' in (20/21) requires a different set of possibilities. Here, the observer is meddling with the open alternatives by supposing that Jones had made the correct first choice. We therefore want the set of possibilities to be $\{w_1, w_2\}$. For this set of possibilities, $\bigcirc L_2$ is true, even after Jones' first choice at w_3. This is because $f(w_3, \{w_1, w_2\}) \subseteq \{w_1, w_2\}$ and w_1, which is a L_2 world, is better than w_2.

The main point may have been lost in these formal details. It's this: a context, in the form of a set of background possibilities, contributes to the interpretation of an 'ought'. For practical oughts (and this is the default), these are the possibilities that vary according to exogenous chance factors and the agent's choice of an action. But 'ought' can also be used impractically, with respect to a counterfactual set of possibilities; such usages are often associated with the verb

---

[7] For some reason, usages with 'be' seem to go better with impractical contexts. This seems related to the difference that philosophers like Castañeda (See [18]) have noted between 'ought to be' and 'ought to do'.

'be'. The truth of an 'ought' statement will depend, among other things, on the context that is used to interpret it.

In our examples from the two-choice version of the game, (19) is practical and (20/21) are impractical. The contradiction between the two is therefore only apparent, since the appropriate contexts for them are different. In fact, although $\bigcirc R_2$ is true relative to a context $\{w_3, w_4\}$ of alternatives, and $\bigcirc L_2$ is true relative to a context $\{w_1, w_2\}$ of alternatives, the two formulas and the deontic contradiction $\bigcirc \bot$ that they entail are never true relative to the same context set of alternatives. So, there is no "paradox" here.

6.2   The Chisholm Quartet

What we concluded about (19) and (20/21)—apparently contradictory, but perfectly compatible if you take the context shift into account—is, directly analogous to the main issue in Chisholm's Paradox. Consider this *Heads Up* paradox:

(22a) It ought to be that Jones chooses left initially.
(22b) It ought to be that if he chooses left initially he chooses left next.
(22c) If Jones does not choose left initially, he ought not to choose left next.
(22d) Jones does not choose left initially.

We propose to formalize these as follows.

(23a) $\bigcirc L_1$
(23b) $L_1 > \bigcirc L_2$.
(23c) $\neg L_1 > \bigcirc \neg L_2$
(23d) $\neg L_1$.[8]

Again we imagine our bystander uttering (22a–d). This time, she speaks as Jones is about to make his first choice, and in the world w_4. We will evaluate (23a–d) from this standpoint. As we saw above, we also must take context into account, in the form of a set of alternatives. Many sets could be in play at this point. Let's consider two possibilities.

Context 1. Total Ignorance. Suppose our observer uses a context set in which all possibilities are open, $X = \{w_1, w_2, w_3, w_4\}$. Here, (23a) is true, since L_1 is true in w_1 and w_1 is the best alternative to w_4. In fact, this context/world combination renders all four premises true. However, as we discussed in section 4.2, a context like total ignorance, which doesn't alter the alternatives, is not a felicitous context for a Chisholm-style premise set.[9]

---

[8] Chisholm may well have intended a wide-scope formalization of (22b), but, as we have argued, such formulations are implausible, and are better treated as generics. We take (22b) to be a plausible formulation, however, given the foregoing discussion.

[9] There are also reasonable interpretations of $\bigcirc$ that make (23a) false in this context, but that issue is beyond the scope of the present paper.

Context 2. Optimal Second Choice. Suppose the observer believes that Jones may lapse in making his first choice, but will regain his senses in the second round. (Perhaps he appears to be temporarily distracted.) Then, the second choice will be optimal, meaning that her context is $\{w_1, w_4\}$.

Relative to this context, (23a) is true, since w_1 is better than w_4 and w_1 is an L_1 world. Furthermore, (23b) is true, but vacuously so since w_1 is the only alternative under which we can evaluate $\bigcirc L_2$ under the antecedent and this context set. This suggests that $\{w_1, w_4\}$ is not a felicitous context for that formula. This leaves (23c), which also is vacuously true.

Context 3. Bold First Choice. Assuming that the first choice is bold, keeping the best possibility available, gives us the set $\{w_1, w_2\}$. Here, (23a) is true; $\mathfrak{M}, \{w_1, w_2\}, w_4 \models \bigcirc L_1$. It's vacuously true, however, since the context assumes he will choose left. (23b) is also true, since w_1 is better than w_2. (23c) is also true, but it's vacuous because the antecedent of the conditional is inconsistent with the context set. Finally, (23d) is true as well, since the world is w_4.

6.3 Plausibility of the Chisholm Premise Set

Each of these contexts satisfies all of the premises of Chisholm's Paradox without sacrificing consistency. Part of the problem presented by the paradox was that these premises are supposed to imply $\bigcirc \bot$. To that extent, we've shown that the Chisholm premise set is not paradoxical on our account. In general, $\bigcirc p$, $p > \bigcirc q$, $\neg p > \bigcirc \neg q$, and $\neg p$ do not imply $\bigcirc \bot$. We lost this implication at the point where we added contexts to the interpretation of $\bigcirc$ and the conditional. While this is an interesting feature of the contextualized satisfaction conditions, it isn't really a solution to the paradox. Contexts two and three both employ vacuous oughts, which signify an inappropriate context. This is a problem because the four terms of Chisholm's Paradox not only seem true, but are meant to seem natural.

We obtain a more satisfactory solution if we say that each of (23a–d) is satisfied and appropriate in some context, but there there is no single context that satisfies them all appropriately. There are at least two linguistically well-motivated ways we might account for this context change.

Accommodation. We know from pragmatics that an utterance attracts an appropriate context. When a sentence is uttered (within limits) a context for interpreting it is selected that makes it a sensible thing to say. Following [19], this phenomenon is called "accommodation."

Aside from general rules, such as "Try to make the utterance true," some special rules seem to apply to the interpretation of 'ought'.

(i) All things equal, prefer an indicative or practical use, in which what is beyond the agent's control is supposed, but what depends on actions under the agent's control is allowed to vary.

(ii) Vacuous cases are to be avoided, and in particular, in interpreting $\bigcirc \phi$, context sets that entail ϕ or $\neg \phi$ are to be avoided.

These rules make $\{w_1, w_3\}$ (Context 2, the optimal second-choice context) the most plausible context for (22a). At any rate, the totally ignorant context falsifies (22a), and contexts that determine the first choice make (22a) vacuous. But this context makes (23b) and (23c) vacuous.

On the other hand, $W = \{w_1, w_2, w_3, w_4\}$ (Context 1, the total ignorance context) is the most plausible for (22b) and (22c). This context makes both conditionals true, and it entails neither the antecedents of the conditionals nor their negations (though it does, on this account of $\bigcirc$, entail (23a)).

Together, these different preferences for contexts may help to explain the plausibility of Chisholm-style paradoxes. The premises seem true and felicitous because, when they're accommodated to their respective appropriate contexts, each *is* true and felicitous. But these different contexts cannot be unified into a single one that makes all the premises true and felicitous.

Assertion. It wasn't necessary to say anything about the last premise, (22d), because any context is compatible with its truth. But things are different if we imagine (22d) to have been asserted. Dynamic theories of assertion, such as the one presented in [20], take assertion to add context to the context. When the initial context for an assertion of ϕ is X, the subsequent context is $X \cap [\![\phi]\!]$.

If we take (20d) to have been asserted, then, we get quite a different picture of the premisses (22a–d). The order of the premises in Chisholm's formulation, of course, doesn't invite this interpretation, since (1d) is his last premise. But the interpretation is still available, we think, even with Chisholm's order, and his wording of (1a) and (1b) actually encourages this interpretation.

To make the case where (22d) has been asserted salient, let's revise the order of premises as follows.

(24a) Jones does not choose left initially.

(24b) It ought to be that Jones chooses left at first.

(24c) It ought to be that if he chooses left initially he chooses left next.

(24d) If Jones does not choose left initially, he ought not to choose left next.

The assertion of (24a), restricts the context to $\{w_3, w_4\}$. This forces a *counterfactual* interpretation of (24b) and (24c), in which this restriction is temporarily suspended and replaced with the totally ignorant context. Chisholm's wording, with 'ought to be' in both (24b) and (24c), encourages this subjunctive interpretation. With a return to the restricted context at the last premise, FDD can be applied, enabling the conclusion 'Jones ought not to choose left next'.

This provides another natural, non-paradoxical interpretation of the Chisholm premises, in which (1a) and (1b) are taken to be subjunctive.

7   Conclusions

Chisholm's paradox is not merely the byproduct of a naive theory of conditional obligation. We have shown that, by integrating linguistic ideas, such as contextual effects and their interaction with assertion and accommodation, we are able to solve the paradox's *unconditional*, equally troublesome variation. Further, by combining this operator with factual deontic detachment, we rendered the Chisholm set consistent. Suggesting that some deontic constructions are generic, we also offered an account of standing obligations within this framework. Finally, citing the vacuous satisfaction of certain premises in contexts allowing all four Chisholm premises to be true, we provided an account of the naturalness of this paradox. Beyond solving the paradox, these ideas provide a theory of deontic operators and conditionals that is linguistically motivated and intuitive.

References

1. Chisholm, R.M.: Contrary-to-duty imperatives and deontic logic. Analysis 24, 33–36 (1963)
2. Benthem, J., Grossi, D., Liu, F.: Priority structures in deontic logic. Theoria 80 (2013)
3. Cantwell, J.: Changing the modal context. Theoria 74, 331–351 (2008)
4. Mott, P.L.: On Chisholm's paradox. Journal of Philosophical Logic 2, 197–211 (1973)
5. Sellars, W.: Reflections on contrary-to-duty imperatives. Noûs 1, 303–344 (1967)
6. Kratzer, A.: The notional category of modality. In: Eikmeyer, H.J., Rieser, H. (eds.) Words, Worlds and Contexts, pp. 38–74. Walter de Gruyter, Berlin (1981)
7. Lewis, D.K.: Counterfactuals. Harvard University Press, Cambridge (1973)
8. Lewis, D.K.: Counterfactuals and comparative possibility. Journal of Philosophical Logic 2, 418–446 (1973)
9. Kolodny, N., MacFarlane, J.: Ifs and oughts. The Journal of Philosophy 107, 115–143 (2010)
10. Stalnaker, R.C.: Indicative conditionals. Philosophia 5, 269–286 (1975)
11. Portner, P.: Modality. Oxford University Press, Oxford (2009)
12. Kratzer, A.: Modals and Conditionals. Oxford University Press, Oxford (2012)
13. Jones, A.J., Carmo, J.: Deontic logic and contrary-to-duties. In: Gabbay, D.M., Guenthner, F. (eds.) Handbook of Philosophical Logic, 2nd edn., vol. VIII, pp. 265–344. Kluwer Academic Publishers, Amsterdam (2002)
14. Thomason, R.H.: The semantics of conditional modality (2012), http://web.eecs.umich.edu/~rthomaso/documents/modal-logic/lkminf.pdf
15. Carlson, G.N., Pelletier, F.J. (eds.): The Generic Book. Chicago University Press, Chicago (1995)
16. Hacquard, V.: Aspects of Modality. Ph.d. dissertation, Linguistics Department, Massachusetts Institute of Technology, Cambridge, Massachusetts (2006)
17. Lerner, A., Leslie, S.J.: Generics, generalism, and reflective equilibrium: Implications for moral theorizing from the study of language. In: Hawthorne, J., Turner, J. (eds.) Philosophical Perspectives 27: Philosophy of Language, pp. 366–403. Wiley Periodicals, Malden (2013)
18. Castañeda, H.N.: Paradoxes of moral reparation: Deontic foci versus circumstances. Philosophical Studies 57, 1–21 (1989)
19. Lewis, D.K.: Scorekeeping in a language game. Journal of Philosophical Logic 8, 339–359 (1979)
20. Stalnaker, R.C.: Assertion. In: Cole, P. (ed.) Syntax and Semantics 9: Pragmatics. Academic Press, New York (1981)

Deontic Reasoning Across Contexts

Justin Snedegar

University of St Andrews, St Andrews, Scotland
js280@st-andrews.ac.uk

Abstract. Contrastivism about 'ought' holds that 'ought' claims are relativized, at least implicitly, to sets of mutually exclusive but not necessarily jointly exhaustive alternatives. This kind of theory can solve puzzles that face other linguistic theories of 'ought', via the rejection or severe restriction of principles that let us make inferences between 'ought' claims. By rejecting or restricting these principles, however, the contrastivist takes on a burden of recapturing acceptable inferences that these principles let us make. This paper investigates the extent to which a contrastivist can do this.

Keywords: deontic modals, contrastivism, context-sensitivity, reasoning.

This paper is about deontic reasoning, or reasoning with *oughts*. This is very plausibly the kind of reasoning deontic logicians aim to formalize. But I am not here directly concerned with formal logical systems. Rather, I am interested in natural language inferences we can make using the English word 'ought'. In particular, I am concerned with whether a particular kind of theory of the meaning of 'ought' can explain some intuitive deontic inferences, which I take to be data that such a theory needs to explain.

Insights from deontic logic are relevant for developing and evaluating accounts of natural language deontic reasoning, and vice versa. Thus, a next step in this project would be to explore proposals in deontic logic, to see what they can teach us about analogous issues in natural language deontic reasoning. Unfortunately, though, all I can do here is flag places where this strategy, of drawing on lessons from deontic logic, seems promising. Near the end of the paper, I will suggest that recent work in preference-based deontic logic, in particular, offers an interesting avenue for future development of the picture I develop.

1 Cross-Context Deontic Reasoning

A simple, standard semantics for 'ought' holds that what you ought to do is what you do in all the best worlds. This simple picture leads to well-known puzzles of deontic reasoning, which have led to interesting complications of the semantics (and of the linguistic theory more generally) of 'ought'. For example, the orthodox Kratzerian semantics for modals, developed in [1], is motivated in part by its ability to solve some of these puzzles. To simplify somewhat, Kratzer's theory relativizes 'ought' (and other modal) claims to contextual parameters,

F. Cariani et al. (Eds.): DEON 2014, LNAI 8554, pp. 208–223, 2014.

including bodies of information and standards, and holds that the puzzles arise from ignoring shifts in these parameters. So what appear to be paradoxes really involve some kind of equivocation.

A different kind of complication is distinctive of what we can call *contrastivism* about 'ought'—the thesis that 'ought' claims are always at least implicitly relativized to sets of alternatives. So to say that you ought to A is really, according to contrastivism, to say that you ought to A out of some set of alternatives.[1] The contrastivist holds that various puzzles about 'ought' arise from ignoring the contrast-sensitivity of 'ought' claims; so again, these puzzles, according to the contrastivist, are due to a kind of equivocation.

Contrastivism, and contextualist theories more generally, provide nice solutions to a variety of puzzles involving 'ought'. But it is easy to overgenerate fallacious equivocations. Some deontic reasoning takes place across contexts. As the context shifts, the content of the relevant 'ought' claims will also shift, according to these contextualist theories. So the concern is that even deontic reasoning which seems unobjectionable will turn out to be fallacious. This is a general statement of the problem I take up in this paper.

Another way to bring out this point, and also a way to preview discussion later in the paper, is to focus on a particular inference rule that simple, non-contextualist theories of 'ought' validate, but that contextualist theories, including contrastivism, seem to invalidate.[2]

Inheritance: If p entails q, then if it ought to be that p, it ought to be that q.

This kind of inference is involved in many of the puzzles of deontic reasoning.

The trouble is that **Inheritance** also explains lots of good instances of deontic reasoning. Consider the following simple example:

(1) I ought to buy milk.
(2) Buying milk entails buying a dairy product.
(3) So I ought to buy a dairy product.

(1)-(3) looks like an example of good deontic reasoning. If we accept **Inheritance**, we have an easy explanation for why it is good: it is just an instance of **Inheritance** (plus *modus ponens*). Once we reject this principle, though, we seem to lack an explanation.

The obvious move for a contextualist is to point out that she does not *reject* **Inheritance** but only *restricts* its application. Specifically, she restricts its application to deontic reasoning that takes place in a single context. So as long as we remain in the same context—as long as the relevant contextual parameters don't shift—throughout the reasoning process encoded in (1)-(3), we can apply

---

[1] For different versions of contrastivism, see [2–5].

[2] An analogous axiom in deontic logic is often called 'inheritance' or 'necessitation': If $\vdash p \rightarrow q$, then $\vdash Op \rightarrow Oq$. The principle I call 'Inheritance' below, though, is rather a schema for explaining natural language inferences using 'ought'.

this restricted version of **Inheritance** to explain why this is good reasoning. The hope is then that there are not single-context versions of the puzzles.[3]

The trouble is that lots of deontic reasoning takes place across contexts. For example, if we follow Kratzer and hold that 'ought' claims are relativized to, among other things, bodies of information, we can imagine a situation in which premise (1) is accepted relative to one body of information, then some information is added (including, if the reasoner is sufficiently ignorant about milk, premise (2)), and finally the conclusion is accepted relative to this updated body of information. Cases like this can still be instances of good deontic reasoning, but they are not explained by the restricted version of **Inheritance**.[4]

Though I think that recapturing good cross-context inferences is a general issue that any contextualist theory should address, my focus here will be on contrastivism. Once we focus on contrastivism, the problem of recapturing good cross-context reasoning becomes the problem of explaining inferences between 'ought' claims that are relativized to different sets of alternatives. After introducing contrastivism in more detail in section 2, I will show, in section 3, that it is relatively easy to secure some of these inferences. In the remainder of the paper, though, I return to **Inheritance**-like inferences. More work must be done to secure these. The only attempt of which I am aware, from Cariani in [5], faces some problems. I will show that a contrastivist picture that is actually simpler in some ways than Cariani's picture can avoid these problems while still securing some attractive inferences.

2 Contrastivism

Contrastivism about 'ought' holds that 'ought' claims are relativized, at least implicitly, to sets of alternatives. As the alternatives shift, the truth of the 'ought' claims can shift, as well. According to the contrastivist, there will be some actions such that, whether you ought to perform them can vary with the alternatives. This is to say more than that the *availability* of other alternatives can affect whether you ought to perform a given action—nearly everyone would accept that. The distinctive contrastivist claim is that the particular alternatives to which we are comparing the action can affect whether or not you ought to perform the action. Here is a simple illustration:

(4) You ought to take the bus rather than drive your SUV.
(5) But it's not the case that you ought to take the bus rather than ride your bike.

---

[3] In deontic logic, lots of work has been done to give restricted versions of the analogous principle, sometimes called 'necessitation', in order to avoid paradoxes of deontic logic. This literature is very large; [6] surveys attempts with an eye specifically to allowing for deontic conflicts. See also [7–9]. Later in the paper I will point out some similarities between contrastivist proposals and some proposals from this formal work.

[4] See [10] for a similar complaint against contextualist formulations of an epistemic closure principle.

Putting things explicitly in terms of sets of alternatives, instead of 'rather than' claims:

(4') You ought to take the bus out of {take the bus, drive your SUV}.
(5') It's not the case that you ought to take the bus out of {take the bus, ride your bike}.

The contrastivist holds that both of these claims may be true, even if all three actions are available.

The alternatives may be provided explicitly, as in the 'rather than' ascriptions (4) and (5) above. More often, they will be provided implicitly, by the *question under discussion*—in particular by the *deliberative* question under discussion. Questions are sets of alternatives, with each alternative corresponding to a potential answer to the question. A *deliberative* question is a special kind of question—a question of what to do—with each alternative corresponding to some action that the relevant agent can perform.[5]

Contrastivism can solve puzzles about 'ought'. Jackson, in [3], uses the following example to motivate his contrastive view of 'ought'. Imagine the following dialogue taking place:

A: It ought to be that Lucretia used less painful poisons.
B: Oh no, it ought to be that she used painless poisons.
C: Oh no, it ought to be that she used political means rather than poison to obtain her ends.
D: Oh no, it ought to be that she never existed at all.
E: Oh no, it ought to be that she existed but made people happy.
F: Oh no, it ought to be that everyone was *already* happy.
And so on.

As Jackson points out, each of these claims seems perfectly appropriate, when uttered. But according to the principle **Inheritance**, they are inconsistent. To focus on just one, suppose that B's claim is true. Then by **Inheritance**, it ought to be that Lucretia existed, since it ought to be that she used painless poisons, and her using painless poisons entails that she existed. But this contradicts D's claim.

Jackson's solution is to hold that "our everyday judgments of what ought to be are all relative to sets of (mutually exclusive but not necessarily jointly exhaustive) alternatives" (p. 180). As we move through the dialogue about Lucretia, the alternatives shift. A's claim is relative to a set like {she uses more painful poisons, she uses less painful poisons}, B's is relative to {she uses (less) painful poisons, she uses painless poisons}, C's is relative to {she uses (painless) poisons, she uses political means}, and so on. Since these claims are relativized to different sets of alternatives, there is in fact no inconsistency. Of course, if we

---

[5] See [11, 12] for classic discussions of this conception of questions. See [13] and the references there for discussion of questions under discussion and their role in theories of communication. See [5] for a contrastivist theory of 'ought' developed in terms of deliberative questions.

were to evaluate an earlier claim relative to a later set of alternatives, we would get inconsistencies. But the data is that the claims seem perfectly appropriate *when made*. Once we consider, say, C's claim, B's claim no longer seems true. The contrastivist explanation is that once we consider C, we have in mind a set of alternatives relative to which B's claim is actually false.

This same kind of move provides a solution to the famous Good Samaritan paradox. This can be seen clearly by focusing on claim E above. 'It ought to be that she existed and made everyone happy' seems true, but she could only make everyone happy if some people were unhappy to begin with (let's suppose). So by **Inheritance**, the truth of E's claim would seem to imply the truth of 'It ought to be that some people were unhappy to begin with', which seems false. But again, by relativizing to sets of alternatives, we can block this inference. E's claim is only true relative to a set like {she existed and made people miserable, she didn't exist and some people were miserable, she existed and made everyone happy}. Relative to this set, it is not true that it ought to be that some people were unhappy to begin with. So the Good Samaritan inference doesn't seem to go through.

According to Jackson's contrastivist theory, it ought to be that p out of Q just in case p is the best out of Q. Or, to put things in agential terms (though Jackson does emphasize that he means to be talking about the 'ought to be' rather than the 'ought to do'—I will try to gloss over this issue here), you ought to A out of Q just in case A is the best alternative in Q.[6] Other contrastivist theories offer more or less significant variations from this simple picture. So the contrastivist reconciliation of the claims above can be put like this: an action which is the best out of one set of alternatives need not be best out of a different set of alternatives.

3   Simple Inferences

One way to think of the contrastivist solutions is as blocking problematic inferences between 'ought' claims. For example, we cannot infer from A's claim one that would be inconsistent with B's, since this would involve a shift in the set of alternatives, and so an equivocation. This description of the solution fits with the broader contextualist move I discussed at the beginning of the paper. But it also highlights the problem of accounting for good cross-context deontic reasoning. It seems that we *can* make some inferences between 'ought' claims which are (according to the contrastivist) relativized to different sets of alternatives.

For example, suppose I decide that I ought to go to church out of {go to church, go to the bar, go to the office}. If going to the bar becomes irrelevant for some reason—maybe the person I wanted to meet there isn't going, or maybe the bar doesn't open on Sundays—I should be able to infer straightforwardly that I ought to go to church out of the new set of alternatives, {go to church,

---

[6] See [14] for a contrastivist theory that attempts to maintain a unified theory of the 'ought to do' and the 'ought to be'.

go to the office}. The removal of going to the bar as a relevant alternative, when it wasn't what I ought to do anyway, shouldn't make a difference.

For another example, if I decide that we ought to go to church out of {go to church, go to the bar, go to the office} and you decide that we ought to go to church out of {go to church, go to the game, stay home}, then when we get together to talk about what to do, we should be able to straightforwardly infer that we ought to go to church out of {go to church, go to the bar, go to the office, go to the game, stay home}. If we come to the same conclusion about what we ought to do, even if we're considering different alternatives, then we should be able to conclude that we ought to do that thing when we consider all of the alternatives together.

Finally, suppose I decide both that I ought to go to church rather than go to the office, and that I ought to go to the office rather than go to the bar. That is, I ought to go to church out of {go to church, go to the office} and I ought to go to the office out of {go to the office, go to the bar}. Now suppose I am wondering what I ought to do out of {go to church, go to the bar}. It seems that it should just *follow* from what I've already decided that I ought to go to church rather than go to the bar: the 'ought to ... rather than ...' relation should turn out to be transitive.

But if we focus on the fact that according to contrastivism, what we ought to do can vary as the set of alternatives varies, it may be initially puzzling how we could make inferences like this. It looks as if they will be just as equivocal as the problematic inferences contrastivism blocks, which serves as the primary motivation for the theory in the first place. An 'ought' claim that is relativized to one set of alternatives wouldn't seem to tell us anything about 'ought' claims relativized to other sets of alternatives. This is how contrastivism solved the puzzles I discussed above, after all.

Fortunately, as I will now argue, the contrastivist can easily capture these inferences. Recall Jackson's claim that you ought to A out of Q just in case A is the *best* alternative in Q. The picture here is that there is a contrast-invariant ranking of actions, and relative to a set of alternatives, you ought to perform the one that ranks the highest in this ranking. With this picture in mind, we can easily get the following inferences:

Subsets: If you ought to A out of Q, then you ought to A out of subsets of Q that contain A.

Unions: If you ought to A out of Q and you ought to A out of R, then you ought to A out of $Q \cup R$, provided the members of $Q \cup R$ are mutually exclusive.

Transitivity: If you ought to A out of $\{A, B\}$ and you ought to B out of $\{B, C\}$, then you ought to A out of $\{A, C\}$.[7]

The proofs of these principles are straightforward. If A is highest-ranked in Q, then, since the ranking is contrast-invariant, it will also be highest-ranked in

---

[7] Arguments for the rationality of intransitive preferences will be relevant for assessing this inference schema. See, for example, [15] and [16]. Unfortunately, I do not have the space to discuss these challenges here. For arguments for the intransitivity of 'better than', which may be more directly relevant, see [17].

subsets of Q of which it is an element; so we get **Subsets**.[8] If A is highest-ranked in Q and highest-ranked in R, then there won't be any alternative in the union of the two which ranks higher than A; so we get **Unions**.[9] Finally, if you ought to A rather than B and ought to B rather than C, then A is higher ranked than B, and B is higher ranked than C. Since this ranking is contrast-invariant, it just follows that A is higher ranked than C; so we get **Transitivity**.[10]

So we have seen that, though contrastivism does face the challenge of accounting for cross-context deontic reasoning, the theory also has resources to explain some of them; the theory has an important kind of *structure*.

4   Inheritance-Like Inferences

The inferences I discussed in the last section have to do with one kind of variation that sets of alternatives can display, which is the kind that Jackson focuses on in arguing for contrastivism. This is what I have elsewhere called *non-exhaustivity*.[11] Sets of alternatives can vary in terms of which possibilities they include. A subset of Q (generally) covers fewer possibilities than Q, and the union of Q and R (generally) covers more possibilities than either Q or R. What we've seen, then, is that contrastivism can account for some attractive cross-context inferences in which the sets of alternatives vary along the dimension of non-exhaustivity.

There is a second way in which sets of alternatives can vary, however. This is what we can, following [5, 21], call *resolution-sensitivity*. The sets of alternatives to which 'ought' claims are relativized can lump together the possibilities in more or less fine-grained ways, or at higher and lower resolutions. For example, the set {go to church, stay home} and the set {drive to church, take the bus to church, stay home and clean, stay home and watch football} vary in this way—the second makes distinctions between possibilities that the former does not; it divides up the relevant possibilities at a higher resolution.

This is a very important feature of contrastivism. First, in terms of general motivations for the theory, this is the feature that Cariani focuses on in developing his contrastivist semantics. Second, in terms of the purposes of this paper, some important cross-context **Inheritance**-like reasoning depends on making

---

[8] It follows as a corollary of **Subsets** that if you ought to A out of Q and you ought to A out of R, then you ought to A out of the intersection, $Q \cap R$, since this will be a subset of Q that contains A. I motivated the **Unions** inference by appealing to joint deontic reasoning; you might think these kinds of consideration would instead (or sometimes) motivate this intersections principle, instead.

[9] The restriction on **Unions**—that the members of $Q \cup R$ must be mutually exclusive—screens off cases in which there are members of Q and R that are each inconsistent with A, but not with each other (e.g., 'buy milk' and 'buy a liquid', where A is 'buy a banana'). These cases will be more complicated, but I will not discuss them here.

[10] Compare the discussion of choice functions in [16] and [18]. The **Subsets** inference is similar to "basic contraction consistency" (Property α), while the **Unions** property is similar to "basic expansion consistency" (Property γ).

[11] See [19, 20].

inferences between 'ought' claims that are relativized to sets of alternatives that vary along this dimension.

4.1   Contrastivism and Inheritance

It will be helpful to see exactly why **Inheritance** is problematic on a contrastivist theory. A naive contrastivist construal of this principle is the following:

Contrastivist Inheritance: If p entails q, then if it ought to be that p out of Q, it ought to be that q out of Q.

The problem here is that this doesn't make sense on a contrastive theory, at least not for one of the simple sort we have discussed so far. This is because the alternatives in a set of alternatives are *mutually exclusive*. But if p entails q, then obviously p and q are not mutually exclusive; hence, they cannot be members of the same set of alternatives. So if it ought to be that p out of Q, then we know that q is not even in Q, so q cannot be the *best* in the set, so it cannot be true that it ought to be that q out of Q.

4.2   Nice Results

This failure of **Inheritance** on a contrastivist theory delivers some nice results. First, we can avoid Ross's paradox (see [22]). From 'You ought to mail the letter', it would follow by **Inheritance** that 'You ought to mail the letter or burn it', and this latter claim seems false, even when the first is true. But 'mail the letter' and 'mail the letter or burn it' cannot be members of the same set of alternatives, since they are not mutually exclusive. Thus, Ross's inference must involve a shift in the set of alternatives, and so an equivocation.

Second, we can get attractive results in Jackson and Pargetter's Professor Procrastinate case.[12] Professor Procrastinate is asked to review a book, since she is the most qualified person to write it. However, she is a terrible procrastinator, so if she accepts, she is unlikely to actually write it—it isn't impossible that she write it, she just probably won't. If she accepts and does not write it, then the book will go unreviewed, which would be the worst outcome. If she declines, someone else will review it. The review won't be as good as it would be if she did write it, but it will definitely get done.

In this case, both of the following claims seem true:

(6) Procrastinate ought to accept and write.
(7) It's not the case that Procrastinate ought to accept.

(6) seems true, since Procrastinate is the most qualified person to write the review; (7) seems true since she would probably not actually get around to writing it, were she to accept.

The problem is that accepting and writing entails accepting, so by **Inheritance**, (6) entails the negation of (7). The contrastivist avoids this contradiction

---

[12] The case is originally presented in [3], and was the focal point of [23].

by blocking this inference. 'Accept and write' and 'accept' cannot be members of the same set of alternatives, since they are not mutually exclusive. Rather, they are members of two different sets of alternatives that differ in resolution. So to infer the negation of (7), properly relativized, from (6), properly relativized, is to equivocate.

The contrastivist solution can also explain why both (6) and (7) are true, not just why they are actually consistent: (6) is relativized to {accept and write, accept and not write, not accept} and (7) to {accept, not accept}. Out of the first set, accepting and writing is the best alternative; out of the second, accepting is not the best alternative, since it would most likely lead to the worst outcome.

Note that the solutions to these puzzles depend on the resolution-sensitivity of 'ought'. {mail the letter, burn the letter, leave it on the table} and {mail the letter or burn the letter, leave it on the table} vary in resolution—the first makes distinctions that the second does not. Similarly for {accept and write, accept and not write, not accept} and {accept, not accept}. A fine-grained 'ought' claim can be true while a corresponding coarse-grained 'ought' claim—one in which the relevant action is a coarse-grained action that "subsumes" the original fine-grained action—is false. This is perfectly fine for the contrastivist, since the 'ought' claims are relativized to different sets of alternatives.

4.3   Coarsening Inferences

The failure of **Inheritance** on the contrastive theory looks like a nice feature, since it provides solutions to Ross's paradox and the Professor Procrastinate puzzle. But as should be clear by now, this also creates problems. Some **Inheritance**-like inferences look like good deontic reasoning, so we need to recapture them.

Cariani (in [5]) calls these **Inheritance**-like inferences *coarsening* inferences, since they involve moving from a more fine-grained 'ought' claim to a more coarse-grained 'ought' claim. Ross's paradox and the Professor Procrastinate puzzle are cases in which coarsening an alternative leads to a false 'ought' claim. Though 'You ought to mail the letter' is true, coarsening to 'mail the letter or burn it' leads to falsity: it is not true that you ought to mail the letter or burn it. Similarly, though 'Procrastinate ought to accept and write' is true, coarsening to 'accept' leads to falsity.

But lots of coarsening inferences look like good deontic reasoning. Consider Cariani's example:

(8) You ought to feed your pets sufficient amounts of non-poisonous food.
(9) You ought to feed your pets.

This looks like a fine inference to make; often coarse-grained 'ought' claims like (9) are perfectly acceptable, and are in some sense supported by more fine-grained 'ought' claims like (8). This can be so even when there are impermissible ways of carrying out the coarse-grained option, incompatible with the more fine-grained option you ought to perform. So even though there are impermissible ways of feeding your pets—ways that are incompatible with feeding them sufficient amounts of non-poisonous food—(9) can still be true. And **Inheritance** offers a straightforward explanation of this: (8) is true, and by **Inheritance**, it entails (9).

So here is where we stand. We want to allow some coarsening inferences, like that between (8) and (9), but not too many, as in Ross's paradox and the Professor Procrastinate puzzle. We have seen that **Inheritance** could offer an explanation of good coarsening inferences, but not without allowing bad coarsening inferences. And we have seen that contrastivism blocks the bad coarsening inferences; the danger is that it will also block the good coarsening inferences. In the next section I will introduce a sophisticated contrastivist account recently developed by Cariani that is designed to solve this problem.

5   Cariani's Semantics

Cariani develops a version of contrastivism that is meant to walk this fine line, blocking the bad **Inheritance**-like inferences while allowing the good ones. Besides the deliberative question, which provides the alternatives, he posits two pieces of contextually-provided machinery:

- A ranking of the alternatives in the set
- A threshold in the ranking that distinguishes permissible from impermissible options; this is the *benchmark*

We then say that the agent ought to perform some action A relative to the set of alternatives Q just in case (i) all the top-ranked options are ways of A-ing, and (ii) all the ways of A-ing are permissible, i.e. above the benchmark.

The diagnosis of Ross's paradox and the Professor Procrastinate puzzle is that there are relevant impermissible ways of performing the coarse-grained action, but not the fine-grained action. So though there are not relevant impermissible ways of mailing the letter, there are relevant impermissible ways of mailing the letter or burning it—namely, burning it. And though there are not relevant impermissible ways of accepting and writing, there are relevant impermissible ways of accepting—namely, accepting and not writing. So the coarse-grained 'ought' claims come out false on Cariani's semantics.

So the problematic coarsening inferences do not go through, on Cariani's picture. On the other hand, when we have a coarsening inference that looks like good deontic reasoning, this is because there are *not* relevant impermissible ways of performing the coarse-grained alternative. So, for (8) and (9), we can imagine that the relevant set of alternatives is {feed your pets sufficient amounts of non-poisonous food, don't feed your pets}. Relative to this set, (8) is true, since feeding your pets sufficient amounts of non-poisonous food is the best option. (9) is also true, since (i) all the top-ranked options are ways of feeding your pets, and (ii) all the relevant ways of feeding your pets are above the threshold.

Cariani's contrastivist picture resembles a proposal for restricting the **Inheritance** principle that Goble considers, but rejects.[13] This restriction is to say

---

[13] [6], note 49.

that, if A entails B *and B is permissible*, if it ought to be that A, it ought to be that B. Goble rejects this principle because, when paired with the possibility of deontic conflicts, it leads to problematic results. But Cariani is not interested in accommodating deontic conflicts, and in fact they seem to be ruled out by condition (i) of his semantics, plus the assumption that the alternatives in Q are mutually exclusive.[14]

Though Cariani's semantics delivers nice results in an elegant way in these cases, it also delivers some unintuitive results. For a catalogue of such problems, see [24]. Here I will just walk through two of Dowell and Bronfman's examples involving the Professor Procrastinate case. The first targets Cariani's claim that all the top-ranked options must be ways of A-ing, for A-ing to be what you ought to do. The second targets the claim that no relevant ways of A-ing can fall below the permissibility threshold, for A-ing to be what you ought to do.

First, in the Professor Procrastinate case, we not only want both (6) and (7) to be true. We also think that (10) is true:

(10) Procrastinate ought to not accept.

But Cariani's semantics cannot deliver this result, since it is not true that all the top-ranked options in are ways of not accepting.

Second, consider a variation of the Procrastinate case in which Procrastinate (despite her name) is actually very reliable, and if she accepts she is very likely to write. But, crucially, it is still possible that she won't. And given that it is pretty important that the review is written, accepting and not writing, unlikely as it is, is still plausibly a relevant alternative. Nevertheless, given that she is very likely to write, intuitively (7) is false, in this variation—Procrastinate ought to accept. But Cariani's semantics cannot deliver this result, since there is a relevant way of accepting, namely accepting and not writing, which falls below the permissibility threshold.

In the next section I will return to the simpler contrastivist picture I introduced in sections 2 and 3 above, and show that it can deliver the right results in all of these cases.

6   A Different Picture

Return to the simpler contrastive theory I introduced above, according to which you ought to A out of Q just in case A is the highest-ranked (or best) alternative in Q. If we are interested in coarsening inferences, we need to think about the relationship between fine-grained alternatives, like 'drive to the store', and more

---

[14] Goble himself accepts a different restriction: If A entails B *and A is permissible*, then if it ought to be that A, it ought to be that B. As Goble shows, this principle has several nice features in a logic that allows for deontic conflicts. But notice that it also generates the **Inheritance**-based puzzles I am concerned with in this paper. For example, accepting and writing is permissible, for Professor Procrastinate, but what we want to avoid is the conclusion that since she ought to accept and write, she ought to accept. Goble's principle validates this inference.

coarse-grained alternatives, like 'go to the store'. One way to think about this is as the relationship between an option and a disjunction of which it is a disjunct. The disjunction corresponds to a coarse-grained alternative that lumps together several relevant fine-grained ways of carrying out that alternative. So 'go to the store' may be identified with 'drive to the store or walk to the store or take the bus to the store', if those are the relevant fine-grained ways of going to the store.

So we need to think about the relationships between the value of a disjunct and the value of the disjunction of which it is a part; on the picture I'm developing here, this is the relationship between the place in the ranking of the disjuncts and the place in the ranking of the disjunction. As I will show below, there are various options here. But there are constraints that any choice will have to meet. To see some of these, consider the following overwhelmingly plausible principle:

Disjunctions: If B is better than C, then $B \vee C$ is no better than B and no worse than C—$B \vee C$ is ranked somewhere between (inclusive) B and C.[15]

Fortunately, this principle is just what we need to get some attractive inferences between 'ought' claims that are relativized to sets that differ in resolution.

6.1 Disjunctions-Supported Inferences

First, **Disjunctions** gives us the following claims about the value of alternatives:

Better Dis: If A is better than both B and C, then A is better than $B \vee C$.
Worse Dis: If A is worse than both B and C, then A is worse than $B \vee C$.

Assuming the simple contrastivist claim that you ought to A out of Q just in case A is the best (or highest ranked) alternative in Q, these give us several deontic inferences, including the following three:

Dis 1: If you ought to A out of $\{A, B, C\}$, then you ought to A out of $\{A, B \vee C\}$.
Dis 2: If you ought to B out of $\{A, B\}$ and you ought to C out of $\{A, C\}$, then you ought to $B \vee C$ out of $\{A, B \vee C\}$.
Dis 3: If you ought to B out of $\{B, E, F\}$ and you ought to C out of $\{C, E, F\}$, then you ought to $B \vee C$ out of $\{B \vee C, E \vee F\}$.

Dis 1 tells us that as long as A-ing is better than all the fine-grained ways of D-ing, you ought to A rather than D. **Dis 2** tells us that as long as all the fine-grained ways of D-ing are better than A-ing, you ought to D rather than A. And **Dis 3** tells us that as long as all the fine-grained ways of D-ing are better than all the fine-grained ways of G-ing, you ought to D rather than G. The proofs of these are straightforward. These are simple, intuitive inferences. So it is important that the contrastivist theory can capture them.

Next I will show that these principles do let us make attractive coarsening inferences, but do not license the problematic inferences in Ross's paradox and the Professor Procrastinate puzzle.

---

[15] That is, the betterness ranking is *interpolative*; see the discussion of preference rankings in [9], p. 482.

6.2 Application to Cases

Consider the following case, which Cariani has discussed in other work ([25]). Suppose the speed limit on this road is 50 mph. Thus, you ought to drive at or under 50 mph. As Cariani points out, we should be able to infer that you ought to drive under 100 mph; this is a coarsening inference, since driving at or under 50 is a way of driving under 100. Cariani's semantics has trouble here, since there are impermissible ways of driving under 100—namely, driving between 50 and 100.

We want to be able to infer (12) from (11):

(11) You ought to drive at or under 50 out of {drive at or under 50, drive between 50 and 100, drive over 100}.
(12) You ought to drive under 100 out of {drive under 100, drive over 100}.

In this case, 'drive under 100' in the second set of alternatives is just the disjunction of two alternatives from the first set, 'drive at or under 50' and 'drive between 50 and 100'. I assume the alternatives are ranked as follows: drive at or under 50 > drive between 50 and 100 > drive over 100. From (11), we can infer that you ought to drive at or under 50 out of {drive at or under 50, drive between 50 and 100}, just given the contrastivist idea that what you ought to do out of a set of alternatives is the best thing in that set and the principle I called **Subsets** above. And from the assumption about the ranking of the alternatives I've just made, we can infer that you ought to drive between 50 and 100 out of {drive between 50 and 100, drive over 100}. Now we can just apply **Dis 2** to get (12). So, though (12) did not follow *directly* from (11), given the contrastive theory I've developed here, it does follow given that theory and the assumption about the ordinal ranking of the alternatives.

A strength of Cariani's semantics is that it is able to give the right results in the Ross's paradox case and in the Professor Procrastinate case. This theory can also deliver those results. We cannot infer (14) from (13):

(13) You ought to mail the letter out of {mail the letter, burn the letter, leave the letter on the table}.
(14) You ought to mail the letter or burn it out of {mail the letter or burn it, leave it on the table}.

The inference fails because there are fine-grained ways of mailing the letter or burning it—namely, burning it—that are ranked below 'leave it on the table'. So we cannot use the deontic inferences generated by **Disjunctions**.

If, for some reason, the other alternatives are even more disastrous than burning the letter, things will be different. Consider, for example, {mail the letter, burn it, poison the water supply}. In this case, it may well be true that you ought to mail the letter or burn it out of {mail the letter or burn it, poison the water supply}. But that is as it should be. An advantage of this theory over Cariani's is that it can block Ross's inference without predicting that 'You ought to mail the letter or burn it' will always be false, simply because burning the letter is impermissible. The truth of this claim may depend on how disastrous the *other* alternatives are. This point will be even more important in the Professor Procrastinate case, as we will see shortly.

The Professor Procrastinate case is similar: there are fine-grained ways of accepting—namely, accepting and not writing—that are ranked below 'not accept', so the **Disjunctions**-supported inferences do not let us infer that Procrastinate ought to accept from the claim that Procrastinate ought to accept and write.

Cariani's semantics mistakenly predicted, in the original Procrastinate case, that 'Procrastinate ought to not accept' is always false. This is because not all of the top-ranked alternatives in {accept and write, accept and not write, not accept} were ways of not accepting. The theory here, on the other hand, can allow that this claim is true. All that has to be true is that 'not accept' is ranked above 'accept'. The theory does not predict that this will always be the case, of course. But that is as it should be. After all, in the modified Procrastinate case borrowed from Dowell and Bronfman above, in which Procrastinate is very likely to write if she accepts, we want 'Procrastinate ought to accept' to be true. Again, Cariani's semantics predicts that this claim will always be false, as long as 'accept and not write' is a relevant alternative. According to the theory here, though, the truth of this claim, and the truth of 'Procrastinate ought to not accept', will depend on how 'accept' and 'not accept' rank, relative to one another. And how 'accept' ranks will depend on how 'accept and write or accept and not write' ranks, since it is just equivalent to this disjunction. We have some options for determining the ranking of this disjunction, as I will now illustrate.

6.3   Options for Other Inferences

Any proposal that is consistent with the contrastivist machinery I have developed so far will have to include **Dis 1–3**, and other **Disjunctions**-supported inferences. But this leaves a great deal of leeway in either embracing or rejecting many other inferences between sets that differ in resolution. Which inferences we accept will depend on how we determine the ranking of a disjunction from the ranking of its disjuncts.

The question of how to determine the ranking of a disjunction from the rankings of its disjuncts parallels a question that arises in Hansson's preference-based deontic logic, of how to determine the preference ranking of a sentence from the preference ranking of "holistic alternatives" in which the sentence is true.[16] Holistic alternatives, for Hansson, are complete specifications of a course of action open to an agent. We can think of these, on the contrastive picture, as maximally fine-grained actions. Hansson's sentences correspond to relatively coarse-grained actions, on the contrastive picture. Here are three options.

First, we may let the value of $A \lor B$ be a *weighted average* of the values of A and B. The value of each disjunct is weighted by its likelihood. For example, if 'accept and write' is much more likely than 'accept and not write', as in the reliable Procrastinate variation, the value of the disjunction will be relatively

---

[16] See [9], section 6.

high; if 'accept and not write' is much more likely, as in the original case, the value of the disjunction will be relatively low (holding fixed the value of each disjunct). This weighted average approach delivers the verdicts I have suggested are the intuitive ones in both the original and the modified Procrastinate cases: it is not the case that unreliable Procrastinate ought to accept, and in fact is the case that she ought *not* accept, while reliable Procrastinate ought to accept. These are "actualist" intuitions; see [23].

Second, we can adopt a *pessimistic* approach, or in more friendly terms, a *cautious* approach, and let the value of the disjunction be the value of its lowest-ranked disjunct. Then no matter how likely Procrastinate is to write, 'accept' is going to be ranked with 'accept and not write'.

Finally, we can adopt a more *optimistic* approach, and let the value of the disjunction be the value of its highest-ranked disjunct. Then no matter how unlikely it is that Procrastinate writes, 'accept' will be ranked with 'accept and write'. This optimistic approach delivers the "possibilist" intuition that even unreliable Procrastinate ought to accept.

In fact, this optimistic approach delivers a coherent contrastivist version of **Inheritance**. Here is a simplified version of it, where the set of alternatives has just three members:

Optimistic Inheritance: If you ought to A out of $\{A, B, C\}$, then (i) you ought to $A \vee B$ out of $\{A \vee B, C\}$ and (ii) you ought to $A \vee C$ out of $\{A \vee C, B\}$.

This principle will generate the same kinds of puzzles as **Inheritance**, of course. But, though contrastivism is motivated in large part by its ability to give a principled rejection of **Inheritance**, we see now that even if you do not share the intuitions that supported this rejection, this is no reason not to be a contrastivist. Settling debates like the one between actualists and possibilists requires settling on how to determine the ranking of a coarse-grained alternative from the ranking of the fine-grained alternatives it subsumes. As I have just shown, contrastivism is in principle compatible with various options here. This kind of neutrality on substantive ethical issues is generally taken to be a good thing in a theory of the meaning of 'ought'.

As I said above, these three options parallel options Hansson considers in developing his preference-based deontic logic. Translating between Hansson's logics and this contrastive picture of the meaning of 'ought' will not be completely straightforward, but given these similarities, it does seem to me a fruitful direction for future research.

Acknowledgements. Thanks to the referees for DEON 2014, an audience at Leeds, and Noah Friedman-Biglin for helpful comments and discussion. Thanks to Shyam Nair, Mark Schroeder, and Walter Sinnott-Armstrong for discussion of earlier work that informed this paper.

References

1. Kratzer, A.: The notional category of modality. In: Eikmeyer, H.J., Reiser, H. (eds.) Words, Worlds, and Contexts, pp. 38–74. de Gruyter, Berlin (1981)
2. Sloman, A.: 'Ought' and 'better'. Mind 79(315), 385–394 (1970)
3. Jackson, F.: On the semantics and logic of obligation. Mind 94(374), 177–195 (1985)
4. Snedegar, J.: Contrastive semantics for deontic modals. In: Blaauw, M. (ed.) Contrastivism in Philosophy, Routledge (2012)
5. Cariani, F.: 'Ought' and resolution semantics. Noûs 47(3), 534–558 (2013)
6. Goble, L.: Normative conflicts and the logic of 'ought'. Noûs 43(3), 450–489 (2009)
7. Goble, L.: A logic of good, should, and would: Part I. Journal of Philosophical Logic 19(2), 169–199 (1990)
8. van der Torre, L.: Reasoning about Obligations: Defeasibility in Preference-based Deontic Logic. Dissertation, Tinbergen institute research series, Erasmus University Rotterdam (1997)
9. Hansson, S.O.: Alternative semantics for deontic logic. In: Gabbay, D., Horty, J., van der Meyden, R., van der Torre, L. (eds.) Handbook of Deontic Logic and Normative Systems, pp. 445–497. College Publications (2013)
10. Schaffer, J.: Closure, contrast, and answer. Philosophical Studies 133(2), 233–255 (2007)
11. Hamblin, C.L.: Questions. Australasian Journal of Philosophy 36, 159–168 (1958)
12. Groenendijk, J., Stokhof, M.: Questions. In: van Benthem, J., ter Meulen, A. (eds.) Handbook of Logic, pp. 1055–1124. Elsevier Science Publishers (1997)
13. Roberts, C.: Information structure in discourse: Towards an integrated formal theory of pragmatics. Semantics and Pragmatics 5, 1–16 (2012)
14. Finlay, S., Snedegar, J.: One ought too many. In: Philosophy and Phenomenological Research (forthcoming)
15. Anand, P.: The philosophy of intransitive preference. The Economic Journal 103(417), 337–346 (1993)
16. Sen, A.: Internal consistency of choice. Econometrica 61(3), 495–521 (1993)
17. Temkin, L.: Rethinking the Good. Oxford University Press, Oxford (2012)
18. Sen, A.: Choice functions and revealed preference. The Review of Economic Studies 38(3), 307–317 (1971)
19. Snedegar, J.: Reason claims and contrastivism about reasons. Philosophical Studies 166(2), 231–242 (2013)
20. Snedegar, J.: Contrastive reasons and promotion. Ethics (forthcoming)
21. Yalcin, S.: Nonfactualism about epistemic modality. In: Egan, A., Weatherson, B. (eds.) Epistemic Modality. Oxford University Press, Oxford (2011)
22. Ross, A.: Imperatives and logic. Theoria 7, 53–71 (1941)
23. Jackson, F., Pargetter, R.: Oughts, options, and actualism. Philosophical Review 95(2), 233–255 (1986)
24. Dowell, J., Bronfman, A.: The language of reasons and oughts. In: Star, D. (ed.) Oxford Handbook of Reasons. Oxford University Press, Oxford (forthcoming)
25. Cariani, F.: The Semantics of 'Ought' and the Unity of Modal Discourse. PhD Thesis, University of California, Berkeley (2009)

Sequent-Based Argumentation for Normative Reasoning

Christian Straßer[1] and Ofer Arieli[2]

[1] Centre for Logic and Philosophy of Science, Ghent University, Belgium
christian.strasser@UGent.be
[2] School of Computer Science, The Academic College of Tel-Aviv, Israel
oarieli@mta.ac.il

Abstract. In this paper we present an argumentative approach to normative reasoning. Special attention is paid to deontic conflicts, contrary-to-duty and specificity cases. These are modeled by means of argumentative attacks. For this, we adopt a recently proposed framework for logical argumentation in which arguments are generated by a sequent calculus of a given base logic (see [1]), and use standard deontic logic as our base logic. Argumentative attacks are realized by elimination rules that allow to discharge specific sequents. We demonstrate our system by means of various well-known benchmark examples.

1 Introduction

Normative reasoning concerns reasoning with and about notions such as obligations, permissions, etc. A paradigmatic instance is so-called factual detachment which says that if φ holds and there is a commitment to ψ conditional on φ, then there is a commitment to ψ. Another instance is aggregation: if there is an obligation to bring about φ and another obligation to bring about ψ then there should be an obligation to bring about $\varphi \wedge \psi$. Allowing for unrestricted factual detachment or unrestricted aggregation is problematic in cases of normative conflicts [2]. For instance, aggregating two conflicting obligations leads to an obligation that commits us to do the impossible. Other problematic cases concern specificity: sometimes more specific obligations or permissions override more general ones. In such cases we want to block factual detachment from the overridden obligations or permissions. Logical accounts of normative reasoning that is tolerant with respect to normative conflicts and/or specificity cases have been shown to be challenging. This has given rise to a variety of approaches (e.g., [3–8]).

In this paper we model normative reasoning by means of logical argumentation. Given a set of facts and a set of possibly conflicting and interdependent conditional obligations or permissions we will demonstrate how this model helps us to identify conflict-free sets that are apt to guide the actions of a user. Furthermore, we will show how it offers an elegant tool to deal with specificity cases. It follows that the entailment relations that are obtained offer conflict-handling

F. Cariani et al. (Eds.): DEON 2014, LNAI 8554, pp. 224–240, 2014.

mechanisms for various types of conflicts, and as such they are adaptive to different application contexts.

Our starting point in modeling normative reasoning is concerned with Dung's well-known abstract argumentation frameworks [9]. These frameworks consist of a set of abstract objects (the 'arguments') and an attack relation between them. Their role is to serve as a tool to analyze and reason with arguments. Various procedures for selecting accepted arguments have been proposed, based on the dialectical relationships between the arguments. Usually, these methods avoid selecting arguments that conflict with each other and allow to respond to every possible attack on the argumentative stance with a counter-argument.

For formalizing normative reasoning we need to enhance abstract argumentation in order to model the structure of arguments. There are various ways of doing so (e.g., [10, 11]). In this paper, we settle for the representation in terms of *sequents* [1]. One advantage of this approach is that it immediately equips us with dynamic proof procedures in the style of adaptive logics [12, 13] that allow for automated reasoning [14]. Another advantage is that we can plug in any Tarskian logic that comes with an adequate sequent calculus as a base logic that produces our arguments.

In this paper we use SDL (standard deontic logic) as our base logic (see Section 2). In this context, the modality O is used to model obligations and permissions are modeled by P, defined by $\neg O \neg$. Accordingly, arguments are (proofs of) derivable sequents $\Gamma \Rightarrow \phi$ (for some finite set of formulas Γ and a formula ψ) in a sequent calculus for SDL, based on Gentzen's LK proof system [15]. Attacks between arguments are represented by attack rules that allow to derive elimination sequents of the form $\Gamma \nRightarrow \phi$, whose effect is the canceling or uncharging of $\Gamma \Rightarrow \phi$ (see [1]).

The following example illustrates (still on the intuitive level) how the sequent-based argumentation framework described above models normative reasoning.

Example 1. Consider the following example by Horty [16]:

- When served a meal you ought to not eat with your fingers.
- However, if the meal is asparagus you ought to eat with your fingers.

The statements above may be represented, respectively, by the formulas $m \supset O\neg f$ and $(m \wedge a) \supset Of$. Now, in case we are indeed served asparagus $(m \wedge a)$ we expect to derive the (unconditional) obligation to eat with your fingers (Of) rather than to not eat with our fingers $(O\neg f)$. This is a paradigmatic case of *specificity*: a more specific obligation cancels (or overrides) a less specific one. In our setting this will be handled by an attack rule advocating specificity (see Example 2 below), according to which the argument $\{m \wedge a, (m \wedge a) \supset Of\} \Rightarrow Of$ attacks the argument $\{m, m \supset O\neg f\} \Rightarrow O\neg f$, and as a consequence Of will be inferable in this case while $O\neg f$ will not.

2   The Base Logic SDL

The base logic that we shall use in this paper is SDL (standard deontic logic, i.e., the normal modal logic KD). The underlying language $\mathcal{L}_{\mathsf{SDL}}$ consists of a propositional constant $\bot$ (representing falsity), the standard operators for conjunction $\wedge$, disjunction $\vee$, and implication $\supset$, and the modal operator O representing obligations. Thus, for instance, the conditional obligation $\phi \supset \mathsf{O}\psi$ may be intuitively understood as "ϕ commits to bring about ψ".

We shall denote formulas in $\mathcal{L}_{\mathsf{SDL}}$ by the lower Greek letter ψ, ϕ, and sets of formulas by the upper Greek letters Γ, Δ, Σ. As usual, we incorporate the modality P for representing permissions, where $\mathsf{P}\psi$ is defined by $\neg\mathsf{O}\neg\psi$. Also, we shall abbreviate the formula $\bot \supset \bot$ by $\top$, write $\mathsf{O}\Gamma$ for the set $\{\mathsf{O}\psi \mid \psi \in \Gamma\}$, and denote $\bigwedge\Gamma$ for the conjunction of the formulas in a finite set Γ.

Reasoning with SDL is done by $\mathcal{L}_{\mathsf{SDL}}$-*sequents* (or just sequents, for short), that is: expressions of the form $\Gamma \Rightarrow \psi$, where Γ is a finite set of $\mathcal{L}$-formulas and $\Rightarrow$ is a symbol that does not appear in $\mathcal{L}_{\mathsf{SDL}}$. We shall denote $\mathsf{Prem}(\Gamma \Rightarrow \psi) = \Gamma$.

Given a set Σ of formulas in $\mathcal{L}_{\mathsf{SDL}}$, we say that a formula ψ *follows* from Σ (in SDL), and denote this by $\Sigma \vdash_{\mathsf{SDL}} \psi$, if there is a subset $\Gamma \subseteq \Sigma$, such that the $\mathcal{L}_{\mathsf{SDL}}$-sequent $\Gamma \Rightarrow \psi$ is provable in the sequent calculus $\mathcal{C}_{\mathsf{SDL}}$ shown in Figure 1. It is easy to verify that $\vdash_{\mathsf{SDL}}$ is a Tarskian consequence relation (that is, reflexive, monotonic and transitive).

Axioms: $\psi \Rightarrow \psi$

Structural Rules:

Weakening: $\dfrac{\Gamma \Rightarrow \Delta}{\Gamma, \Gamma' \Rightarrow \Delta, \Delta'}$ Cut: $\dfrac{\Gamma_1 \Rightarrow \Delta_1, \psi \quad \Gamma_2, \psi \Rightarrow \Delta_2}{\Gamma_1, \Gamma_2 \Rightarrow \Delta_1, \Delta_2}$

Logical Rules:

$[\wedge\Rightarrow]\ \dfrac{\Gamma, \psi, \varphi \Rightarrow \Delta}{\Gamma, \psi \wedge \varphi \Rightarrow \Delta}$ $[\Rightarrow\wedge]\ \dfrac{\Gamma \Rightarrow \Delta, \psi \quad \Gamma \Rightarrow \Delta, \varphi}{\Gamma \Rightarrow \Delta, \psi \wedge \varphi}$

$[\vee\Rightarrow]\ \dfrac{\Gamma, \psi \Rightarrow \Delta \quad \Gamma, \varphi \Rightarrow \Delta}{\Gamma, \psi \vee \varphi \Rightarrow \Delta}$ $[\Rightarrow\vee]\ \dfrac{\Gamma \Rightarrow \Delta, \psi, \varphi}{\Gamma \Rightarrow \Delta, \psi \vee \varphi}$

MP: $\dfrac{}{\Gamma, \phi, \phi \supset \psi \Rightarrow \psi}$ $[\Rightarrow\supset]\ \dfrac{\Gamma, \psi \Rightarrow \varphi, \Delta}{\Gamma \Rightarrow \psi \supset \varphi, \Delta}$

$[\neg\Rightarrow]\ \dfrac{\Gamma \Rightarrow \Delta, \psi}{\Gamma, \neg\psi \Rightarrow \Delta}$ $[\Rightarrow\neg]\ \dfrac{\Gamma, \psi \Rightarrow \Delta}{\Gamma \Rightarrow \Delta, \neg\psi}$

KR: $\dfrac{\Gamma \Rightarrow \phi}{\mathsf{O}\Gamma \Rightarrow \mathsf{O}\phi}$ DR: $\dfrac{\Gamma \Rightarrow \phi}{\mathsf{O}\Gamma \Rightarrow \neg\mathsf{O}\neg\phi}$

Fig. 1. The proof system $\mathcal{C}_{\mathsf{SDL}}$

Note 1. The proof system $\mathcal{C}_{\mathsf{SDL}}$ is equivalent to Gentzen's well-known sequent calculus LK for classical propositional logic, extended with rules for the modal operator O [17]. In particular, in $\mathcal{C}_{\mathsf{SDL}}$ the rule [MP] is primitive and the rule

$$[\supset\Rightarrow]\quad \frac{\Gamma \Rightarrow \psi, \Delta \quad \Gamma, \varphi \Rightarrow \Delta}{\Gamma, \psi \supset \varphi \Rightarrow \Delta}$$

is admissible (i.e., it is derivable from the rules of C_{SDL}), while in LK it is the other way around. We switched between [MP] and [⊃⇒] since this allows to simplify some of the formalities to be developed in the sequel (namely, it allows for a more straightforward formulation of some sequent elimination rules in Section 3).

Note 2. SDL has the usual problems or 'paradoxes' that are associated with a material account of implication (such as $\phi \supset (\psi \supset \phi)$ which in terms of conditional obligations becomes $\mathsf{O}\phi \supset (\psi \supset \mathsf{O}\phi)$). Furthermore, straightforward accounts of modeling conditional obligations with material implication is plagued with consistency problems for various types of conditional obligations (e.g., contrary-to-duty and specificity cases). For instance, applying SDL to Forrester's example of the gentle murderer (see [18] and Example 6 below) results in triviality. These problems are usually taken to be the death sentence for an account of conditional obligations based on SDL and material implication. The system developed in this paper may be seen as a sign of caution: although the 'paradoxes' of material implication are here to stay, we can give an account which solves the consistency problems and gives intuitive results for contrary-to-duty and specificity cases (as will be demonstrated below with various examples).

Note 3. Instead of basing SDL on classical propositional logic and LK, our framework also allows for other variants, such as intuitionistic logic and Gentzen's LJ. This may be justified by the fact that, e.g., legal or medical systems often involve uncertainties, thus excluded middle is sometimes rejected in them, and proofs are required to be constructive. To keep things as simple as possible, we will proceed the discussion in terms of classical logic.

3 Logical Argumentation for Normative Reasoning

In what has become the orthodox approach based on Dung's representation [9], formal argumentation is studied on the basis of so-called argumentation frameworks. An argumentation framework in its most abstract form is a directed graph, where the nodes present (abstract) arguments and the arrows present argumentative attacks.

Definition 1. *An* (abstract) argumentation framework *is a pair* $\langle Args, Attack \rangle$, *where Args is an enumerable set of elements, called (abstract)* arguments, *and Attack is a relation between arguments whose instances are called* attacks.

When it comes to specific applications of formal argumentation it is often useful to provide an *instantiation* of (abstract) argumentation frameworks. Instantiations provide a specific account of the structure of arguments, and the concrete nature of argumentative attacks. There are various formal accounts available that provide frameworks for instantiating abstract argumentation such as assumption-based argumentation [10], ASPIC [11], etc. Here we settle for a recently proposed account based on sequent-based calculi [1].

The basic idea behind our instantiation is that arguments are C_{SDL}-proofs.

Definition 2. $\mathrm{Arg}(\Sigma)$ *is the set of* $\mathcal{C}_{\mathsf{SDL}}$-*proofs of sequents of the form* $\Gamma \Rightarrow \psi$ *for some* $\Gamma \subseteq \Sigma$.

For specifying the attack relation we complement $\mathcal{C}_{\mathsf{SDL}}$ with *sequent elimination rules*. Unlike the inference (or, sequent introduction) rules of $\mathcal{C}_{\mathsf{SDL}}$, the conclusions of sequent elimination rules are of the form $\Gamma \not\Rightarrow \psi$, and their intuitive meaning is the discharging of the sequent $\Gamma \Rightarrow \psi$.

We use attacks to model normative conflicts as well as conflict resolution by rules such as specificity (e.g., 'lex specialis' in legal contexts). Normative conflicts occur in cases in which we can construct arguments for conflicting obligations (and permissions).

Example 2. Consider the following sequent elimination rule:

$$\text{SPEC} \quad \frac{\Gamma, \phi \supset \psi \Rightarrow \psi \quad \Gamma \Rightarrow \phi \quad \Gamma' \Rightarrow \phi' \quad \phi \Rightarrow \phi' \quad \psi \Rightarrow \neg\psi' \quad \Gamma', \phi' \supset \psi' \Rightarrow \psi'}{\Gamma', \phi' \supset \psi' \not\Rightarrow \psi'}$$

This rule aims at formalizing the principle of specificity. It states that when two sequents $\Gamma' \Rightarrow \psi'$ and $\Gamma \Rightarrow \psi$ are conflicting, the one which is more specific gets higher precedence, and so the other one is discarded. Thus, in Example 1 for instance, SPEC allows to discharge the sequent $m, m \supset \mathsf{O}\neg f \Rightarrow \mathsf{O}\neg f$ in light of the more specific sequent $m \wedge a, (m \wedge a) \supset \mathsf{O}f \Rightarrow \mathsf{O}f$. We also say that the latter sequent *attacks* the former.

Some variations of SPEC are given below (where $\mathsf{NN}' \in \{\mathsf{OO}, \mathsf{OP}, \mathsf{PO}\}$):[1,2]

$$\text{NN'SPEC} \quad \frac{\begin{array}{c}\Gamma, \phi \supset \mathsf{N}\psi \\ \Rightarrow \mathsf{N}\psi\end{array} \quad \Gamma \Rightarrow \phi \quad \Gamma' \Rightarrow \phi' \quad \phi \Rightarrow \phi' \quad \psi \Rightarrow \neg\psi' \quad \begin{array}{c}\Gamma', \phi' \supset \mathsf{N}'\psi' \\ \Rightarrow \mathsf{N}'\psi'\end{array}}{\Gamma', \phi' \supset \mathsf{N}'\psi' \not\Rightarrow \mathsf{N}'\psi'}$$

For instance, POSPEC models permission as derogation [19]: a permission may suspend a more general obligation.

Example 3. In order to illustrate a conflict for which there is no overriding principle such as specificity that resolves it, suppose we have two triggered conditional obligations that conflict: $\Sigma = \{a, b, a \supset \mathsf{O}c, b \supset \mathsf{O}\neg c\}$. One could imagine an argumentative context in which one proponent presents an argument for $\mathsf{O}c$ by proving $a, a \supset \mathsf{O}c \Rightarrow \mathsf{O}c$. The opponent may rebut this argument for $\mathsf{O}\neg c$ by proving $b, b \supset \mathsf{O}\neg c \Rightarrow \mathsf{O}\neg c$. In a unilateral context this may be considerations and counter-considerations of a single reasoner. Such argumentative attacks may be modeled by sequent elimination rules as (where $\mathsf{NN}' \in \{\mathsf{OO}, \mathsf{OP}, \mathsf{PO}\}$):

$$\text{NN'CONF} \quad \frac{\Gamma \Rightarrow \mathsf{N}\psi \quad \psi \Rightarrow \neg\psi' \quad \Gamma' \Rightarrow \mathsf{N}'\psi'}{\Gamma' \not\Rightarrow \mathsf{N}'\psi'}$$

Some further sequent elimination rules for handling conflicting sequents are listed in Figure 2. We will not further discuss them here but we will come back to them in Section 4.

---

[1] In this and the following attack rules we also intend to capture unconditional obligations such as $\mathsf{O}\psi$. E.g., that $\phi, \phi \supset \mathsf{O}\psi \Rightarrow \mathsf{O}\psi$ OOSPEC-attacks $\mathsf{O}\neg\psi \Rightarrow \mathsf{O}\neg\psi$.

[2] Note that a 'PPSPEC'-variant would not be sensible since permissions with incompatible content do not conflict in any intuitive sense.

$$\text{CON} \quad \frac{\Rightarrow \neg\bigwedge\Gamma \quad \Gamma,\Gamma' \Rightarrow \psi}{\Gamma,\Gamma' \not\Rightarrow \psi} \qquad\qquad \text{NIC} \quad \frac{\Gamma \Rightarrow \neg\phi \quad \Gamma' \Rightarrow \mathsf{N}\phi}{\Gamma' \not\Rightarrow \mathsf{N}\phi}$$

$$\text{NN'CONFU} \quad \frac{\Gamma,\phi \supset \mathsf{N}\psi \Rightarrow \mathsf{N}\psi \quad \Gamma \Rightarrow \phi \quad \psi \Rightarrow \neg\psi' \quad \Gamma',\phi' \supset \mathsf{N}'\psi' \Rightarrow \psi''}{\Gamma',\phi' \supset \mathsf{N}'\psi' \not\Rightarrow \psi''}$$

$$\text{NCONFU'} \quad \frac{\Gamma \Rightarrow \neg(\phi \supset \mathsf{N}\psi) \quad \Gamma',\phi \supset \mathsf{N}\psi \Rightarrow \psi'}{\Gamma',\phi \supset \mathsf{N}\psi \not\Rightarrow \psi'}$$

$$\text{NCTD} \quad \frac{\begin{array}{c}\Gamma,\phi \supset \mathsf{N}\psi\\\Rightarrow \mathsf{N}\psi\end{array} \quad \Gamma \Rightarrow \phi \quad \Gamma' \Rightarrow \phi' \quad \phi \Rightarrow \phi' \quad \psi \Rightarrow \neg\psi' \quad \begin{array}{c}\Gamma',\phi' \supset \mathsf{O}\psi'\\\Rightarrow \mathsf{O}\psi'\end{array}}{\Gamma',\phi' \supset \mathsf{O}\psi' \not\Rightarrow \mathsf{O}\psi'}$$

$$\text{NN'SPECU} \quad \frac{\begin{array}{c}\Gamma,\phi \supset \mathsf{N}\psi\\\Rightarrow \neg(\phi' \supset \mathsf{N}'\psi')\end{array} \quad \Gamma \Rightarrow \phi \quad \phi \Rightarrow \phi' \quad \psi \Rightarrow \neg\psi' \quad \begin{array}{c}\Gamma',\phi' \supset \mathsf{N}'\psi'\\\Rightarrow \psi''\end{array}}{\Gamma',\phi' \supset \mathsf{N}'\psi' \not\Rightarrow \psi''}$$

$$\text{FCONF} \quad \frac{\Gamma \Rightarrow \neg\bigwedge_{i=1}^{n}\phi_i \supset \mathsf{N}_i\psi_i \quad \Gamma',\phi_1 \supset \mathsf{N}_1\psi_1,\ldots,\phi_n \supset \mathsf{N}_n\psi_n \Rightarrow \psi}{\Gamma',\phi_1 \supset \mathsf{N}_1\psi_1,\ldots,\phi_n \supset \mathsf{N}_n\psi_n \not\Rightarrow \psi}$$

Fig. 2. Some more sequent elimination rules for normative reasoning (where $\mathsf{NN'} \in \{\mathsf{OO},\mathsf{OP},\mathsf{PO}\}$ and $\mathsf{N} \in \{\mathsf{O},\mathsf{P}\}$)

Attacks between arguments are defined with reference to some $A \in \mathsf{Arg}(\Sigma)$ as follows:

- $\hat{A}$ denotes the top sequent in the proof A;[3]
- We say that a sequent $\Gamma \Rightarrow \psi$ is a *subsequent of* A if it is contained in A, and $\mathsf{Prem}(\hat{A}) \vdash_{\mathsf{SDL}} \bigwedge\Gamma$ (or, equivalently, if $\mathsf{Prem}(\hat{A}) \Rightarrow \bigwedge\Gamma \in \mathsf{Arg}(\Sigma)$).[4]

According to the next definition, an argument is attacked in some of its subsequents (including its top sequent).

Definition 3. *Let* $R = \frac{\Gamma_1 \Rightarrow \phi_1 \ldots \Gamma_n \Rightarrow \phi_n}{\Gamma_n \not\Rightarrow \phi_n}$ *be a sequent elimination rule in Figure 2, and let* $\mathcal{R}$ *be a set of such elimination rules.*

- *A sequent* $\mathfrak{s}$ *R-attacks a sequent* $\mathfrak{s}'$, *if there is an* $\mathcal{L}_{\mathsf{SDL}}$-*substitution* θ *such that* $\mathfrak{s} = \theta(\Gamma_1) \Rightarrow \theta(\phi_1)$ *and* $\mathfrak{s}' = \theta(\Gamma_n) \Rightarrow \theta(\phi_n)$. *We say that* $\mathfrak{s}$ $\mathcal{R}$-*attacks* $\mathfrak{s}'$ *if* $\mathfrak{s}$ *R-attacks* $\mathfrak{s}'$ *for some* $R \in \mathcal{R}$.
- *An argument* $A \in \mathsf{Arg}(\Sigma)$ *R-attacks an argument* $B \in \mathsf{Arg}(\Sigma)$ *if* $\hat{A}$ *R-attacks some subsequent of* B. *Similarly,* A $\mathcal{R}$-*attacks* B *if* A *R-attacks* B *for some* $R \in \mathcal{R}$.

---

[3] The top sequent is the top of the proof tree A if we conceive of proofs as trees, or the last line in the proof A if proofs are considered to be lists of lines.

[4] Intuitively speaking, the second condition warrants that the subsequents of a proof A of $\mathfrak{s} = \Gamma \Rightarrow \psi$ are only those sequents whose premises are charged in the proof of $\mathfrak{s}$. Take for instance the proof of $\Rightarrow \phi \supset \phi$ from $\phi \Rightarrow \phi$ by $[\Rightarrow\supset]$. This prevents for instance attacks on A by $\neg\phi \Rightarrow \neg\phi$.

Definition 4. *A normative argumentation framework induced by a set of elimination rules* $\mathcal{R}$ *is the logical argumentation framework* $\mathcal{AF}_{\mathcal{R}}(\Sigma) = \langle \mathsf{Arg}(\Sigma),$ *Attack* $\rangle$ *in which* $(A, B) \in$ *Attack iff* A $\mathcal{R}$*-attacks* B.

Normative Entailments Induced by Argumentation Frameworks

We are ready now to use (normative) argumentation frameworks for normative reasoning. As usual in the context of abstract argumentation, we do so by incorporating Dung's notion of extension [9], defined next.

Definition 5. *Let* $\mathcal{AF} = \langle Args, Attack \rangle$ *be an argumentation framework, and let* $\mathcal{E} \subseteq Args$. *We say that* $\mathcal{E}$ *attacks an argument* A *if there is an argument* $B \in \mathcal{E}$ *that attacks* A *(i.e.,* $(B, A) \in$ *Attack). The set of arguments that are attacked by* $\mathcal{E}$ *is denoted* $\mathcal{E}^+$. *We say that* $\mathcal{E}$ *defends* A *if* $\mathcal{E}$ *attacks every argument* B *that attacks* A. *The set* $\mathcal{E}$ *is called* conflict-free *if it does not attack any of its elements (i.e.,* $\mathcal{E}^+ \cap \mathcal{E} = \emptyset$*),* $\mathcal{E}$ *is called* admissible *if it is conflict-free and defends all of its elements, and* $\mathcal{E}$ *is* complete *if it is admissible and contains all the arguments that it defends. The minimal complete subset of Args is called the* grounded extension *of* $\mathcal{AF}$, *and a maximal complete subset of Args is called a* preferred extension *of* $\mathcal{AF}$.

Let $\mathcal{AF}_{\mathcal{R}}(\Sigma) = \langle \mathsf{Arg}(\Sigma), Attack \rangle$ be a normative argumentation framework.

- $\Sigma \mathbin{\vdash\mkern-10mu\sim}_{\mathsf{gr}} \psi$ if there is $A \in \mathsf{Arg}(\Sigma)$ in the grounded extension of $\mathcal{AF}_{\mathcal{R}}(\Sigma)$ such that $\hat{A} = \Gamma \Rightarrow \psi$.[5]
- $\Sigma \mathbin{\vdash\mkern-10mu\sim}_{\mathsf{pr}}^{\sqcap} \psi$ $[\Sigma \mathbin{\vdash\mkern-10mu\sim}_{\mathsf{pr}}^{\sqcup} \psi]$ if in every [some] preferred extension of $\mathcal{AF}_{\mathcal{R}}(\Sigma)$ there is $A \in \mathsf{Arg}(\Sigma)$ with $\hat{A} = \Gamma \Rightarrow \psi$.[6,7]

We will use the notation $\mathbin{\vdash\mkern-10mu\sim}$ whenever a statement applies to each of the defined consequence relations.

4   Some Examples

In this section we will demonstrate our argumentative model for normative reasoning by means of various examples.

Example 4. Let us recall Example 1, where $\Sigma_1 = \{m, a, m \supset \mathsf{O}\neg f, (m \wedge a) \supset \mathsf{O}f\}$. Some arguments in $\mathsf{Arg}(\Sigma_1)$ are listed in Figure 3 (right). We do not spell out the very simple proofs given by each argument but only list the top sequents and subsequent relationships. For instance, arguments A, B, C, D and E are one-liner proofs, argument F is obtained from B and C by weakening, etc. Figure 3 (left) shows an attack diagram where the only attack rule is OOSPECU.

---

[5] Recall that by the definition of $\mathsf{Arg}(\Sigma)$, this implies that $\Gamma \subseteq \Sigma$.

[6] A more cautious approach is to define: $\Sigma \mathbin{\vdash\mkern-10mu\sim}_{\mathsf{pr}}^{\sqcap} \psi$ $[\Sigma \mathbin{\vdash\mkern-10mu\sim}_{\mathsf{pr}}^{\sqcup} \psi]$ if there is an $A \in \mathsf{Arg}(\Sigma)$ with $\hat{A} = \Gamma \Rightarrow \psi$ that is in every [some] preferred extension of $\mathcal{AF}_{\mathcal{R}}(\Sigma)$.

[7] Similar entailment relations may of-course be defined for other semantics of abstract argumentation such as [semi-]stable semantics, ideal semantics, etc.

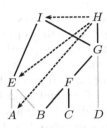

$$\hat{A} = m \supset O\neg f \Rightarrow m \supset O\neg f$$
$$\hat{B} = m \Rightarrow m$$
$$\hat{C} = a \Rightarrow a$$
$$\hat{D} = (m \wedge a) \supset Of \Rightarrow (m \wedge a) \supset Of$$
$$\hat{E} = m, m \supset O\neg f \Rightarrow O\neg f$$
$$\hat{F} = m, a \Rightarrow m \wedge a$$
$$\hat{G} = m, a, (m \wedge a) \supset Of \Rightarrow Of$$
$$\hat{H} = m, a, (m \wedge a) \supset Of \Rightarrow \neg(m \supset O\neg f)$$
$$\hat{I} = m, a, m \supset O\neg f, (m \wedge a) \supset Of \Rightarrow O\bot$$

Fig. 3. (Part of) the normative argumentation framework of Example 4: dashed arrows are OOSPECU-attacks, solid black lines indicate subsequents (the top sequents of lower arguments are subsequents of higher ones) and the gray line merely helps the reader to see which sequents share premises

We observe that H OOSPECU-attacks A and E, and since $\hat{E}$ is a subsequent of I, the latter is also attacked by H. It follows that, as expected, we have the following deductions:

- $\Sigma_1 \not\hspace{-0.3em}\sim O\neg f$. Indeed, one cannot derive $O\neg f$ since the application of MP to $m \supset O\neg f$ (depicted by argument E) gets attacked by H.[8]

- $\Sigma_1 \hspace{0.1em}\sim\hspace{-0.6em}\mid\hspace{0.1em} Of$. Indeed, G is not OOSPECU-attackable by an argument in $\mathsf{Arg}(\Sigma)$, thus it is part of every grounded and preferred extension of the underlying normative argumentation framework, and so its descendent follows from Σ_1.[9]

Example 5. Caminada [20] gives the following example for a deontic conflict that is not resolved by a resolution principle such as specificity: snoring is a misbehavior ($s \supset m$), it is allowed to remove misbehaving people from the library ($m \supset Pr$), it is not allowed to remove a professor from the library ($p \supset \neg Pr$), people who misbehave are subject to a fine ($m \supset Of$). Now suppose we have a snoring professor resulting in the following set: $\Sigma_{\mathsf{pro}} = \{s, p, s \supset m, p \supset \neg Pr, m \supset Pr, m \supset Of\}$. We can proof the two sequents $\mathfrak{s}_1 = p, p \supset O\neg r \Rightarrow O\neg r$ and $\mathfrak{s}_2 = s, s \supset m, m \supset Pr \Rightarrow Pr$ which NN'CONF-attack each other (where NN' $\in \{OP, PO\}$).

Caminada uses this example to illustrate what is sometimes considered a shortcoming of deductive approaches to defeasible reasoning [20, 21]. Given two conflicting inference steps described schematically by $\phi \rightsquigarrow \neg\theta$ and $\psi_1 \rightsquigarrow \psi_2 \rightsquigarrow \theta$. When "$\rightsquigarrow$" is contrapositive, we get $\phi \rightsquigarrow \neg\theta \rightsquigarrow \neg\psi_2$. This schematic

---

[8] Note that $m \supset O\neg f$ cannot be derived either, due to the attack of H on A.

[9] It is important to note that G *is* OOSPECU-attackable by SDL-derivable arguments, but none of them is in $\mathsf{Arg}(\Sigma_1)$. For instance, since material implication allows for strengthening of antecedents ($\phi \supset \psi \Rightarrow (\phi \wedge \phi') \supset \psi$), we have that $m \supset O\neg f \Rightarrow (m \wedge a) \supset O\neg f$ is SDL-derivable, and so G is attackable by an argument with, say, the SDL-derivable top sequent $m, m \supset O\neg f, m, a, (m \wedge a) \supset O\neg f \Rightarrow \neg((m \wedge a) \supset O\neg f)$. Yet, since $m \wedge a \supset O\neg f \notin \Sigma_1$, this argument is not in $\mathsf{Arg}(\Sigma_1)$. We note, further, that the sequent $a, m, m \supset O\neg f \Rightarrow \neg((m \wedge a) \supset Of)$ is derivable, but it does not OOSPECU-attack $\hat{G}$ and $\hat{H}$ though it is attacked by $\hat{H}$.

representation applied to our example yields $p \rightsquigarrow O\neg r \rightsquigarrow \neg Pr \rightsquigarrow \neg m$. Thus, the sequent $\mathfrak{s}_4 = p, p \supset O\neg r, m \supset Pr \Rightarrow \neg m$ is provable and conflicts with $\mathfrak{s}_3 = s, s \supset m \Rightarrow m$. Caminada argues that this violates the principle to keep conflicts as local as possible and so deontic conditionals are not to be contrapositive.

In our case, although contraposition holds in SDL, we can 'undercut' $\mathfrak{s}_4$ by attack rules such as FCONF as follows: In order to construct an argument for $\neg m$ we need the two conflicting conditional obligations $p \supset O\neg r$ and $m \supset Pr$ as premises. The fact that they are both triggered and conflicting in view of the given factual information s, p and $s \supset m$ can be formally expressed by the derivable sequent $\mathfrak{s}_7 = s, p, s \supset m \Rightarrow \neg((m \supset Pr) \wedge (p \supset O\neg r))$. Now, $\mathfrak{s}_7$ attacks all sequents that have both $p \supset O\neg r$ and $m \supset Pr$ as premises, such as $\mathfrak{s}_4$.

In Figure 4 we depict an excerpt of an attack diagram for Σ_{pro} with the attack rules FCONF and NN'CONF. We get, e.g., $\Sigma_{\mathsf{pro}} \hspace{1pt}\vdash_{\mathsf{gr}} Of$ and $\Sigma_{\mathsf{pro}} \not\vdash_{\mathsf{gr}} Pr$. If we use only FCONF, we also get $\Sigma_{\mathsf{pro}} \hspace{1pt}\vdash_{\mathsf{gr}} Pr$ and $\Sigma_{\mathsf{pro}} \hspace{1pt}\vdash_{\mathsf{gr}} O\neg r$ but we are still not able to accept arguments with both conflicting conditional obligations as premise (e.g., A_4 and A_6) since such arguments are FCONF-attacked by A_7.[10]

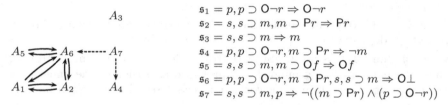

$$\mathfrak{s}_1 = p, p \supset O\neg r \Rightarrow O\neg r$$
$$\mathfrak{s}_2 = s, s \supset m, m \supset Pr \Rightarrow Pr$$
$$\mathfrak{s}_3 = s, s \supset m \Rightarrow m$$
$$\mathfrak{s}_4 = p, p \supset O\neg r, m \supset Pr \Rightarrow \neg m$$
$$\mathfrak{s}_5 = s, s \supset m, m \supset Of \Rightarrow Of$$
$$\mathfrak{s}_6 = p, p \supset O\neg r, m \supset Pr, s, s \supset m \Rightarrow O\bot$$
$$\mathfrak{s}_7 = s, s \supset m, p \Rightarrow \neg((m \supset Pr) \wedge (p \supset O\neg r))$$

Fig. 4. An attack diagram for Example 5 where $\hat{A}_i = \mathfrak{s}_i$. FCONF attacks are dashed, NN'CONF-attacks are solid $(N, N' \in \{O, P\})$.

Example 6. In the next example we take a look at contrary-to-duty (in short, CTD) obligations. A paradigmatic example is Forrester's Gentle Murderer scenario [18]: generally, one ought not to kill $(\top \supset O\neg f)$. However, upon killing, this should be done gently $(k \supset O(k \wedge g))$. Let $\Sigma_2 = \{k, \top \supset O\neg k, k \supset O(k \wedge g)\}$.

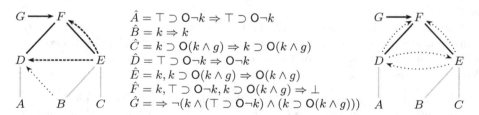

$$\hat{A} = \top \supset O\neg k \Rightarrow \top \supset O\neg k$$
$$\hat{B} = k \Rightarrow k$$
$$\hat{C} = k \supset O(k \wedge g) \Rightarrow k \supset O(k \wedge g)$$
$$\hat{D} = \top \supset O\neg k \Rightarrow O\neg k$$
$$\hat{E} = k, k \supset O(k \wedge g) \Rightarrow O(k \wedge g)$$
$$\hat{F} = k, \top \supset O\neg k, k \supset O(k \wedge g) \Rightarrow \bot$$
$$\hat{G} = \Rightarrow \neg(k \wedge (\top \supset O\neg k) \wedge (k \supset O(k \wedge g)))$$

Fig. 5. Two modelings of Forrester's Gentle Murderer

This is in line with Goble's [6, p. 27/28] analysis of an enriched version of Horty's Smith argument [7]: given $\{O\neg f, O(f \vee s), O\neg s\}$ (f is fighting in the army, s is performing civil service) he advocates to let both Os and $O\neg s$ be derivable without aggregating them to $O(s \wedge \neg s)$.

Van der Torre and Tan [22] distinguish CTD-obligations from cases of specificity. In the former the general obligations are not canceled or overridden but have still normative force (despite the fact that they are violated), while in cases of specificity the more general conditional obligations are canceled and thus deprived of normative force. There are various ways in which in our framework this distinction can be taken into account. One way of doing so is as follows. Instead of using strong rules such as OOSPECU in Example 4 that 'destroy' overridden conditional obligations in the sense that they do not appear in the consequence set, we can make use of rules such as OCTD (Figure 2) that preserve 'overshadowed' conditional CTD obligations despite the fact that detachment is blocked, or incorporate OIC that blocks detachment from violated obligations. This is illustrated in Figure 5 (left) with the attack rules OCTD (dashed arrow), OIC (dotted arrow) and CON (solid arrow). Alternatively, we could model overshadowing by means of OOCONF instead of OCTD. This is illustrated in Figure 5 (right) with attack rules OOCONF (dotted arrows) and CON (solid arrow). Where $\Xi = \{A, B, C, G\}$, we have two preferred extensions: $\Xi \cup \{D\}$ and $\Xi \cup \{E\}$. Hence, $\Sigma_2 \hspace{0.1em}\vdash^{\cup}_{\mathsf{pr}} \mathsf{O}\neg k$ and $\Sigma_2 \hspace{0.1em}\vdash^{\cup}_{\mathsf{pr}} \mathsf{O}(k \wedge g)$. In the skeptical approach we get $\Sigma_2 \hspace{0.1em}\vdash^{\cap}_{\mathsf{pr}} \mathsf{O}(\neg k \vee (k \wedge g))$ and $\Sigma_2 \hspace{0.1em}\vdash^{\cap}_{\mathsf{pr}} \mathsf{O}\neg k \vee \mathsf{O}(k \wedge g)$. Yet another option is to use a very liberal approach with CON only. This will block arguments with inconsistent premises such as F but otherwise allows e.g., to derive both $\mathsf{O}\neg k$ and $\mathsf{O}(g \wedge k)$ even via the grounded approach: $\Sigma_2 \hspace{0.1em}\vdash_{\mathsf{gr}} \mathsf{O}\neg k$ and $\Sigma_2 \hspace{0.1em}\vdash_{\mathsf{gr}} \mathsf{O}(k \wedge g)$.

Example 7. Consider the next paradigmatic CTD-case (Chisholm paradox, [23]):
  - It ought to be that Jones visits his neighbors.
  - It ought to be that if Jones goes, he tells them that he is coming.
  - If Jones doesn't go, then he ought not to tell them that he is coming.
  - Jones doesn't visit his neighbors.

In the modeling of this configuration, specific requirements have been posed. First, the logical model should not trivialize the set. Second, the formal representation of the four sentences should be rendered logically independent. It is obvious that by modeling conditional obligations via $\phi \supset \mathsf{O}\psi$ we will fail to meet the second requirement since with material implication we have $\neg\phi \supset (\phi \supset \psi)$ and hence we get $\neg g \supset (g \supset \mathsf{O}t)$ (where g is going to the neighbors and t is telling them). Since $\{\neg g, g \supset \mathsf{O}t, \neg g \supset \mathsf{O}\neg t, \mathsf{O}g\}$ is SDL-consistent, argumentation frameworks based on this set and based on the previously discussed attack rules are conflict-free. Hence, the first criterion is met.[11]

Example 8. Let us consider a variant of Example 4. Suppose that beside the obligation not to eat with your fingers we have the permission to do so in case asparagus is served, but it is considered impolite to eat asparagus with your

---

[11] An alternative modeling of 2 by $\mathsf{O}(g \supset t)$ is not appropriate here, since it would be ad hoc to model some conditional obligations by $\phi \supset \mathsf{O}\psi$ and others by $\mathsf{O}(\phi \supset \psi)$ whenever we run into problems with logical dependency. Moreover, given $\mathsf{O}g$ and $\mathsf{O}(g \supset t)$ we would be able to derive $\mathsf{O}t$ although g is not derivable, i.e., although the conditional obligation is not triggered.

fingers if there is a guest who considers this rude. The enriched set of premises may look as follows: $\Sigma_3 = \{a, m, c, m \supset O\neg f, (m \wedge a) \supset Pf, (m \wedge a \wedge c) \supset O\neg f\}$. The situation is depicted in Figure 6, with the attack rules OPSPECU (dotted arrows) and POSPECU (dashed arrows).

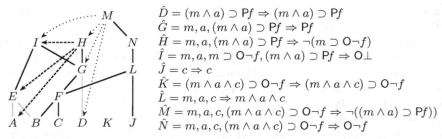

$\hat{D} = (m \wedge a) \supset Pf \Rightarrow (m \wedge a) \supset Pf$
$\hat{G} = m, a, (m \wedge a) \supset Pf \Rightarrow Pf$
$\hat{H} = m, a, (m \wedge a) \supset Pf \Rightarrow \neg(m \supset O\neg f)$
$\hat{I} = m, a, m \supset O\neg f, (m \wedge a) \supset Pf \Rightarrow O\bot$
$\hat{J} = c \Rightarrow c$
$\hat{K} = (m \wedge a \wedge c) \supset O\neg f \Rightarrow (m \wedge a \wedge c) \supset O\neg f$
$\hat{L} = m, a, c \Rightarrow m \wedge a \wedge c$
$\hat{M} = m, a, c, (m \wedge a \wedge c) \supset O\neg f \Rightarrow \neg((m \wedge a) \supset Pf))$
$\hat{N} = m, a, c, (m \wedge a \wedge c) \supset O\neg f \Rightarrow O\neg f$

Fig. 6. A normative argumentation framework for Example 8 (arguments A, B, C, E, F are as in Figure 3)

Thus, $\Sigma_3 \mathbin{\vert\!\sim} O\neg f$ (as expected), since N is defended, while G is not. Note that arguments A and E are also defended, since their only attacker H is attacked by the defended M. In argumentation theory A and E are said to be reinstated.

Example 9. Next we take a look at a simple conflict that is neither a specificity nor a CTD-case. Let $\Sigma_4 = \{a, b, a \supset O(c \wedge d), b \supset O(\neg c \wedge d)\}$. Figure 7 shows the situation for the attack rule OOCONFU (dotted arrows).

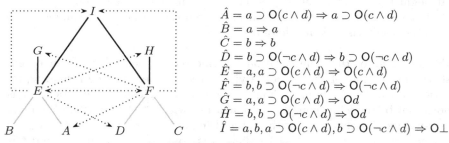

$\hat{A} = a \supset O(c \wedge d) \Rightarrow a \supset O(c \wedge d)$
$\hat{B} = a \Rightarrow a$
$\hat{C} = b \Rightarrow b$
$\hat{D} = b \supset O(\neg c \wedge d) \Rightarrow b \supset O(\neg c \wedge d)$
$\hat{E} = a, a \supset O(c \wedge d) \Rightarrow O(c \wedge d)$
$\hat{F} = b, b \supset O(\neg c \wedge d) \Rightarrow O(\neg c \wedge d)$
$\hat{G} = a, a \supset O(c \wedge d) \Rightarrow Od$
$\hat{H} = b, b \supset O(\neg c \wedge d) \Rightarrow Od$
$\hat{I} = a, b, a \supset O(c \wedge d), b \supset O(\neg c \wedge d) \Rightarrow O\bot$

Fig. 7. A simple conflict

We have the following preferred extensions: $\{A, B, E, G\}$ and $\{C, D, F, H\}$. Note that we have the 'floating conclusion'[12] $\Sigma_4 \mathbin{\vert\!\sim^{\cap}_{pr}} Od$ since one of G and H is in every preferred extension.

Example 10. The next example illustrates a conflict between three obligations. Let $\Sigma_5 = \{c, c \supset O(a \vee b), c \supset O(a \vee \neg b), c \supset O\neg a\}$. It is interesting to note that modeling this scenario with OOCONFU is problematic. In this case no conflicts are triggered since the triple-conflict is not reducible to a binary conflict that

---

[12] In nonmonotonic reasoning *floating conclusions* are conclusions that are obtained from each of a set of otherwise conflicting arguments.

fits the attack rule OOCONFU. This may be avoided by using OCONFU' instead of OOCONFU, as we get for instance $\Sigma_5 \mathrel{\mid\!\sim}^{\cap}_{\mathsf{pr}} Oa \vee O(\neg a \wedge b) \vee O(\neg a \wedge \neg b)$. This example shows that elimination rules should be carefully chosen.[13]

5 Basic Properties and Relation to Input/Output Logic

We start with two basic observations which can easily be verified by the reader:

1. For any set of attack rules previously defined: whenever Σ is SDL-consistent (i.e., $\Sigma \not\vdash_{\mathsf{SDL}} \perp$) then $\Sigma \vdash_{\mathsf{SDL}} \psi$ iff $\Sigma \mathrel{\mid\!\sim} \psi$. It is easy to verify that in this case all arguments in $\mathsf{Arg}(\Sigma)$ are selected since no argumentative attacks occur.

2. Where CON is part of the attack rules, (i) $\Sigma \mathrel{\mid\!\sim} \phi$ implies that ϕ is SDL-consistent (i.e., $\phi \not\vdash_{\mathsf{SDL}} \perp$) and, consequently, (ii) $\mathrel{\mid\!\sim}$ is strongly paraconsistent (i.e., for all Σ, $\Sigma \mathrel{\not\mid\!\sim} \perp$).

The framework of Input/Output logics [24] represents one of the standard approaches in conditional deontic logic. Many logics devised in this framework come with a simple and intuitive syntactic as well as a semantic characterization. The framework has been extended in order to deal with conflicts among conditionals such as in Contrary-to-Duty obligations [25]. There are in-depth studies concerning the modeling of permissions [19]. Moreover, there are links to other frameworks such as default logic (see [25]), logic programming [26] and adaptive logics [27]. In view of this it is interesting to notice that I/O logics can also be related to our framework, as will be established below (see Theorem 1 and Corollary 1).

In the following we focus on premise sets Σ that consist of non-modal formulas (representing 'facts' or 'input') and formulas of the type $\phi \supset O\psi$ (representing conditional obligations). For this, let Σ_F be a set of non-modal propositional formulas, Σ_O a set of pairs of non-modal formulas (ψ, ϕ) ('I/O-pairs') and $\Sigma_O^* = \{\psi \supset O\phi \mid (\psi, \phi) \in \Sigma_O\}$. Let also CPL be classical propositional logic, and denote by $\mathrm{Cn}_{\mathsf{CPL}}(\Gamma)$ the transitive closure of Γ with respect to $\vdash_{\mathsf{CPL}}$. The following definitions describe the 'out' and the 'out$_2$'-function in [24]:

Definition 6. $\mathsf{out}(\Sigma_F, \Sigma_O) = \{\psi \mid (\phi, \psi) \in \Sigma_O, \Sigma_F \vdash_{\mathsf{CPL}} \phi\}$.

Definition 7. $\phi \in \mathsf{out}_2(\Sigma_F, \Sigma_O)$ *iff* $\phi \in \mathrm{Cn}_{\mathsf{CPL}}(\mathsf{out}(\Xi, \Sigma_O))$ *for all CPL-maximal consistent extensions Ξ of Σ_F. In the degenerated case in which Σ_F is CPL-inconsistent, we define* $\mathsf{out}_2(\Sigma_F, \Sigma_O)$ *to be* $\mathrm{Cn}_{\mathsf{CPL}}(\{\psi \mid (\psi', \psi) \in \Sigma_O\})$.

The following is a corollary of Observation 4 in [24]:[14]

Lemma 1. $\Sigma_F \cup \Sigma_O^* \vdash_{\mathsf{SDL}} O\phi$ *iff* $\phi \in \mathsf{out}_2(\Sigma_F, \Sigma_O)$.

---

[13] In Section 5 we will prove that OCONFU' is rather well-behaved and can be used to give an argumentative account of a specific Input/Output logic.

[14] In [24] the authors show the correspondence for all normal modal logics L, for which $\vdash_{\mathsf{K}} \subseteq \vdash_{\mathsf{L}} \subseteq \vdash_{\mathsf{K45}}$.

In order to deal with situations in which $\mathsf{out}(\Sigma_F, \Sigma_O)$ is inconsistent, Makinson and Van Der Torre [25] 'contextualize' their output-functions to maximal sets of conditionals that are consistent with Σ_F, so-called maxfamilies:[15]

Definition 8. *We consider the following sets:*

- $\Gamma_O \in \mathsf{maxfamily}(\Sigma_F, \Sigma_O)$ *iff* $\mathsf{out}_2(\Sigma_F, \Gamma_O)$ *is* CPL-*consistent and for all* $(\psi, \phi) \in \Sigma_O \setminus \Gamma_O$, $\mathsf{out}_2(\Sigma_F, \Gamma_O \cup \{(\psi, \phi)\})$ *is not* CPL-*consistent.*
- $\psi \in \mathsf{out}_2^{\cup}(\Sigma_F, \Sigma_O)$ *iff* $\psi \in \bigcup_{\Gamma_O \in \mathsf{maxfamily}(\Sigma_F, \Sigma_O)} \mathsf{out}_2(\Sigma_F, \Gamma_O)$.
- $\psi \in \mathsf{out}_2^{\cap}(\Sigma_F, \Sigma_O)$ *iff* $\psi \in \bigcap_{\Gamma_O \in \mathsf{maxfamily}(\Sigma_F, \Sigma_O)} \mathsf{out}_2(\Sigma_F, \Gamma_O)$.

We now show that in our argumentative approach the Input/Output logics in Definition 8 can be characterized by means of the attack rule OCONFU'.

Theorem 1. *If* Σ_F *is* CPL-*consistent, then the set of all the preferred extensions of* $\mathcal{AF}_{\mathsf{OCONFU'}}(\Sigma_F \cup \Sigma_O^*)$ *is* $\{\mathsf{Arg}(\Sigma_F \cup \Gamma_O^*) \mid \Gamma_O \in \mathsf{maxfamily}(\Sigma_F, \Sigma_O)\}$.

Proof (Sketch). Let $\Gamma_O \in \mathsf{maxfamily}(\Sigma_F, \Sigma_O)$. By Lemma 1, $\Sigma_F \cup \Gamma_O^*$ is SDL-consistent and hence $\mathsf{Arg}(\Sigma_F \cup \Gamma_O^*)$ is conflict-free. Thus, each argument A attacking any argument in $\mathsf{Arg}(\Sigma_F \cup \Gamma_O^*)$ is such that $A \notin \mathsf{Arg}(\Sigma_F \cup \Gamma_O^*)$. Let $A \in \mathsf{Arg}(\Sigma_F \cup \Sigma_O^*) \setminus \mathsf{Arg}(\Sigma_F \cup \Gamma_O^*)$. This means that there is a $\psi \supset \mathsf{O}\phi \in \mathsf{Prem}(\hat{A}) \cap (\Sigma_O^* \setminus \Gamma_O^*)$. Since $\mathsf{out}_2(\Sigma_F, \Gamma_O \cup \{(\psi, \phi)\})$ is CPL-inconsistent we have by Lemma 1 that $\Sigma_F \cup \Gamma_O^* \cup \{\psi \supset \mathsf{O}\phi\}$ is SDL-inconsistent. Thus, there is a finite $\Theta \subseteq \Sigma_F \cup \Gamma_O^*$ such that $\Theta, \psi \supset \mathsf{O}\phi \Rightarrow \bot$ is $\mathcal{C}_{\mathsf{SDL}}$-provable. By $[\Rightarrow\supset]$, we derive $\mathfrak{s} = \Theta \Rightarrow \neg(\psi \supset \mathsf{O}\phi)$. Let C be the corresponding proof with $\hat{C} = \mathfrak{s}$. Then $C \in \mathsf{Arg}(\Sigma_F \cup \Gamma_O^*)$ and C OCONFU'-attacks A. We have shown that $\mathsf{Arg}(\Sigma_F \cup \Gamma_O^*)$ is defended and that it is maximally so.

Now assume there is an admissible extension Ξ of $\mathcal{AF}_{\mathsf{OCONFU'}}(\Sigma_F \cup \Sigma_O^*)$ such that there is no $\Gamma_O \in \mathsf{maxfamily}(\Sigma_F, \Sigma_O)$ for which $\Xi \subseteq \mathsf{Arg}(\Sigma_F \cup \Gamma_O^*)$. Hence, there is no $\Gamma_O \in \mathsf{maxfamily}(\Sigma_F, \Sigma_O)$ for which $\Gamma_\Xi = \bigcup_{A \in \Xi}\{(\psi, \phi) \mid \psi \supset \mathsf{O}\phi \in \mathsf{Prem}(\hat{A})\} \subseteq \Gamma_O$. This means $\mathsf{out}_2(\Sigma_F, \Gamma_\Xi)$ is CPL-inconsistent. By Lemma 1, $\Sigma_F \cup \Gamma_\Xi^*$ is SDL-inconsistent. Hence, there are finite $\Theta_F \subseteq \Sigma_F$ and $\Theta_O^* \subseteq \Gamma_\Xi^*$ such that $\Theta_F, \Theta_O^* \Rightarrow \bot$ is $\mathcal{C}_{\mathsf{SDL}}$-derivable. Since Σ_F is CPL-consistent, $\Theta_O^* \neq \emptyset$. With Weakening and $[\Rightarrow\supset]$ we have an argument C, with $\hat{C} = \Theta_F, \Theta_O^* \setminus \{\psi \supset \mathsf{O}\phi\} \Rightarrow \neg(\psi \supset \mathsf{O}\phi)$ where $\psi \supset \mathsf{O}\phi \in \Theta_O^*$. By the definition of Γ_Ξ, there is an $A \in \Xi$ for which $\psi \supset \mathsf{O}\phi \in \mathsf{Prem}(\hat{A})$. By the subformula property of SDL (see [17]) we can suppose that:

(†) for all $\gamma \supset \mathsf{O}\gamma'$ that occur in subsequents of C, $(\gamma, \gamma') \in \Theta_O$.

Then C OCONFU'-attacks A. Also, by (†), the only way to attack C leads to an attack on Ξ as well. Thus, Ξ cannot be defended from C. □

Corollary 1. *Where the only attack rule is* OCONFU', *for every* $\lambda \in \{\cup, \cap\}$ *it holds that* $\psi \in \mathsf{out}_2^{\lambda}(\Sigma_F, \Sigma_O)$ *iff* $\Sigma_F \cup \Sigma_O^* \mathrel{\vdash_{\mathsf{pr}}^{\lambda}} \mathsf{O}\psi$.

---

[15] The approach in [25] is more general since it takes into account sets of additional constraints beside our requirement of consistency.

Example 11. Suppose that $\Sigma_O = \{(p_1, q_1 \wedge q_2), (p_2, \neg q_1 \wedge q_2)\}$ and $\Sigma_F = \{p_1, p_2\}$. We have $\mathsf{maxfamily}(\Sigma_F, \Sigma_O) = \{\{(p_1, q_1 \wedge q_2)\}, \{(p_2, \neg q_1 \wedge q_2)\}\}$. Since $q_2 \in \mathsf{out}_2(\Sigma_F, \{(p_1, q_1 \wedge q_2)\}) \cap \mathsf{out}_2(\Sigma_F, \{(p_2, \neg q_1 \wedge q_2)\})$, also $q_2 \in \mathsf{out}_2^{\cap}(\Sigma_F, \Sigma_O)$.

In the normative argumentation framework $\mathcal{AF}_{\mathsf{OCONFU}'}(\Sigma_F, \Sigma_O^*)$ we have two preferred extensions: one with e.g. arguments with top sequents $p_1, p_2 \supset O(q_1 \wedge q_2) \Rightarrow \neg(p_2 \supset O(\neg q_1 \wedge q_2))$, $p_1, p_1 \supset O(q_1 \wedge q_2) \Rightarrow O(q_1 \wedge q_2)$, and $p_1, p_1 \supset O(q_1 \wedge q_2) \Rightarrow Oq_2$, and another one with e.g. arguments with top sequents $p_1, p_2 \supset O(\neg q_1 \wedge q_2) \Rightarrow \neg(p_1 \supset O(q_1 \wedge q_2))$, $p_2, p_2 \supset O(\neg q_1 \wedge q_2) \Rightarrow O(\neg q_1 \wedge q_2)$, and $p_2, p_2 \supset O(\neg q_1 \wedge q_2) \Rightarrow Oq_2$. Thus, $\Sigma_F \cup \Sigma_O^* \hspace{0.2em}\vdash_{\mathsf{pr}}^{\cap} Oq_2$.

Further investigations of entailment relations resulting from the application of attack rules other than OCONFU' will be considered in a future work.

6   Discussion and Outlook

The idea to use argumentation and abstract argumentation in particular to model normative reasoning is not new. Two examples are [28, 29]. The approach in [28] is based on bipolar abstract argumentation frameworks: beside an attack arrow a support arrow is used to express conditional obligations. Also in [29] Dung's framework is enhanced by a support relation this time signifying evidential support. Prolog-like predicates are used to encode argument schemes of normative reasoning and an algorithm is provided to translate them into an argumentation framework. One of the main differences in our approach based on logical argumentation is that we use a base logic (SDL) that generates all the given arguments (on the basis of a premise set). As a consequence an additional support relation is not needed since argumentative support is intrinsically modeled by considering arguments as proofs in SDL. A by-product of this is that our approach is more tightly linked to deontic logic.

Deontic logicians mainly agree that modeling conditional obligations on the basis of SDL and material implication is futile due to problems with CTD-obligations and specificity [2]. Therefore more research interest has been directed towards bi-conditionals. Specificity cases for instance call for weakened principles of strengthening the antecedent which are still strong enough to support many intuitively valid inferences. E.g., the principle of Rational Monotonicity has been challenged in [30] and replaced by a weakened version which itself has been criticized in [31]. In contrast, our base logic uses the standard implication of CPL to model conditional obligations and allows for full strengthening of the antecedent. Unwanted applications of the latter are avoided by means of argumentative attacks that are triggered e.g. in cases of specificity. As a consequence, our consequence relations are non-monotonic. There are other non-monotonic accounts of normative reasoning such as [7] based on default logic, Input/Output logic [25], or adaptive logics [3, 5, 6, 32]. Due to space restrictions we postpone a more elaborate comparison with these frameworks to future work.

In future work we also plan to investigate ways to combine and prioritize among attack rules, to distinguish preferences/priorities among obligations and

238   C. Straßer and O. Arieli

permissions, and to relate our work to different accounts of permission [19, 33], as demonstrated next.

Example 12. Let us add the facts r_1, r_2 and the conditional obligations $r_1 \supset \mathsf{O}\neg s$ and $(r_1 \wedge r_2) \supset \mathsf{O}s$ to Σ_F and Σ_O (respectively) in Example 11. This results in the premise set $\Sigma = \{p_1, p_2, r_1, r_2, p_1 \supset \mathsf{O}(q_1 \wedge q_2), p_2 \supset \mathsf{O}(\neg q_1 \wedge q_2), r_1 \supset \mathsf{O}\neg s, (r_1 \wedge r_2) \supset \mathsf{O}s\}$. Let us also add the attack rule OOSPECU to the previously used OCONFU'. One would expect that we get the consequence $\mathsf{O}s$ since arguments for the conflicting $\mathsf{O}\neg s$ such as proofs with the top sequent $r_1, r_1 \supset \mathsf{O}\neg s \Rightarrow \mathsf{O}\neg s$ are OOSPECU-attacked by arguments with the top sequent $r_1, r_2, (r_1 \wedge r_2) \supset \mathsf{O}s \Rightarrow \neg(r_1 \supset \mathsf{O}\neg s)$. However, the latter arguments are OCONFU'-attacked by arguments with head $r_1, r_2, r_1 \supset \mathsf{O}\neg s \Rightarrow \neg((r_1 \wedge r_2) \supset \mathsf{O}s)$. More generally, adding OOSPECU to OCONFU' doesn't alter the semantic selections. Hence, in order to model configurations in which arbitrary deontic conflicts occur together with specificity cases, we may need to prioritize OOSPECU-attacks over OCONFU'. The details of this are left for future research.

There are various other resolution principles for conflicts besides specificity and the latter does not apply in all cases or may be in conflict with other principles. For instance, "lex posterior derogat legi priori" may apply expressing that more recent laws override older ones. In order to model this we need to express temporal information and hence enhance our language.

Finally, we plan to investigate whether other nonmonotonic approaches and non truth-functional logics can be expressed in our framework.[16] Also, we shall examine base logics that are obtained from SDL by removing some of the inference rules in $\mathcal{C}_{\mathsf{SDL}}$, and so such logics may not have deterministic matrices. There is also forthcoming work on dynamic proofs for sequent-based argumentation [14], which may be useful to automatize normative reasoning as modeled in this paper.

References

1. Arieli, O.: A sequent-based representation of logical argumentation. In: Leite, J., Son, T.C., Torroni, P., van der Torre, L., Woltran, S. (eds.) CLIMA XIV 2013. LNCS, vol. 8143, pp. 69–85. Springer, Heidelberg (2013)
2. Aqvist, L.: Deontic logic. In: Gabbay, D.M., Guenthner, F. (eds.) Handbook of Philosophical Logic, 2nd edn., vol. 8, pp. 147–264. Kluwer (2002)
3. Beirlaen, M., Straßer, C., Meheus, J.: An inconsistency-adaptive deontic logic for normative conflicts. Journal of Philosophical Logic 2(42), 285–315 (2013)
4. Carmo, J., Jones, A.J.I.: Deontic logic and contrary-to-duties. In: Gabbay, D.M., Guenthner, F. (eds.) Handbook of Philosophical Logic, 2nd edn., vol. 8, pp. 265–343. Kluwer (2002)
5. Straßer, C.: A deontic logic framework allowing for factual detachment. Journal of Applied Logic 9(1), 61–80 (2010)

---

[16] For instance, we will check whether out_4 from [25] may be expressible by using $\Gamma_O^\star = \Gamma_O^\star \cup \{\mathsf{O}\phi \supset \mathsf{O}\psi \mid \phi \supset \psi \in \Gamma_O\}$ instead of $\Gamma_O^\star$ as in Corollary 1.

6. Goble, L.: Deontic logic (adapted) for normative conflicts. Logic Journal of the IGPL (2013)
7. Horty, J.F.: Reasoning with moral conflicts. Noûs 37(4), 557–605 (2003)
8. Straßer, C., Beirlaen, M.: An Andersonian deontic logic with contextualized sanctions. In: Ågotnes, T., Broersen, J., Elgesem, D. (eds.) DEON 2012. LNCS, vol. 7393, pp. 151–169. Springer, Heidelberg (2012)
9. Dung, P.M.: On the acceptability of arguments and its fundamental role in non-monotonic reasoning, logic programming and n-person games. Artifical Intelligence 77, 321–358 (1995)
10. Dung, P.M., Kowalski, R.A., Toni, F.: Assumption-based argumentation. Argumentation in Artificial Intelligence, 199–218 (2009)
11. Prakken, H.: An abstract framework for argumentation with structured arguments. Argument and Computation 1(2), 93–124 (2011)
12. Batens, D.: A universal logic approach to adaptive logics. Logica Universalis 1, 221–242 (2007)
13. Straßer, C.: Adaptive Logic and Defeasible Reasoning. Applications in Argumentation, Normative Reasoning and Default Reasoning. Trends in Logic, vol. 38. Springer (2014)
14. Arieli, O., Straßer, C.: Dynamic derivations for sequent-based logical argumentation. To appear in the Proceedings of COMMA 2014. IOS Press (2014)
15. Gentzen, G.: Investigations into logical deduction (1934) An English translation appears in Szabo, M.E. (ed.): The Collected Works of Gerhard Gentzen. North-Holland (1969) (in German)
16. Horty, J.F.: Moral dilemmas and nonmonotonic logic. Journal of Philosophical Logic 23, 35–65 (1994)
17. Valentini, S.: The sequent calculus for the modal logic D. Unione Matematica Italiana. Bollettino. A. Serie 7(7), 3 (1993)
18. Forrester, J.: Gentle murder, or the adverbial samaritan. Journal of Philosophy 81, 193–197 (1984)
19. Stolpe, A.: A theory of permission based on the notion of derogation. Journal of Applied Logic 8(1), 97–113 (2010)
20. Caminada, M.: On the issue of contraposition of defeasible rules. Frontiers in Artificial Intelligence and Applications 172, 109–115 (2008)
21. Prakken, H.: Some reflections on two current trends in formal argumentation. In: Logic Programs, Norms and Action, pp. 249–272 (2012)
22. Van Der Torre, L., Tan, Y.H.: Cancelling and overshadowing: two types of defeasibility in defeasible deontic logic. In: Proc. IJCAI 1995, pp. 1525–1532 (1995)
23. Chisholm, R.M.: Contrary-to-duty imperatives and deontic logic. Analysis 24, 33–36 (1963)
24. Makinson, D., Van Der Torre, L.: Input/Output logics. Journal of Philosophical Logic 29, 383–408 (2000)
25. Makinson, D., Van Der Torre, L.: Constraints for Input/Output logics. Journal of Philosophical Logic 30(2), 155–185 (2001)
26. Gonçalves, R., Alferes, J.J.: An embedding of input-output logic in deontic logic programs. In: Ågotnes, T., Broersen, J., Elgesem, D. (eds.) DEON 2012. LNCS, vol. 7393, pp. 61–75. Springer, Heidelberg (2012)
27. Straßer, C., Beirlaen, M., Putte, F.V.D.: Dynamic proof theories for input/output logic. Under review (2014)
28. Gabbay, D.: Bipolar argumentation frames and contrary to duty obligations, preliminary report. In: Fisher, M., van der Torre, L., Dastani, M., Governatori, G. (eds.) CLIMA XIII 2012. LNCS, vol. 7486, pp. 1–24. Springer, Heidelberg (2012)

29. Oren, N., Luck, M., Miles, S., Norman, T.: An argumentation inspired heuristic for resolving normative conflict. In: Proc. AAMAS 2008, pp. 41–56 (2008)
30. Goble, L.: A proposal for dealing with deontic dilemmas. In: Lomuscio, A., Nute, D. (eds.) DEON 2004. LNCS (LNAI), vol. 3065, pp. 74–113. Springer, Heidelberg (2004)
31. Straßer, C.: An adaptive logic framework for conditional obligations and deontic dilemmas. Logic and Logical Philosophy 19(1-2), 95–128 (2010)
32. Meheus, J., Beirlaen, M., Van De Putte, F.: Avoiding deontic explosion by contextually restricting aggregation. In: Governatori, G., Sartor, G. (eds.) DEON 2010. LNCS, vol. 6181, pp. 148–165. Springer, Heidelberg (2010)
33. Makinson, D., Van Der Torre, L.: Permission from an input/output perspective. Journal of Philosophical Logic 32(4), 391–416 (2003)

Combining Constitutive and Regulative Norms in Input/Output Logic

Xin Sun and Leendert van der Torre

Computer Science and Communication, University of Luxembourg

Abstract. In this paper we study three semantics to combine constitutive and regulative norms. In the first semantics, called the simple-minded semantics, the output of the constitutive norms are intermediate facts used as input for the regulative norms. The second method is called throughput, and adds the input of the constitutive norms to the intermediate facts. The third method is called reusable throughput, because it reuses the output of the regulative norms in the input of the constitutive norms. In addition, we refine these three so-called abstract semantics such that the obligations are labeled with the intermediate facts used to derive them. These explanations in the labels can be used for norm change, interpretation or defeasible argumentation. We present complete axiomatisations for the abstract and refined versions of the three semantics.

Keywords: deontic logic, input/output logic, constitutive norms.

1   Introduction

Constitutive norms are normative reasoning discussed in the handbook of deontic logic and normative systems [6], besides permissive norms, prima facie norms, and normative positions. They are usually contrasted with norms regulating the behavior of human beings by indicating which behaviors are obligatory, permitted and forbidden. Constitutive norms do not regulate actions or states-of-affairs, but rather they define new possible actions or states of affairs. In this paper we have little to say about constitutive norms themselves, and we refer the reader to the overview chapter of Grossi and Jones in the handbook [9]. Instead, we are interested in the *combination* of constitutive and regulative norms. As there are various ways to combine these two kinds of norms, and we believe none of them is perfect, this raises a new question. Besides choosing a logic for constitutive norms and a logic for regulative norms, the new question is:

Research Question. Which combination method to choose for combining constitutive and regulative norms?

Our approach in this paper to answer this question is to define three ways of combining these two kinds of norms, and to axiomatize these combinations. As always, the axiomatization presents the characteristic properties of the combination methods, which can be used to choose the method appropriate for a particular application. Moreover, we make as little commitments as possible about the representation of the norms. For example, constitutive norms are represented by count-as conditionals "X counts as

F. Cariani et al. (Eds.): DEON 2014, LNAI 8554, pp. 241–257, 2014.

Y in the context Z" [24,12,10], but there does not seem to be a consensus on the representation of the context. We therefore follow Lindahl and Odelstad [15] and Boella and van der Torre [1,2] and abstract away from the context. We thus represent constitutive norms as rules "X counts as Y." We use the general input/output logic approach [17,18] to represent both constitutive and regulative norms, and in particular we use a 'minimal' input/output logic recently introduced by Parent and van der Torre [20,21].

The research question breaks down into the following four subquestions. First, how to axiomatize the simple-minded combination method, as used by Lindahl and Odelstad [15], visualized below in Figure 1? Here C and R are the set of constitutive and regulative norms respectively. A is a set of formulas representing the facts. I is another set of formulas representing the intermediate concepts derived by the fact A and constitutive norms C. These intermediate facts are the input of regulative norms R. O is the output of intermediate facts I together with regulative norms R. If we write $I(C,A)$ for the intermediate facts derived from the facts A using constitutive norms C, and $\bigcirc(R,I)$ for the obligations derived from the intermediate facts I using the regulative norms R, then we can represent their simple-minded combination method as $\ominus^*(R,C,A) = \bigcirc(R,I(C,A))$.

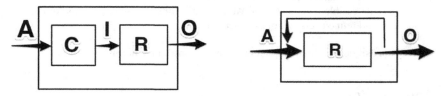

Fig. 1. Lindahl & Odelstad's combination   **Fig. 2.** Reusable input/output logic

Second, how to relate obligations explicitly to their intermediate facts, such that these can be used for explanation tasks in norm change, interpretation, and defeasible argumentation? For example, assume that a piece of paper counts as a contract, represented by the constitutive norm (*paper, contract*), and that the contract obliges us to pay money, represented by the regulative norm (*contract, pay*). From these two norms we want to derive the intermediate fact *contract* from the fact *paper*, and the obligation *pay*$_{contract}$, which means we are obliged to pay because there is a contract. For example, in argumentation, we can present a rebutting argument there is no obligation to pay, or an undercutting argument that there is no contract. As another example, consider the well known story of tû-tû discussed by Ross [23]. Suppose (*eat, tû-tû*) represents "If a person has eaten the chief's food, then she is tû-tû", and (*tû-tû, purification*) represents "If a person is tû-tû, then she is obligatory to be subjected to a ceremony of purification." Given the fact *eat*, from these two norms we can derive the institutional fact of *tû-tû* and the obligation *purification*$_{tû-tû}$, which means the person should be subjected to a ceremony of purification because she is tû-tû. Likewise, an obligation *congratulate*$_{checkmate}$ says that you have to congratulate your opponent because you are checkmate in chess, and the obligation *takecarefamily*$_{married}$ says that you have to take care of your family because you are married. Ross argues that similar examples can be found everywhere in the law. Taxation law, for instance, provides an abundance of examples: it usually stipulates who is to count as a taxable subject, and it also provides numerous examples of mandatory norms predicated on the basis of these classifications in the form of tax liabilities.

Third, how to introduce the assumption that brute facts can be used to detach obligations from regulative norms too? Tosatto et al. [26] discuss the difference between $\Theta^*(R,C,A) = \bigcirc(R,I(C,A))$ and $\Theta(R,C,A) = \bigcirc(R,A \cup I(C,A))$. For example, given $R = \{(a,x),(p,y)\}$, $C = \{(b,p)\}$ and $A = \{a,b\}$, we have $\Theta^*(R,C,A) = \{y\}$ because $I(C,A) = \{p\}$ and $\bigcirc(R,\{p\}) = \{y\}$, and $\Theta(R,C,A) = \{x,y\}$ because $A \cup I(C,A) = \{a,b,p\}$ and $\bigcirc(R,\{a,b,p\}) = \{x,y\}$.

Fourth, how to ensure that the combined system has the same properties as the individual systems? For example, Parent and van der Torre [21] argue for a new form of deontic detachment, called aggregative deontic detachment. The corresponding rule for aggregative deontic detachment is called *aggregative cumulative transitivity* (ACT):

$$(\text{ACT})\frac{x \in \bigcirc(R,a) \qquad y \in \bigcirc(R,a \wedge x)}{x \wedge y \in \bigcirc(R,a)}$$

In this paper we use aggregative input/output logic to represent constitutive and regulative norms. The two combinations mentioned by Tosatto et al. [26] are therefore represented by $\bigcirc(R,\bigcirc(C,A))$ and $\bigcirc(R,A \cup \bigcirc(C,A))$, where the operator $\bigcirc$ is defined by Parent and van der Torre [21]. We show in this paper that even though aggregative input/output logic satisfies ACT, both $\bigcirc(R,\bigcirc(C,A))$ and $\bigcirc(R,A \cup \bigcirc(C,A))$ do not. We therefore define a third combination of constitutive and regulative norms by forcing $\bigcirc(R,A \cup \bigcirc(C,A))$ to satisfy ACT. We call these three combinations simpleminded, throughput and reusable throughput combination respectively. For each of the three combinations, we further distinguish an *abstract combination*, where the output of the combination is a set of obligations, and a *detailed combination*, where the output of the combination is a set of obligations together with institutional facts.

Inspired by the input/output terminology, we use the following notation. We use $\circledcirc_1$, $\circledcirc_1^+$ and $\circledcirc_3^+$ for the semantics for the simple-minded, throughput and reusable throughput abstract combination respectively. The corresponding derivation system are represented by $\rhd_1$, $\rhd_1^+$ and $\rhd_3^+$. For the detailed combinations, we use $\odot_1$, $\odot_1^+$ and $\odot_3^+$ for semantics and $\blacktriangleright_1$, $\blacktriangleright_1^+$ and $\blacktriangleright_3^+$ for derivation system.

The layout of this paper is as follows. We survey aggregative input/output logic in Section 2. Then we introduce simple-minded, throughput and reusable combinations in Section 3 to 5 respectively. In Section 6 and 7 we discuss related work and future research.

2   Aggregative Input/Output Logic

Parent and van der Torre [20,21] introduce aggregative input/output logic, based on the following ideas. On the one hand, deontic detachment (DD) or cumulative transitivity (CT) is fully in line with the tradition in deontic logic. For instance, the Danielsson-Hansson-Lewis semantics [5,11,14] for conditional obligation validates such a law. On the other hand, they also observe that potential counterexamples to DD may be found in the literature. Parent and van der Torre illustrate this with the following example, due to Broome [4, §7.4]:

> You ought to exercise hard everyday
>
> If you exercise hard everyday, you ought to eat heartily
>
> ?* You ought to eat heartily

Intuitively, the obligation to eat heartily no longer holds if you take no exercise. Like the others, Parent and van der Torre claim that this counterexample suggests an alternative (they call it "aggregative") form of detachment, which keeps track of what has been detached. They therefore reject the CT rule, and they accept the weaker ACT rule. As a consequence, and following an established tradition in the literature [7,27,25], weakening the output is no longer accepted either. In Parent and van der Torre [21], ACT is motivated as follows: "the counterexample usually given to CT in the literature no longer work when ACT is used in place of CT. This is because they all rely on the intuition that the obligation y ceases to hold when the obligation of (a, x) is violated". Essentially the same argument is reproduced here by the example above. For another example consider the following:

Example 1. Consider the following situation:

- $(order, deliver)$ When you receive a purchase order from a customer, you have to deliver the good to the customer.
- $(order \wedge deliver, pay)$ If a customer sends a purchase order and the good are delivered to the customers, then the customer has the obligation to pay.
- $order \wedge \neg deliver$ The goods are ordered but not delivered.

Given the above situation, applying CT we derive $(order, pay)$. Then using factual detachment, which is a basic mechanism of input/output logic, we detach pay as our obligation. This conclusion is problematic, because intuitively we do not have to pay if the good is not delivered. On the contrary, applying ACT we can derive the argument $(order, deliver \wedge pay)$ but not $(order, pay)$. The problematic conclusion has disappeared.

Let $\mathbb{P} = \{p_0, p_1, \ldots\}$ be a set of propositional letters and L be the propositional language built upon $\mathbb{P}$. We write $\phi \dashv\vdash \psi$ for logical equivalence in the logic L. Let R be a set of ordered pairs of formulas of L. A pair $(a, x) \in R$, call it a regulative norm, is read as "given a, it ought to be x". Let $A \subseteq L$, $R(A) = \{x \in L | (a, x) \in R, a \in A\}$ be set theoretically understood as the image of A under function R. The semantics of aggregative input/output logic is defined as following:

Definition 1 (Aggregative input/output logic [21]). *For every $R \subseteq L \times L$, $A \subseteq L$, $x \in O(R, A)$ iff there is finite $R' \subseteq R$ with $R'(A) \neq \varnothing$ such that $\forall B = Cn(B)$, if $A \cup R'(B) \subseteq B$ then $x \dashv\vdash \bigwedge R'(B)$.*

The above definition is partially visualized by Figure 2. In the definition there is a qualification over a logically closed set B, which represents the input of R. B is required to extend $A \cup R'(B)$ because the left arrow is labeled by A and there is influx from the right arrow (representing $R'(B)$) to the left arrow.

At a first glance it seems that such a reusable approach conflates the distinction between "p is the case" and "p is obligatory." This is true in the sense that both facts and detached obligations are used as input to detach new obligations. Of course "p is the case" and "p is obligatory" are semantically different, but we believe such difference is not a sufficient reason to reject aggregative deontic detachment. For defense of aggregative deontic detachment the readers are suggested to consult the companying paper of Parent and van der Torre [21].

The following example illustrates aggregative deontic detachment.

Example 2. Let regulative norms be $R = \{(a, x), (a \wedge x, y)\}$ and input $A = \{a\}$. We have $O(R, A) = \{x, x \wedge y, \ldots\}$. The table below illustrates how to calculate $O(R, A)$, where B^* is the smallest set of formulas such that we have $B^* = Cn(B^*)$ and moreover $A \cup R'(B^*) \subseteq B^*$. For $R = R'$ we can then derive the obligation for $x \wedge y$, but not an obligation for y, illustrating that ACT holds in aggregative input/output logic, but CT does not.

A	R'	B^*	$R'(B^*)$	$\bigwedge R'(B^*)$
$\{a\}$	$\{(a, x)\}$	$Cn(\{a, x\})$	$\{x\}$	$\{x, \ldots\}$
$\{a\}$	R	$Cn(\{a, x, y\})$	$\{x, y\}$	$\{x \wedge y, \ldots\}$

The proof system contains three rules: strengthening of the antecedent (SI), output equivalence (OEQ) and aggregative cumulative transitivity (ACT).

Definition 2 (Proof system of aggregative input/output logic [21]). *Let $D(R)$ be the smallest set of arguments such that $R \subseteq D(R)$ and $D(R)$ is closed under the following rules:*

- *SI: from (a, x) and $b \vdash a$ to (b, x)*
- *OEQ: from (a, x) and $x \dashv\vdash y$ to (a, y)*
- *ACT: from (a, x) and $(a \wedge x, y)$ to $(a, x \wedge y)$.*

The rule AND is derivable in aggregative input/output logic.

- AND: from (a, x) and (a, y) to $(a, x \wedge y)$

Parent and van der Torre define $x \in D(R, A)$ iff there exist $a_1, \ldots, a_n \in A$ such that $(a_1 \wedge \ldots \wedge a_n, x) \in D(R)$. The following completeness result is proved [21].

Theorem 1 (Completeness of aggregative input/output logic [21]). *Given an arbitrary normative system R and a set A of formulas, $D(R, A) = O(N, A)$.*

In this paper we use a variant of aggregative input/output logic: we delete the restriction of $R'(A) \neq \emptyset$ in Definition 1. Correspondingly, we require $(\top, \top)$ to be included in the derivation system to ensure the soundness and completeness result.

Definition 3 (Variant of aggregative input/output logic). *For every $R \subseteq L \times L$, $A \subseteq L$, $x \in \bigcirc(R, A)$ iff there is finite $R' \subseteq R$ such that $\forall B = Cn(B)$, if $A \cup R'(B) \subseteq B$ then $x \dashv\vdash \bigwedge R'(B)$. $\triangleright(R)$ is the smallest set of arguments such that $\{(\top, \top)\} \cup R \subseteq \triangleright(R)$ and $\triangleright(R)$ is closed under the rules SI, OEQ and ACT.*

Soundness and completeness of $\bigcirc$ and $\triangleright$ follows the same lines as the soundness and completeness proofs of O and D.

3   Simple-Minded Combination

We use $\bigcirc$ not only for regulative norms, but also for constitutive norms. However, as constitutive norms cannot be violated, the issue raised by Parent and van der Torre does not occur for constitutive norms. In other words, we could have taken an input/output logic like reusable output or out_3 as well. The reason we choose to use $\bigcirc$ is uniformity, and in the detailed combination the derived obligations are more informative in the sense that the subscript contains *all* the institutional facts needed to derive the obligation.

The idea of simple-minded combination is illustrated by Figure 3. There is a set of constitutive norms $C \subseteq L \times L$ and a set of regulative norms $R \subseteq L \times L$. The input A is a set of formulas representing facts. $I = \bigcirc(C, A)$ is the output produced by the semantics of aggregative input/output logic given C and A. I is understood as the intermediate facts and used as the input to regulative norms R to generate obligations $O = \bigcirc(R, I)$.

We use aggregative input/output logic as our tool to analyze both constitutive and regulative norms. Since aggregative input/output logic is reusable in the sense its output can be reused as input, we represent simple-minded combination by Figure 3 with arrows representing reusability.

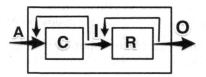

Fig. 3. Simple-minded combination

3.1   Simple-Minded Abstract Combination

Simple-minded abstract combination can be built straightforwardly by a composition of two aggregative input/output logics.

Definition 4 (Semantics of simple-minded abstract combination). *Let C, R be two sets of constitutive and regulative norms respectively, the semantics of simple-minded abstract combination is:*

$$\circledcirc_1(C, R, A) = \bigcirc(R, \bigcirc(C, A)).$$

Example 3. Let $A = \{eat\}$, $C = \{(eat, t\hat{u}\text{-}t\hat{u})\}$, $R = \{(t\hat{u}\text{-}t\hat{u}, purification), (eat, sorry)\}$. Here *(eat, sorry)* means "if a person has eaten the chief's food, then she should say sorry." Then we have $\bigcirc(C, A) = \{t\hat{u}\text{-}t\hat{u}, \dots\}$, $\circledcirc_1(C, R, A) = \{purification, \dots\}$. Note we do not have *sorry* $\in \circledcirc_1(C, R, A)$ because $eat \notin \bigcirc(C, A)$.

The proof system of simple-minded abstract combination is base on the derivation system $\triangleright$ for constitutive and regulative norms, with an additional composition.

Definition 5 (Derivation system of simple-minded abstract combination). *Let C, R be two sets of constitutive and regulative norms respectively, the proof system of simple-minded abstract combination is defined as follows:*

$$\triangleright_1(C,R) = \{(a,x) \mid \text{there is } p \in L \text{ such that } (a,p) \in \triangleright(C) \text{ and } (p,x) \in \triangleright(R)\}$$

We call the rule to derive $(a,x) \in \triangleright_1(C,R)$ *from* $(a,p) \in \triangleright(C)$ *and* $(p,x) \in \triangleright(R)$ *abstract constitutive regulative transitivity (ACRT).*

Example 4. From $C = \{(a,x),(a \wedge x,y)\}$, $R = \{(y,z)\}$ we can derive $(a,z) \in \triangleright_1(C,R)$ as following:

$$\cfrac{\cfrac{(a,x) \in C \quad (a \wedge x,y) \in C}{(a,x \wedge y) \in \triangleright(C)}\ (ACT) \quad \cfrac{(y,z) \in R}{(x \wedge y,z) \in \triangleright(R)}\ (SI)}{(a,z) \in \triangleright_1(C,R)}\ (DCRT)$$

Like in the proof theory of aggregative input/output logic, we let $x \in \triangleright(C,R,A)$ iff there exist $a_1, \ldots, a_n \in A$ such that $(a_1 \wedge \ldots \wedge a_n, x) \in \triangleright(C,R)$. The semantics and proof theory of simple-minded abstract combination are connected by the following completeness result:

Theorem 2 (Completeness of simple-minded abstract combination). *Let C, R be two sets of constitutive and regulative norms respectively, and a a formula, we have*

$$x \in \bigodot_1(C,R,A) \text{ iff } x \in \triangleright_1(C,R,A).$$

The reader may be surprised about our choice of calling $\triangleright_1$ a proof system, whereas it has a semantical flavor, in the sense that it is defined by the composition of binary relations—which is a set-theoretic and not a syntactic operation. Consequently, the coincidence is not a classical completeness result in the sense of connecting a calculus with a set-theoretic construction. However, $\triangleright_1$ still is a kind of very simple proof system, building on derivations in the underlying logic $\triangleright$, and it is in spirit similar to the other proof systems used in the input/output logic framework and in this paper. We therefore prefer to call Theorem 2 an completeness rather than a representation result.

The following proposition shows some basic properties of simple-minded abstract combination.

Proposition 1. $\triangleright_1(C,R)$ *validates SI, OEQ and AND, but not ACT.*

3.2 Simple-Minded Detailed Combination

In the semantics of aggregative input/output logic (Definition 2), we pick a set R' of the norms and qualify over a set of formulas B, which is closed under logical consequence. In the semantics of simple-minded detailed combination, we pick two sets C' and R', and we qualify over two sets of formulas B_1, B_2, which are both closed under logical consequence. The set B_1 is the input for C'. As visualized in Figure 3, we require it to extend $A \cup C'(B_1)$ because there is an arrow labelled A inject to C' and there is another arrow, the arrow from I to A, also inject to C'. Here note that I is $C'(B_1)$. Similarly, the set B_2 is the input for R'. We require it to contain $C'(B_1) \cup R'(B_2)$ because there is an arrow labeled I inject to R and there is another arrow, the arrow from O to I, also inject to R. Here note that O is $R'(B_2)$.

For the detailed combinations, we want to produce not only an obligation, but an obligation together with institutional facts. Formally, the output for the detailed combination is of the form x_p, where $x, p \in L$. As far as we know, such mixed output has not been defined in the input/output logic literature yet. Technically, the semantics of simple-minded detailed combination is defined as follows, in a flavor similar to aggregative input/output logic:

Definition 6 (Semantics of simple-minded detailed combination). *Let C, R be two sets of constitutive and regulative norms respectively, we define $x_p \in \bigodot_1(C, R, A)$ iff there is finite $C' \subseteq C, R' \subseteq R$ such that for all $B_1 = Cn(B_1), B_2 = Cn(B_2)$, if $A \cup C'(B_1) \subseteq B_1$ then $p \dashv \vdash \bigwedge C'(B_1)$, if $C'(B_1) \cup R'(B_2) \subseteq B_2$, then $x \dashv \vdash \bigwedge R'(B_2)$.*

Definition 7 (Proof theory of simple-minded detailed combination). *Let C, R be two sets of constitutive and regulative norms respectively, the proof theory of simple-minded detailed combination is:*

$$\blacktriangleright_1(C, R) = \{(a, x_p) \mid \text{there is } p \in L \text{ such that } (a, p) \in \triangleright(C) \text{ and } (p, x) \in \triangleright(R)\}$$

We call the rule deriving $(a, x_p) \in \blacktriangleright_1(C, R)$ from $(a, p) \in \triangleright(C)$ and $(p, x) \in \triangleright(R)$ detailed constitutive regulative transitivity (DCRT).

Example 5 illustrates that the subscript contains *all* the institutional facts needed to derive the obligation.

Example 5 (continued). From $C = \{(a, x), (a \wedge x, y)\}$, $R = \{(y, z)\}$ we can derive $(a, z_{x \wedge y}) \in \blacktriangleright_1(C, R)$ as follows:

$$\dfrac{\dfrac{(a, x) \in C \quad (a \wedge x, y) \in C}{(a, x \wedge y) \in \triangleright(C)} (ACT) \quad \dfrac{(y, z) \in R}{(x \wedge y, z) \in \triangleright(R)} (SI)}{(a, z_{x \wedge y}) \in \blacktriangleright_1(C, R)} (DCRT)$$

In other words, to derive the obligation for z we need the institutional facts x and y.

With the proof theory and semantics as defined, we have the following completeness result.

Theorem 3 (Completeness of simple-minded detailed combination). *Let C, R be two sets of constitutive and regulative norms respectively,*
$$x_p \in \blacktriangleright_1(C, R, A) \text{ iff } x_p \in \bigodot_1(C, R, A).$$

The above proof theory relies heavily on the proof theory of aggregative input/output logic. Moreover, it works separately on constitutive and regulative norms and combines them together at the last step by the DCRT rule. We alternatively define an equivalent proof theory more directly on expressions of the form (a, x_p).

Definition 8. *[Alternative proof theory of simple-minded detailed combination] Given C, R, let $\blacktriangleright'_1(C, R)$ be the smallest set of arguments such that $(\top, \top_\top) \in \blacktriangleright'_1(C, R)$, $\{(a, \top_p) \mid (a, p) \in C\} \subseteq \blacktriangleright'_1(C, R)$ and $\blacktriangleright'_1(C, R)$ is closed under the following rules:*

– *SI: strengthening of the input: from (a, x_p) to (b, x_p) whenever $b \vdash a$*

- *IOEQ: intermediate and output equivalence: from* (a, x_p) *to* (a, y_q) *if* $p \dashv\vdash q$ *and* $x \dashv\vdash y$
- *ACTI: aggregative cumulative transitivity for the intermediate: from* (a, x_p) *and* $(a \wedge p, x_q)$ *to* $(a, x_{p \wedge q})$
- *ACTO: aggregative cumulative transitivity for output: from* (a, x_p) *and* $(a \wedge x, y_p)$ *to* $(a, x \wedge y_p)$,

and the following indexed constitutive/regulative transitivity (ICRT) rule:

- *if* $(a, \top_p) \in \blacktriangleright_1'(C, R)$ *and* $(p, x) \in \triangleright(R)$ *then* $(a, x_p) \in \blacktriangleright_1'(C, R)$.

Example 6. (continued) Given $C = \{(a, x), (a \wedge x, y)\}$, $R = \{(y, z)\}$, we first derive $(a, \top_x)$ and $(a \wedge x, \top_y)$, then we derive $(a, z_{x \wedge y}) \in \blacktriangleright_1'(C, R)$ as follows:

$$\cfrac{\cfrac{(a, \top_x) \in \blacktriangleright_1'(C, R) \quad (a \wedge x, \top_y) \in \blacktriangleright_1'(C, R)}{(a, \top_{x \wedge y}) \in \blacktriangleright_1'(C, R)} \text{(ACTI)} \quad \cfrac{(y, z) \in R}{(x \wedge y, z) \in \triangleright(R)} \text{(SI)}}{(a, z_{x \wedge y}) \in \blacktriangleright_1'(C, R)} \text{(ICRT)}$$

The proof theory $\blacktriangleright_1'(C, R)$ may look unusual at first glance, but it resembles the proof theory of aggregative input/output logic. They both contain rules like strengthening of the input, output equivalence and aggregative cumulative transitivity. The two derivation systems $\blacktriangleright_1(C, R)$ and $\blacktriangleright_1'(C, R)$ are equivalent.

Proposition 2. *Let* C, R *be two sets of constitutive and regulative norms respectively,* $\blacktriangleright_1(C, R) = \blacktriangleright_1'(C, R)$.

4   Throughput Combination

In this section we strengthen simple-minded combination to throughput combination such that the input A can directly be used by regulative norms R, see Figure 4.

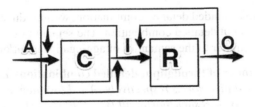

Fig. 4. Throughput combination

4.1   Throughput Abstract Combination

Throughput abstract combination is visualized by Figure 4, where both A and the output $\bigcirc(C, A)$ are part of the input of $\bigcirc(R, I)$.

Definition 9 (**Semantics of throughput abstract combination**). *Let C, R be two sets of constitutive and regulative norms respectively,* $\odot_1^+(C, R, A) = \bigcirc(R, A \cup \bigcirc(C, A))$.

The following is the proof system, in which the abstract constitutive regulative transitivity rule of simple minded combination is replaced by abstract constitutive regulative cumulative transitivity.

Definition 10 (**Proof system of throughput abstract combination**). *Let C, R be two sets of constitutive and regulative norms respectively, the proof system of throughput abstract combination is:*
$$\triangleright_1^+(C, R) = \{(a, x) \mid \text{there is } p \in L \text{ such that } (a, p) \in \triangleright(C) \text{ and } (a \wedge p, x) \in \triangleright(R)\}$$
We call the rule to derive $(a, x) \in \triangleright_1(C, R)$ from $(a, p) \in \triangleright(C)$, $(a \wedge p, x) \in \triangleright(R)$ abstract constitutive regulative cumulative transitivity (ACRCT).

Example 7. Given $C = \{(a, x), (a \wedge x, y)\}$ and $R = \{(a \wedge x \wedge y, z)\}$, we can derive $(a, z) \in \triangleright_1^+(C, R)$ as following:

$$\frac{\dfrac{(a, x) \in C \quad (a \wedge x, y) \in C}{(a, x \wedge y) \in \triangleright(C)} \ (ACT) \qquad (a \wedge x \wedge y, z) \in R}{(a, z) \in \triangleright_1^+(C, R)} \ (ACRT)$$

Here note that $(a, z) \notin \triangleright_1(C, R)$ because $(x \wedge y, z) \notin \triangleright(R)$. It can be further proved that $\triangleright_1(C, R) \subseteq \triangleright_1^+(C, R)$

The semantics and proof theory of throughput abstract combination are connected by the following completeness result.

Theorem 4 (**Completeness of throughput abstract combination**). *Let C, R be two sets of constitutive and regulative norms respectively, we have $x \in \odot_1^+(C, R, a)$ iff $(a, x) \in \triangleright_1^+(C, R)$.*

4.2 Throughput Detailed Combination

In parallel to the simple-minded detailed combination, we introduce the semantics and proof theory of throughput detailed combination. The semantics of simple-minded detailed combination is similar to the semantics of aggregative input/output logic.

Definition 11 (**Semantics of throughput detailed combination**). *Let C, R be two sets of constitutive and regulative norms respectively, $A \subseteq L$, we define $x_p \in \odot_1^+(C, R, A)$ iff there is finite $C' \subseteq C$, $R' \subseteq R$ such that for all $B_1 = Cn(B_1)$, $B_2 = Cn(B_2)$, if we have $A \cup C'(B_1) \subseteq B_1$, then $p \dashv\vdash \bigwedge C'(B_1)$, if $A \cup C'(B_1) \cup R'(B_2) \subseteq B_2$, then we also have $x \dashv\vdash \bigwedge R'(B_2)$.*

Like the semantics of simple-minded detailed combination, here we pick two sets C' and R', and we qualify over two sets of formulas B_1, B_2, which are both closed under logical consequence. The only difference is that for B_2, here we require it to extend A, while in simple-minded detailed combination we do not have such a requirement. The reason of this difference can be visualized by comparing Figure 3 and 4. In Figure 4 there is an arrow from A to I, while in Figure 3 there is not.

Definition 12 (Proof system of throughput detailed combination). *Let* C, R *be two sets of constitutive and regulative norms respectively, the proof system of throughput detailed combination is:*

$$\blacktriangleright_1^+(C,R) = \{(a,x_p) \mid \text{there is } p \in L \text{ such that } (a,p) \in \, \triangleright(C) \text{ and } (a \wedge p, x) \in \, \triangleright(R)\}.$$

We call the rule to derive $(a,x_p) \in \, \blacktriangleright_1^+(C,R)$ *from* $(a,p) \in \, \triangleright(C)$, $(a \wedge p, x) \in \, \triangleright(R)$ *detailed constitutive regulative cumulative transitivity (DCRCT).*

The proof system defined by Definition 12 and semantics are sound and complete.

Theorem 5 (Completeness of throughput detailed combination). *For all set of constitutive norms* C, *and regulative norms* R, $(a,x_p) \in \, \blacktriangleright_1^+(C,R)$ *iff* $x_p \in \, \odot_1^+(C,R,a)$.

Like the proof system $\blacktriangleright_1$, $\blacktriangleright_1^+$ heavily relies on the proof system of aggregative input/output logic. A more independent proof system is defined as follows:

Definition 13 (Alternative proof system of throughput detailed combination). *Let* C, R *be two sets of constitutive and regulative norms respectively, Let* $\blacktriangleright_1^{+'}(C,R)$ *be the smallest set such that* $(\top, \top_\top) \in \, \blacktriangleright_1^{+'}(C,R)$, $\{(a, \top_p) \mid (a,p) \in C\} \subseteq \, \blacktriangleright_1^{+'}(C,R)$, $\{(a, x_\top) \mid (a,x) \in R\} \subseteq \, \blacktriangleright_1^{+'}(C,R)$ *and* $\blacktriangleright_1^{+'}(C,R)$ *is closed under the rules SI, IOEQ, ACTI, ACTO and the following rule:*

- *if* $(a, \top_p) \in \, \blacktriangleright_1^{+'}(C,R)$ *and* $(a \wedge p, x) \in \, \triangleright(R)$ *then* $(a, x_p) \in \, \blacktriangleright_1^{+'}(C,R)$

One difference between $\blacktriangleright_1^{+'}(C,R)$ and $\blacktriangleright_1'(C,R)$ is: for $\blacktriangleright_1^{+'}(C,R)$ we require it to extend $\{(a, x_\top) \mid (a,x) \in R\}$. This feature reveals that regulative arguments can be derived directly in throughput combination. The equivalence of $\blacktriangleright_1^+$ and $\blacktriangleright_1^{+'}$ is stated in the following proposition.

Proposition 3. *For all* $C, R \subseteq L \times L$, $(a,x_p) \in \, \blacktriangleright_1^+(C,R)$ *iff* $(a,x_p) \in \, \blacktriangleright_1^{+'}(C,R)$.

5 Throughput Reusable Combination

Now we turn to throughput reusable combinations. As illustrated by the arrow from O to A in Figure 5, throughput reusable combination is the extension of throughput combination which allowing the output of regulative norms to be reused as input for constitutive norms. In this case the input of C have three resource: the arrow A, the arrow from I to A, and the arrow from O to A. The input of R have exactly the same resource. Therefore we can change Figure 5 to Figure 6 such that C and R have the same input.

5.1 Throughput Reusable Abstract Combination

The fact that throughput reusable combination is the extension of throughput combination which allows the reusability of output from R suggests that the former can be defined as the extension of latter validating the ACT rule. While the proof theory of throughput reusable abstract combination is a straightforward extension of its non-reusable companion, its semantics looks closer to the semantics of aggregative input/output logic.

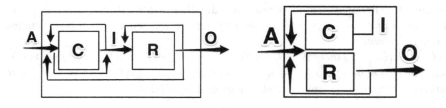

Fig. 5. Reusable combination   **Fig. 6.** Serial visualization

Definition 14 (Semantics of throughput reusable abstract combination). *Let C, R be two sets of constitutive and regulative norms respectively, and $A \subseteq L$, we define $x \in \bigodot_3^+(C, R, A)$ iff there exist finite $C' \subseteq C, R' \subseteq R$ such that for all $B = Cn(B)$: if $A \cup C'(B) \cup R'(B) \subseteq B$ then $x \dashv\vdash \bigwedge R'(B)$.*

Definition 15 (Proof system of throughput reusable abstract combination). *Let C, R be two sets of constitutive and regulative norms respectively, the proof system of throughput reusable abstract combination is defined as follows:*
$\triangleright_3^+(C, R)$ *is the smallest set such that* $\triangleright_1^+(C, R) \subseteq \triangleright_3^+(C, R)$ *and* $\triangleright_3^+(C, R)$ *is closed under the ACT rule.*

The above semantics reflects the ideas illustrated by Figure 6. We qualify over a set B, which is the input for C and R. Such B is an extension of A, $I = C(B)$ and $O = R(B)$. We therefore require $A \cup C'(B) \cup R'(B) \subseteq B$. The semantics and proof theory of throughput reusable abstract combination are connected by the following completeness result:

Theorem 6 (Completeness of throughput reusable abstract combination). *Let C, R be two sets of constitutive and regulative norms respectively, we have $x \in \bigodot_3^+(C, R, a)$ iff $(a, x) \in \triangleright_3^+(C, R)$.*

5.2 Throughput Reusable Detailed Combination

The semantics of throughput reusable detailed combination is an extension of its abstract companion.

Definition 16 (Semantics of throughput reusable detailed combination). *Let C, R be two sets of constitutive and regulative norms respectively, and $A \subseteq L$, we define $x_p \in \bigodot_3^+(C, R, A)$ iff there exist finite $C' \subseteq C, R' \subseteq R$ such that for all $B = Cn(B)$, if $A \cup C'(B) \cup R'(B) \subseteq B$ then $p \dashv\vdash \bigwedge C'(B)$ and $x \dashv\vdash \bigwedge R'(B)$.*

The following variant from an example of Makinson [16] illustrates that the combinations studied in this paper can also be used in other combination problems, not only for combining constitutive and regulative norms.

Example 8. Let $C = \{(A4, \text{size}),(\text{t25}\times15, \text{area})\}$, $R = \{(\text{size}, \text{t25}\times15), (\text{area}, \text{ref10})\}$. Here "A4" means "the paper is an A4 paper", "size" means "the paper is of standard size", "t25$\times$15" means "the text area is 25 by 15 cm", "area" means "the paper is of standard text area", "ref10" means "the font size for the reference is 10 points." In this setting we have $\odot_3^+(C, R, \{A4\}) = \{\top, \text{t25}\times15, \text{t25}\times15 \wedge \text{ref10}, \ldots\}$. For detailed combination we have $\odot_3^+(C, R, \{A4\}) = \{\top_\top, \text{t25}\times15_{\text{size}}, \text{t25}\times15 \wedge \text{ref10}_{\text{size}\wedge\text{area}}, \ldots\}$. The calculation can be illustrated by the following table, in which the column of A represents the input, C' and R' represent the subset of C and R, B^* represents the smallest set such that that $B^* = Cn(B^*)$ and $A \cup C'(B^*) \cup R'(B^*) \subseteq B^*$.

A	C'	R'	B^*	$C'(B^*)$	$R'(B^*)$
A4	$\varnothing$	$\varnothing$	$Cn(\{A4\})$	$\varnothing$	$\varnothing$
A4	$\{(A4, \text{size})\}$	$\{(\text{size}, \text{t25}\times15)\}$	$Cn(\{A4, \text{size}, \text{t25}\times15\})$	$\{\text{size}\}$	$\{\text{t25}\times15\}$
A4	C	R	$Cn(\{A4,\text{size},\text{t25}\times15, \text{area}, \text{ref10}\})$	$\{\text{size}, \text{area}\}$	$\{\text{t25}\times15, \text{ref10}\}$

In the second row of the table, $C'(B^*) = R'(B^*) = \varnothing$. This explains the reason of $\top \in \odot_3^+(C, R, \{A4\})$ and $\top_\top \in \odot_3^+(C, R, \{A4\})$. The third row explains the reason for $\text{t25}\times15 \in \odot_3^+(C, R, \{A4\})$, $\text{t25}\times15_{\text{size}} \in \odot_3^+(C, R, \{A4\})$, and the fourth row for $\text{t25}\times15 \wedge \text{ref10}$.

Definition 17 introduces the proof system of throughput reusable detailed combination $\blacktriangleright_3^+$. It is an extension of $\blacktriangleright_1^{+'}$, but its formation is simpler in the sense we add one rule called ACTIO but delete both ACTI and ACTO. Both ACTI and ACTO are derivable in $\blacktriangleright_3^+$.

Definition 17 (Proof system of reusable detailed combination). *Let $\blacktriangleright_3^+(C, R)$ be the smallest set such that $(\top, \top_\top) \in \blacktriangleright_3^+(C, R)$, $\{(a, x_\top) \mid (a, x) \in R\} \subseteq \blacktriangleright_3^+(C, R)$, $\{(a, \top_p) \mid (a, p) \in C\} \subseteq \blacktriangleright_3^+(C, R)$ and $\blacktriangleright_3^+(C, R)$ is closed under the rules SI, IOEQ, the following rule ACTIO*

– *ACTIO: aggregate cumulative transitivity for the intermediate and output: from (a, x_p) and $(a \wedge p \wedge x, y_q)$ to $(a, x \wedge y_{p \wedge q})$,*

and the following rule

– *if $(a, \top_p) \in \blacktriangleright_3^+(C, R)$ and $(a \wedge p, x) \in \triangleright(R)$ then $(a, x_p) \in \blacktriangleright_3^+(C, R)$.*

Example 9 (continued). From $C = \{(A4, \text{size}), (\text{t25}\times15, \text{area})\}$, $R = \{(\text{size}, \text{t25}\times15), (\text{area}, \text{ref10})\}$ we can derive expressions $(A4, \top_{\text{size}})$, $(\text{t25}\times15, \top_{\text{area}})$, $(\text{size}, \text{t25}\times15_\top)$ and $(\text{area}, \text{ref10}_\top)$. The following is the derivation:

$$\cfrac{(A4, \top_{\text{size}}) \quad \cfrac{(\text{size}, \text{t25}\times15_\top)}{(A4_{\wedge\top\wedge}\text{size}, \text{t25}\times15_\top)}\text{(SI)}}{\cfrac{(A4, \text{t25}\times15_{\wedge\top}\text{size}_{\wedge\top})}{(A4, \text{t25}\times15_{\text{size}})}\text{(IOEQ)}}\text{(ACTIO)}$$

Theorem 7 (Completeness of reusable detailed combination). *Given an arbitrary constitutive normative system C, regulative normative system R and a set A of formulas, $(a, x_p) \in \blacktriangleright_3^+(C, R)$ iff $x_p \in \odot_3^+(C, R, a)$.*

6   Related Work

Grossi and Jones [9] track the distinction between constitutive norms and regulative norms to at least Rawls [22]. Searle [24] uses 'X counts as Y in context C'' as a canonical presentation of constitutive norms. Jones and Sergot [12] formalize the context as an institution. Grossi et al. [10] formalize the context as a set norms.

Grossi and Jones [9] present a classification of approaches to formalize constitutive norms, and we refer to their chapter for further background on the logic of constitutive norms. On the one hand the choice of logic of constitutive norms is orthogonal to the choice of combination method. On the other hand, we adopt aggregative input/output logic as our logic for constitutive norms, which is different from all the approaches summarized by Grossi and Jones [9].

Modalities can be combined using possible world semantics. Boutilier [3] introduces a model $M = (W, \geq_1, \geq_2, V)$ with $\geq_1, \geq_2$ total pre-orders over W, reflecting normality and preference respectively. Each pre-order defines a classical Danielsson-Hansson-Lewis dyadic operator $\bigcirc^i$. Roughly, Boutilier defines a modality $\bigcirc^{12}(B|A)$ as the best of the most normal A worlds satisfy B:

- $M, w \vDash \bigcirc^i(B \mid A)$ iff $opt_{\geq_i}(\|A\|) \subseteq \|B\|$, for $i = 1, 2$
- $M, w \vDash \bigcirc^{12}(B \mid A)$ iff $opt_{\geq_2}(opt_{\geq_1}(\|A\|)) \subseteq \|B\|$

Here $opt_{\geq_i}(S) = \{w \in S | \forall u \in S, w \geq_i u\}$, for every $S \subseteq W$. According to the semantics, the combined modality does not satisfy the combination properties discussed in this paper:

Observation 1. $\{\bigcirc^1(q \mid p), \bigcirc^2(r \mid q)\} \nvDash \bigcirc^{12}(r \mid p)$

Lang and van der Torre [13] define $\bigcirc^{12}(B \mid A)$ in the same models by: the most normal $A \wedge B$ is preferred to the most normal $A \wedge \neg B$, and they compare their definition with Boutilier's. Observation 1 also holds for Lang and van der Torre's combination. A further comparison between these modal logic approaches and our approach is left for further research.

7   Future Research

We use aggregative input/output logic as our basis. Apart from the problem of pragmatic oddity and the irrelevant obligation problem mentioned in Parent and van der Torre [21], how to deal with the case when the output of constitutive norms is inconsistent is worthy of future research. The throughput reusable combination is formed by adding ACT to throughput combination. We can form a weaker version of reusable combination by adding ACT to simple-minded combination. The task is to define a semantics and prove the completeness result.

Makinson and van der Torre [17] developed input/output logic not only for obligations:

"In a range of contexts, one comes across processes resembling inference, but where input propositions are not in general included among outputs, and the operation is not in any way reversible. Examples arise in contexts of conditional obligations, goals, ideals, preferences, actions, and beliefs. Our purpose is to develop a theory of such input/output operations."

Therefore, in future research we want to investigate whether our framework can be used also for combining other modalities. Consider the well known problem of combining beliefs and desires: you may desire to go to the dentist, you may believe that going to the dentist means that you will have pain, but you do not desire to have pain. We can model this in our framework, if C stands for beliefs (or knowledge) and R are desires (or obligations), then we have $C = (dentist, pain)$ and $R = (\top, dentist)$. In all the three abstract combinations we have $dentist \in \bigcirc_{*s}(C, R, \varnothing)$ but not $pain \in \bigcirc_{*s}(C, R, \varnothing)$. It has been argued also that side effects are important, and that we need to avoid only unwanted side effects [8].

Input/output logic can be translated into modal logic. Makinson and ver der Torre have done such a translation in their original paper [17]. In our case, we can translate $(a, p) \in \rhd(C)$ to $I_c(a) \to O_c(p)$, $(p, x) \in \rhd(R)$ to $I_r(p) \to O_r(x)$. Here I_c, I_r, O_c and O_r are all modal operators. Moreover, we can translate $(a, x) \in \blacktriangleright(C, R)$ to $I_c(a) \to O_r(x)$ and $(a, x_p) \in \blacktriangleright(C, R)$ to $I_c(a) \to O_c(p) \wedge O_r(x)$. How to give sound and complete representations of the logics and combination methods discussed in this paper using modal logic are problems to be solved in the future. We generalized input/output logic by considering two sets of norms. It can be further extended to LIONS, as foreseen by Makinson and van der Torre [19]. Moreover, to refer to the special topic of DEON14, it may be a first step towards a Kratzer style semantics of natural language, because Kratzer's semantics can combine various kinds of ordering bases too.

8   Summary

To reason with constitutive and regulative norms, one has to choose a logic for the constitutive norms, a logic for the regulative norms, and a semantics to combine these two logics. In this paper we consider the question which semantics to choose for combining constitutive and regulative norms, a topic which has not raised much attention thus far, without committing ourselves to particular logics for constitutive or regulative norms. To make our analysis general, we use the 'minimal' logic introduced by Parent and van der Torre. Nevertheless, it contains two assumptions. First, strengthening of the input seems to reflect that rules do not have exceptions, whereas both constitutive and regulative norms encountered in practice often do have such exceptions. We do not consider this a limitation, because if we add priorities or a normality relation to reflect prima facie norms, exactly the same analysis can be given. Second, Parent and van der Torre's logic satisfies aggregative deontic detachment. This is only a weak notion of deontic detachment, and some kind of deontic detachment is needed in the logics to be able to define the reusability semantics. We distinguished three semantics to combine constitutive and regulative norms:

The simple-minded combination is the least committed, and thus the safest one to use. It clearly distinguishes the input, intermediate facts and output obligations, there is no possible source for confusion. It may be used, for example, when the input may not be true at the intermediate stage. For example, in legal interpretation the input may contains a bicycle which is also a vehicle, but the bicycle may not count as a vehicle in the legal sense. Using the simple-minded semantics, the intermediate facts may not contain a fact that the bicycle is a vehicle, whereas the input does. The proof system shows that the extension of the semantic is minimal, in the sense that it contains only a transitivity proof rule for the combination (Definition 4 and Theorem 2).

The throughput combination includes the input among the intermediate facts. The proof system shows that the difference with the simple-minded semantics is small, we just have to replace the transitivity axiom by a cumulative transitivity rule (Theorem 4). However, it means that for example in the bicycle case, we need to introduce another concept in the intermediate facts to represent that the bicycle does not count as a vehicle in the legal sense. In many common examples, it seems that the throughput semantics is preferred to the simple-minded semantics.

The throughput reusable combination considers as the intermediate state the facts closed under both the constitutive and regulative norms. This seems very strong, but the proof system shows that this corresponds precisely to the aggregative cumulative transitivity rule for the combined system (Theorem 6). So if this rule is desired, then this semantics has to be chosen. For example, if we start from a system satisfying ACT, and then refining it with systems for constitutive and regulative norms, then we need to refine it in this way.

In this paper we also introduce new detailed logics for combining constitutive and regulative norms, deriving expressions x_p for x is obligatory because of the intermediate concepts p, or simply x meaning x is obligatory without referring to intermediate concept. We have extended each of the three above systems with a proof system for these refined expressions.

References

1. Boella, G., van der Torre, L.: Regulative and constitutive norms in normative multi-agent systems. In: Dubois, D., Welty, C.A., Williams, M.-A. (eds.) Principles of Knowledge Representation and Reasoning, Whistler, Canada, pp. 255–266. AAAI Press (2004)
2. Boella, G., van der Torre, L.: A logical architecture of a normative system. In: Goble, L., Meyer, J.-J.C. (eds.) DEON 2006. LNCS (LNAI), vol. 4048, pp. 24–35. Springer, Heidelberg (2006)
3. Boutilier, C.: Toward a logic for qualitative decision theory. In: Doyle, J., Sandewall, E., Torasso, P. (eds.) Proceedings of the 4th International Conference on Principles of Knowledge Representation and Reasoning, Bonn, Germany, pp. 75–86. Morgan Kaufmann (1994)
4. Broome, J.: Rationality Through Reasoning. Wiley-Blackwell, West Sussex (2013)
5. Danielsson, S.: Preference and Obligation: Studies in the Logic of Ethics. Filosofiska Freningen, Uppsala (1968)
6. Gabbay, D., Horty, J., Parent, X., van der Meyden, R., van der Torre, L. (eds.): Handbook of Deontic Logic and Normative Systems. College Publications, London (2013)
7. Goble, L.: Murder most gentle: the paradox: deepens. Philosophical Studies 64, 217–227 (1991)

8. Governatori, G., Rotolo, A.: Bio logical agents: Norms, beliefs, intentions in defeasible logic. Autonomous Agents and Multi-Agent Systems 17(1), 36–69 (2008)

9. Grossi, D., Jones, A.: Constitutive norms and counts-as conditionals. In: Horty, J., Gabbay, D., Parent, X., van der Meyden, R., van der Torre, L. (eds.) Handbook of Deontic Logic and Normative Systems, pp. 407–441. College Publications, London (2013)

10. Grossi, D., Meyer, J.-J.C., Dignum, F.P.M.: Counts-as: Classification or Constitution? An Answer Using Modal Logic. In: Goble, L., Meyer, J.-J.C. (eds.) DEON 2006. LNCS (LNAI), vol. 4048, pp. 115–130. Springer, Heidelberg (2006)

11. Hansson, B.: An analysis of some deontic logics. Noûs, 373–398 (1969)

12. Jones, A., Sergot, M.: A formal characterisation of institutionalised power. The Logic Journal of IGPL, pp. 427–443 (1996)

13. Lang, J., van der Torre, L.: From belief change to preference change. In: Ghallab, M., Spyropoulos, C.D., Fakotakis, N., Avouris, N. (eds.) Proceedings of the 2008 Conference on ECAI 2008: 18th European Conference on Artificial Intelligence, Amsterdam, pp. 351–355. IOS Press (2008)

14. Lewis, D.K.: Counterfactuals. Blackwell, Oxford (1973)

15. Lindahl, L., Odelstad, J.: Normative systems and their revision: An algebraic approach. Artificial Intelligence and Law 11(2-3), 81–104 (2003)

16. Makinson, D.: On a fundamental problem in deontic logic. In: McNamara, P., Prakken, H. (eds.) Norms, Logics and Information Systems, Amsterdam, pp. 29–54. IOS Press (1999)

17. Makinson, D., van der Torre, L.: Input-output logics. Journal of Philosophical Logic 29, 383–408 (2000)

18. Makinson, D., van der Torre, L.: Constraints for input/output logics. Journal of Philosophical Logic 30(2), 155–185 (2001)

19. Makinson, D., van der Torre, L.: What is input/output logic? In: Lowe, B., Malzkorn, W., Rasch, T. (eds.) Foundations of the Formal Sciences II: Applications of Mathematical Logic in Philosophy and Linguistics, pp. 163–174 (2003)

20. Parent, X., van der Torre, L.: Aggregative deontic detachment for normative reasoning (short paper). In: Proceedings of the 14th International Conference on Principles of Knowledge Representation and Reasoning (KR 2014) (2014)

21. Parent, X., van der Torre, L.: "sing and dance!" input/output logics without weakening. In: 12th International Conference on Deontic logic and Normative Systems (DEON 2014) (2014)

22. Rawls, J.: Two concepts of rules. The Philosophical Review 64(1), 3–32 (1955)

23. Ross, A.: Tû-tû. Harvard Law Review 70, 812–825 (1957)

24. Searle, J.R.: Speech Acts: an Essay in the Philosophy of Language. Cambridge University Press, Cambridge (1969)

25. Stolpe, A.: Normative consequence: The problem of keeping it whilst giving it up. In: van der Meyden, R., van der Torre, L. (eds.) DEON 2008. LNCS (LNAI), vol. 5076, pp. 174–188. Springer, Heidelberg (2008)

26. Tosatto, S., Boella, G., van der Torre, L., Villata, S.: Abstract normative systems: Semantics and proof theory. In: Brewka, G., Eiter, T., McIlraith, S.A. (eds.) Principles of Knowledge Representation and Reasoning, Rome, pp. 358–368. AAAI Press (2012)

27. van der Torre, L., Tan, Y.: Contrary-to-duty reasoning with preference-based dyadic obligations. Ann. Math. Artif. Intell. 27(1-4), 49–78 (1999)

A Deontic Logic of Actions and States

Robert Trypuz and Piotr Kulicki*

John Paul II Catholic University of Lublin,
Faculty of Philosophy
Al. Racławickie 14, 20-950 Lublin, Poland
{kulicki,trypuz}@kul.pl
http://www.kul.pl

Abstract. Deontic considerations are usually conducted in one of the following contexts: (i) general norms expressed in: legal documents, regulations or implicitly present in society in the form of moral or social rules; (ii) specific norms, i.e., duties of particular agents in particular situations. Norms of the two kinds refer to obligatory, permitted and forbidden actions or states. In the paper we start from presenting a simple deontic logic of actions and a simple deontic logic of states. Then we combine them to provide a unified logic of general norms. Finally, we discuss the way in which specific norms can be introduced into the system.

Keywords: deontic action logic, norms on actions and states, general and specific norms, algebra of actions.

Introduction

The distinction between general and individualised norms is the key one for deontic logic. General norms appear in the sources of norms, such as legal documents, agreements, orders, informal social regulations etc. They are external to agents and are usually formed in an abstract way. Individualised norms (sometimes also called subjective normative positions) are connected with a particular situation of an agent. They are the results of applying (by an agent himself or by a judge) all general norms the agent should comply with to the situation in which the agent acts or has acted. The distinction is present in the theory of law [7] and was recently discussed in the context of deontic logic in [2], where the notion of obligation is used instead of the notion of individualised norm.

Among general norms there are those that concern obligatory, recommended, permitted or prohibited actions and those that concern desired, preferred, acceptable or forbidden states (see e.g. [1,16]).

In every case a norm of conduct is a pronouncement which points out for the addressee a more or less generally defined conduct in any, or

---

* The project was funded by the National Science Centre of Poland (DEC-2011/01/D/HS1/04445).

F. Cariani et al. (Eds.): DEON 2014, LNAI 8554, pp. 258–272, 2014.

*under specified, circumstances. Hence, it sets him the duty of undertaking
a specified action, and refraining from any other discordant with it. It
also sometimes happens that a norm points out for an addressee a duty
of bringing about some state of affairs without any indications of the
manner in which this state of affairs is to be attained. [16, chapter viii]*

We shall call the norms of the two types a-norms (for action norms) and
s-norms (for state norms) respectively. Both kinds can be present in the same
normative systems. For example in [1] the authors point out both kinds of norms
in the Spanish constitution.

In deontic logic the two types of norms are not usually present together in formal systems. They are often regarded in deontic literature as linguistic variants
of the same normative reality. We believe that there is a need for a deontic logic
in which we can reason about norms of the two kinds. There are some works that
tackle the problem such as [3,14] and recently [4], but we are not fully pleased
with the solutions present there mostly due to the fact that they do not really
relate deontic properties of actions with the properties of states.

We shall consider general norms formulated as obligations and prohibitions.
The two notions are present in most normative contexts and complement each
other in normative systems. There also exist norms formulated with other notions, but we shall not deal with them in the present paper. We especially omit
permissive norms that seem to be of a different character from the ones we consider – permission usually expresses derogation in a normative system (see e.g.
[10]). We shall only use permission as a counterpart of prohibition (in the sense
that an action or a state is permitted if it is not prohibited).

In our logic by actions we mean action types (not tokens) – see [15]. From the
linguistic point of view we consider general names referring to actions. Names of
actions are arguments of operators of obligation and prohibition. Together with
the deontic operators they make up a-norms. Similarly we deal with states. We
do not refer directly to particular states, but we use propositions to describe
them. Each proposition can be then connected to all states in which it is true.
Propositions are arguments of operators of obligation and prohibition. Together
with the deontic operators they make up s-norms. Thus, collections of actions
or states are forbidden or obligatory. What does it mean for an individual action or state? We believe, following natural language and legal practice, that the
approach to prohibition and obligation should be different. When we prohibit
an action we prohibit the execution of every action denoted by a general action
name and when we prohibit bringing about a state described by a proposition we
prohibit all their concrete realisations. Any sub-action or sub-proposition (action
or proposition referring to a subset of action tokens or states) of a prohibited
action or proposition is also forbidden. In contrast, the obligation concerning an
action name or proposition is fulfilled if any action token or state fulfilling the
specification is realised. However, obligations should not be overgeneralised, i.e.,
the fact that a set of action tokens or states is obligatory does not entail that its

supersets are also obligatory[1]. We would obtain norms that are less useful then the original ones.

We are interested in two kinds of reasoning about norms. One is a derivation of new general norms from the general norms already accepted. In the case of norms expressing obligations, a derived norm is always more specific (referring to a smaller set of action tokens or states) than the norms from which it is derived. That guarantees that a derived norm points to new information about the normative system. We want to be able to combine two a-norms together and two s-norms together, but also a-norms with s-norms.

The other kind of normative reasoning we are interested in is discovering individualised norms (obligations from [2]) for a particular agent and situation in a normative environment. Ignoring a sophisticated ontological distinction between general and individualised norms we attempt to obtain that result by finding the most specific general norm derived from all norms applicable to the situation. That allows us to discuss both kinds of norms (general and individualised) in one formal system.

In section 1 we introduce a model and our notions of norms within that model. In sections 2 and 3 we define a language and introduce a logic. In section 4 we compare our approach with other works.

1   Frames for Deontic Actions, Deontic States and a Bridge between Them

In this section we first introduce two frames for deontic actions and states. We shall see that their sets of required (illegal) actions and states have the same formal properties. The reason for distinguishing between them is ontological (i.e., they correspond to different entities in the normative reality).

1.1   Deontic Frames for Actions and States: $\mathcal{DAF}$ and $\mathcal{DSF}$

Deontic action frame $\mathcal{DAF}$. A deontic action frame with sets of legal and illegal actions was described in Segerberg's [9]. We have systematised the results presented there in [13,12,11] and extended his deontic action frame by a required set of actions (corresponding with obligation in the language). The deontic action frame in question is a structure:

$$\mathcal{DAF} = \langle \mathcal{AF}, \mathcal{ILL}^a, \mathcal{REQ}^a \rangle$$

$\mathcal{AF}$ is an action frame being a triple[2]:

$$\mathcal{AF} = \langle \mathcal{W}, \mathcal{E}, \mathcal{Step} \rangle$$

---

[1] For example Ross paradox is a formula which overgeneralises obligation and causes a loss of information. That is the reason we intend to avoid it in our system.

[2] The action frame presented in this section could be reduced without any loss to a set of events $\mathcal{E}$, since we do not really take advantage of interpreting actions as sets of transitions. Instead, we could think of them as sets of events from $\mathcal{E}$. This point of view was presented and studied in [9,11]. One can also understand the action frame presented in this section as a one-state action frame.

where $\mathcal{W}$ is a nonempty, finite set of states (possible worlds), $\mathcal{E}$ is a nonempty, finite set of basic action types used for a cross-situation identification of actions and $\mathcal{S}tep$ is a set of *action steps*. Every element of $\mathcal{S}tep$ is a triple $\langle w_1, w_2, e \rangle$, where $w_1, w_2 \in \mathcal{W}$ are initial and final states respectively and $e \in \mathcal{E}$ is a label of an action causing the transition from w_1 to w_2. Subsets of $\mathcal{S}tep$ represent *action types*, so we can model a parallel execution of two actions by an intersection of respective sets of action steps and a free choice between two actions by their sum.

We do not impose any restrictions on the frame $\mathcal{AF}$. Thus, it may happen that we have an indeterministic execution of action, e.g., $\langle w_1, w_2, e \rangle \in \mathcal{S}tep$ and $\langle w_1, w_3, e \rangle \in \mathcal{S}tep$ and, on the other hand, that the same transition is a result of the execution of two different actions, e.g., $\langle w_1, w_2, e_1 \rangle \in \mathcal{S}tep$ and $\langle w_1, w_2, e_2 \rangle \in \mathcal{S}tep$.

$\mathcal{ILL}^a(w)$ and $\mathcal{REQ}^a(w)$ are defined as functions from $\mathcal{W}$ to $2^{2^{\mathcal{S}tep}}$, so sets of actions are their values. $\mathcal{ILL}^a(w)$ is a set of illegal (forbidden) actions, whereas $\mathcal{REQ}^a(w)$ – a set of required (obligatory) ones. We assume that each element of $\mathcal{ILL}^a(w)$ and $\mathcal{REQ}^a(w)$ encodes an a-norm which comes from a legal document, social practice, etc. or is inferred from other norms.

Because of the potentially indeterministic character of our theory we assume that whenever $\langle w, w_1, e \rangle$ belongs to some X in $\mathcal{ILL}^a(w)$ or $\mathcal{REQ}^a(w)$, then for any w_2, $\langle w, w_2, e \rangle$ also belongs to X. Other properties of $\mathcal{ILL}^a(s)$ and $\mathcal{REQ}^a(s)$ are characterised below (their detailed description can be found in [13]).

For any $w \in \mathcal{W}$ and $X, Y \in 2^{\mathcal{S}tep}$, $\mathcal{ILL}^a(w)$ satisfies the following conditions making up $\mathcal{ILL}^a(w)$ to be an ideal (in algebraic sense):

$$X \in \mathcal{ILL}^a(w) \ \& \ Y \subseteq X \implies Y \in \mathcal{ILL}^a(w) \tag{1}$$

$$X \in \mathcal{ILL}^a(w) \ \& \ Y \in \mathcal{ILL}^a(w) \implies X \cup Y \in \mathcal{ILL}^a(w) \tag{2}$$

$$\emptyset \in \mathcal{ILL}^a(w) \tag{3}$$

A necessary condition for two elementary transitions to be required is that their intersection should be required (agglomeration principle):

$$X \in \mathcal{REQ}^a(w) \ \& \ Y \in \mathcal{REQ}^a(w) \implies X \cap Y \in \mathcal{REQ}^a(w) \tag{4}$$

We also accept trimming and economy principles, respectively:

$$X \in \mathcal{REQ}^a \ and \ Y \in \mathcal{ILL}^a \implies X \cap -Y \in \mathcal{REQ}^a \tag{5}$$

$$X \in \mathcal{REQ}^a \implies -X \in \mathcal{ILL}^a \tag{6}$$

Deontic state frame $\mathcal{DSF}$. A deontic state frame is a structure

$$\mathcal{DSF} = \langle \mathcal{W}, \mathcal{REQ}_s, \mathcal{ILL}_s \rangle$$

where $\mathcal{W}$ is a set of states (as in $\mathcal{AF}$ above), $\mathcal{REQ}_s$ and $\mathcal{ILL}_s$ are functions: $\mathcal{W} \longrightarrow 2^{2^{\mathcal{W}}}$. $\mathcal{REQ}_s(w)$ and $\mathcal{ILL}_s(w)$ are sets of required and illegal sets of states (propositions), respectively. They represent s-norms and are counterparts of $\mathcal{REQ}^a(w)$ and $\mathcal{ILL}^a(w)$, satisfying the same formal conditions. However, one may also apply different constrains on action and state frames (e.g., consistency of obligation constraint: $\emptyset \notin \mathcal{REQ}_s$).

1.2   Deontic Action and State Frame $\mathcal{DASF}$

A deontic action and state frame is a structure combining the two above frames: $\mathcal{DAF}$ and $\mathcal{DSF}$:

$$\mathcal{DASF} = \langle \mathcal{AF}, \mathcal{ILL}^a, \mathcal{REQ}^a, \mathcal{ILL}_s, \mathcal{REQ}_s \rangle$$

Having two deontic sets concerning actions ($\mathcal{ILL}^a, \mathcal{REQ}^a$) and two sets concerning states ($\mathcal{ILL}_s, \mathcal{REQ}_s$), we intend to link them together to find all the possible combinations of actions and states regulated by norms (see table 1). We intend to infer new norms by taking into account pairs of norms ⟨a-norm, s-norm⟩.

Table 1. Binding a-norms with s-norms

$\times$	$\mathcal{REQ}_s(w)$	$\mathcal{ILL}_s(w)$
$\mathcal{REQ}^a(w)$	(i) $\mathcal{REQ}^a(w) \times \mathcal{REQ}_s(w)$	(ii) $\mathcal{REQ}^a(w) \times \mathcal{ILL}_s(w)$
$\mathcal{ILL}^a(w)$	(iii) $\mathcal{ILL}^a(w) \times \mathcal{REQ}_s(w)$	(iv) $\mathcal{ILL}^a(w) \times \mathcal{ILL}_s(w)$

In cell (i) of table 1 we have all possible connections between obligatory actions and obligatory states, in cell (ii) we bind obligatory actions with forbidden states, etc. On the basis of the connections between a-norms and s-norms we shall make up new definitions of required and illegal actions taking into account norms on states. We will see below that pairs from the sets from (i), (ii), (iii) lead to new binary obligations and prohibitions, whereas pairs from (iv) lead only to prohibitions.

Let us first define a new set $\mathcal{REQ}_s^a$ of required actions in the context of s-norms. In the definition we shall use the function "$means$". A means to bring about states from X starting from w is a set of action steps beginning in w and resulting in any w' in X. Formally:

$$means(w, X) \triangleq \{s \in \mathcal{Step} : \exists\, e \in \mathcal{E}, w' \in \mathcal{W} \text{ s.t. } s = \langle w, w', e \rangle \,\&\, w' \in X\} \quad (7)$$

It is worth noting here that the model may be constructed in such a way that sets of states are not compatible with sets of actions in the sense that for some sets of state X there may be no action in $e \in \mathcal{E}$ such that $means(w, X)$ is a set of all $\langle w, w', e \rangle$. Thus, it may be impossible to express the norms obtained by binding a-norms and s-norms either as a-norms or as s-norms (e.g., in [4] it is always possible; see comparison of our theory with [4] in section 4).

$\mathcal{REQ}_s^a$ is defined as follows:

$Z \in \mathcal{REQ}_s^a(w) \triangleq$
(i) $\exists X, Y \, \langle X, Y \rangle \in \mathcal{REQ}^a(w) \times \mathcal{ILL}_s(w) \,\&\, Z = X \setminus means(w, Y)$ or   (8)
(ii) $\exists X, Y \, \langle X, Y \rangle \in \mathcal{ILL}^a(w) \times \mathcal{REQ}_s(w) \,\&\, Z = means(w, Y) \setminus X$

The former condition in definition (8) states that action Z is required in w if there exists a pair (a-required action X, s-illegal state of affairs Y) such that Z is a set of a-required action steps from X which do not lead to s-illegal states from Y. The latter condition states that Z consists of means to achieve s-required states from Y which are not a-illegal action steps from X.

It is worth noting that from either of the conditions (i) or (ii) from definition (8) it follows that

(iii) $\exists X, Y \; \langle X, Y \rangle \in \mathcal{REQ}^a(w) \times \mathcal{REQ}_s(w) \; \& \; Z = X \cap means(w, Y)$

stating that Z is required in w if there is a pair (a-required action X, s-required state of affairs Y) such that Z is a set of action steps from X being a means to bring about Y (starting from w). This inference is possible because of the economy principle (6), transferring complements of a required action to the set of illegal ones.

Let us also stress that the conditions in (8) do not exclude the fact that actions in $\mathcal{REQ}_s^a(w)$ lead to illegal states in $\mathcal{ILL}_s$ and that required states in $\mathcal{REQ}_s(w)$ are achieved by illegal actions in $\mathcal{REQ}^a(w)$.

A new set $\mathcal{ILL}_s^a$ consists of actions which are a-illegal or are means to bring about s-illegal states. Formally:

$$\mathcal{ILL}_s^a(w) \triangleq \{Z \subseteq Step : \exists X, Y \; \langle X, Y \rangle \in \mathcal{ILL}^a(w) \times \mathcal{ILL}_s(w) \; \& \\ Z = X \cup means(w, Y)\} \qquad (9)$$

Theorem 1. $\mathcal{ILL}_s^a$ *and* $\mathcal{REQ}_s^a$ *satisfy conditions analogous to (1), (2), (3), (4), (5), (6).*

Proof. The proof is straightforward but time and space consuming.

Let us notice here that, for the reasons presented in the introductory section, obligation does not 'go up': for $A, B \subseteq Step$, such that $A \subset B$, it is possible that $A \in \mathcal{REQ}^a(w)$ and $B \notin \mathcal{REQ}^a(w)$. Moreover, for three actions such that $A \subset B \subset C$ we can have: $A \in \mathcal{REQ}^a(w)$, $B \notin \mathcal{REQ}^a(w)$ and $C \in \mathcal{REQ}^a(w)$. We have exactly the same situation for $\mathcal{REQ}_s(w)$. Thus, we can say that $\mathcal{REQ}^a(w)$ and $\mathcal{REQ}_s(w)$ are not extensional. The same fact also concerns $\mathcal{REQ}_s^a$. On the other hand, due to conditions (1) and (2), $\mathcal{ILL}_s^a$ has the formal properties of ideal (cf. [11]).

1.3   Individualised Norms in the Model

As we have mentioned in the introductory section one of the main purposes of the present paper is to derive individualised norms from general a-norms and s-norms. Such individualised norms describe what an agent should and what should not do in a particular situation. We want to achieve that goal by finding the most specific general norm. We presented a similar solution for deontic action logic in [6] and now we extend it for the system including s-norms.

We shall express our most specific norm in the form of the greatest $\mathcal{ILL}_s^a(w)$ set. First let us notice that the set $\mathcal{REQ}_s^a(w)$ may be empty. Then the only norms to be considered are prohibitions. Moreover, in the case of non-empty $\mathcal{REQ}_s^a(w)$, due to the obligation economy principle (6) the most specific (thus the most informative) norm expressed as obligation has its prohibited counterpart. Due to conditions imposed on the set $\mathcal{ILL}_s^a(w)$ the greatest element is also the sum of all elements: $\bigcup \mathcal{ILL}_s^a(w)$.

It may happen that the choice of norms is such that $\bigcup \mathcal{ILL}_s^a(w)$ equals $Step$. In that case the system of norms in inconsistent in the sense that one cannot comply with it. In the opposite case $Step \setminus \bigcup \mathcal{ILL}_s^a(w)$ gives a complete recipe defining what an agent should do.

One of the benefits of the solution is the possibility of defining strong permission (or free choice permission - see [9,14,13]) within the model. As it is argued in [14] the notion is useful since, given a norm expressed with the use of strong permission, an agent can freely choose between the actions regulated by the norm and be sure that, whatever the choice is, it is legal and that no other norms or regulations have to be taken into account.

Since $\bigcup \mathcal{ILL}_s^a(w)$ collects all illegal steps, its complement $Step \setminus \bigcup \mathcal{ILL}_s^a(w)$ collects all legal steps. Thus we can define the set $\mathcal{LEG}_s^a(w)$ analogously to $\mathcal{REQ}_s^a(w)$ and $\mathcal{ILL}_s^a(w)$, collecting norms that can be expressed as strong permissions.

$$\mathcal{LEG}_s^a(w) \triangleq 2^{Step \setminus \bigcup \mathcal{ILL}_s^a(w)} \tag{10}$$

The set $\mathcal{LEG}_s^a(w)$, defined that way, has the same properties as the set $\mathcal{ILL}_s^a(w)$, namely the conditions analogous to (1) and (2) hold (as it is expected for strong permission; cf. [9,13]).

If the set $\mathcal{REQ}_s^a(w)$ is not empty, because of (6) we also have:

$$Step \setminus \bigcup \mathcal{ILL}_s^a(w) = \bigcap \mathcal{REQ}_s^a(w) \tag{11}$$

and we can define the set of legal steps as $\bigcap \mathcal{REQ}_s^a(w)$.

Let us finally notice one more property of the set $\bigcup \mathcal{ILL}_s^a(w)$. Namely, we can define it without the use of $\mathcal{ILL}_s^a(w)$, taking into account the following equation:

$$\bigcup \mathcal{ILL}_s^a(w) = \bigcup \mathcal{ILL}^a(w) \cup \bigcup means(\mathcal{ILL}_s(w)) \tag{12}$$

Thus the sets $\mathcal{ILL}_s^a(w)$ and $\mathcal{REQ}_s^a(w)$ are not indispensable in the model for defining individualised norms.

2 Language for $\mathcal{DASF}$ and Its Interpretation

2.1 Language for $\mathcal{DASF}$

A language for $\mathcal{DASF}$ is defined in Backus-Naur notation in the following way:

$$\alpha \ ::= \ a_i \mid \mathbf{0} \mid \mathbf{1} \mid \overline{\alpha} \mid \alpha \sqcup \alpha \mid \alpha \sqcap \alpha \tag{13}$$

$$\varphi \ ::= \ p_i \mid \top \mid \neg\varphi \mid \varphi \vee \varphi \mid \varphi \wedge \varphi \tag{14}$$

$$\psi \ ::= \ \alpha = \alpha \mid \mathsf{F}^a(\alpha) \mid \mathsf{O}^a(\alpha) \mid \mathsf{F}^s(\varphi) \mid \mathsf{O}^s(\varphi) \mid \neg\psi \mid \psi \wedge \psi \tag{15}$$

where a_i belongs to a finite set of *action generators* Act_0, "**0**" is the impossible action and "**1**" is the universal action, "$\overline{\alpha}$" – not α (complement of α), "$\alpha \sqcup \beta$" – α or β (a free choice between α and β); "$\alpha \sqcap \beta$" – α and β (parallel execution of α and β); "$\alpha = \beta$" means that α is identical with β; "$\mathsf{F}^a(\alpha)$" – α is forbidden, "$\mathsf{O}^a(\alpha)$" – α is obligatory; p_i belongs to a set of atomic propositions Atm, "$\top$" represents truth; "$\mathsf{O}^s(\varphi)$" – the state of affairs φ is obligatory; $\mathsf{F}^s(\varphi)$" – the state of affairs φ is forbidden. For a fixed Act_0, by Act we shall understand the set of formulas defined by (13). Obviously $Act_0 \subseteq Act$. Let us notice that the language is protected from iteration of deontic state operators.

2.2 Interpretation for Actions and Deontic Action Operators

$\mathcal{I}^a : Act \longrightarrow 2^{Step}$ is an interpretation function for $\mathcal{DAF}$ defined as follows:

$$\mathcal{I}^a(a_i) \subseteq Step, \ \text{for } a_i \in Act_0 \tag{16}$$

$$\mathcal{I}^a(\mathbf{0}) = \emptyset \tag{17}$$

$$\mathcal{I}^a(\mathbf{1}) = Step \tag{18}$$

$$\mathcal{I}^a(\alpha \sqcup \beta) = \mathcal{I}^a(\alpha) \cup \mathcal{I}^a(\beta) \tag{19}$$

$$\mathcal{I}^a(\alpha \sqcap \beta) = \mathcal{I}^a(\alpha) \cap \mathcal{I}^a(\beta) \tag{20}$$

$$\mathcal{I}^a(\overline{\alpha}) = Step \setminus \mathcal{I}^a(\alpha) \tag{21}$$

Thus, every action generator is interpreted as a set of labelled transitions, the impossible action has no transitions, the universal action brings about all possible transitions, operations "$\sqcup$", "$\sqcap$" between actions and "$^-$" on a single action are interpreted as set-theoretical operations on interpretations of actions.

$\mathcal{I}^a$ is an interpretation of actions insensitive to their preconditions and takes into account all possible executions of actions in all the states they can be executed. To make our interpretation related to a particular state we introduce $\mathcal{I}^a(w, \alpha)$ being a local interpretation of action (relativised to situation w) and define it as follows:

$$\mathcal{I}^a(w, \alpha) = \mathcal{I}^a(\alpha) \cap exe(w) \tag{22}$$

where

$$exe(w) \triangleq \{\langle w, w', e \rangle : \langle w, w', e \rangle \in Step\} \tag{23}$$

is a set of all action steps executable in state w.

Satisfaction conditions for the action formulas in any model $\mathcal{M} = \langle \mathcal{DAF}, \mathcal{I}^a \rangle$ are defined below:

$$\mathcal{M}, w \models \mathsf{F}^a(\alpha) \iff \mathcal{I}^a(w, \alpha) \in \mathcal{ILL}^a(w)$$
$$\mathcal{M}, w \models \mathsf{O}^a(\alpha) \iff \mathcal{I}^a(w, \alpha) \in \mathcal{REQ}^a(w)$$
$$\mathcal{M}, w \models \alpha = \beta \iff \mathcal{I}^a(\alpha) = \mathcal{I}^a(\beta)$$

2.3   Interpretation for Propositions and Deontic State Operators

$v : Atm \longrightarrow 2^{\mathcal{W}}$ is a standard valuation function that assigns a subset $v(p_i)$ of $\mathcal{W}$ to each proposition in Atm. We shall think of $v(p_i)$ as a semantical representation of proposition p_i in the model, i.e., a set of states in $\mathcal{W}$ where p is true (takes place).

Satisfaction conditions for the state formulas in any model $\mathcal{M} = \langle \mathcal{DSF}, \mathcal{I}^a \rangle$ are defined below:

$$\mathcal{M}, w \models \mathsf{O}^s(\alpha) \iff \|\varphi\|^{\mathcal{M}} \in \mathcal{REQ}_s(w)$$
$$\mathcal{M}, w \models \mathsf{F}^s(\alpha) \iff \|\varphi\|^{\mathcal{M}} \in \mathcal{ILL}_s(w)$$

$\|\varphi\|^{\mathcal{M}}$ is a *truth set* of the sentence φ in the model $\mathcal{M}$, i.e., a set of states at which φ is true. Formally[3]: $\|\varphi\|^{\mathcal{M}} = \{w \in \mathcal{W} : \mathcal{M}, w \models \varphi\}$.

2.4   A Bridge between Deontic Actions and States

Now we introduce deontic operators, O and F, combing actions and results. Both operators have two arguments – an action and a formula being a result of the action. We shall read them in natural language and understand intuitively as follows:

- "$\mathsf{O}(\alpha, \varphi)$" – it is obligatory to execute α in such a way that φ is its result. For "$\mathsf{O}(\alpha, \varphi)$" to be true we require that action α or formula φ have to be obligatory, i.e., there exists an a-norm making α obligatory or there exists an s-norm making φ obligatory.
- "$\mathsf{F}(\alpha, \varphi)$" – it is forbidden to execute α or bring about φ as its result. For a particular behaviour to make it forbidden, it is enough that one out of the two conditions is fulfilled.

Formally we can define the new operators in the following way:

$$\mathsf{O}(\alpha, \varphi) \triangleq (\mathsf{O}^a(\alpha) \wedge \mathsf{F}^s(\neg\varphi)) \vee (\mathsf{O}^s(\varphi) \wedge \mathsf{F}^a(\overline{\alpha})) \tag{24}$$

$$\mathsf{F}(\alpha, \varphi) \triangleq \mathsf{F}^a(\alpha) \vee \mathsf{F}^s(\varphi) \tag{25}$$

We can also define weak permission:

$$\mathsf{P}_{weak}(\alpha, \varphi) \triangleq \neg\mathsf{F}^a(\alpha) \wedge \neg\mathsf{F}^s(\varphi) \tag{26}$$

From the definition we can derive the following formula:

$$\mathsf{P}_{weak}(\alpha, \varphi) \equiv \neg\mathsf{F}(\alpha, \varphi) \tag{27}$$

---

[3] The following facts about a truth set are known:

- $\|p_i\|^{\mathcal{M}} = v(p_i)$, for every $p_i \in Atm$
- $\|\top\|^{\mathcal{M}} = \mathcal{W}$
- $\|\neg\varphi\|^{\mathcal{M}} = \mathcal{W} \setminus \|\varphi\|^{\mathcal{M}} = -\|\varphi\|^{\mathcal{M}}$
- $\|\varphi \wedge \psi\|^{\mathcal{M}} = \|\varphi\|^{\mathcal{M}} \cap \|\psi\|^{\mathcal{M}}$
- $\|\varphi \vee \psi\|^{\mathcal{M}} = \|\varphi\|^{\mathcal{M}} \cup \|\psi\|^{\mathcal{M}}$

Satisfaction conditions for O *and* F. Let us now turn to the satisfaction conditions for O and F. The same set of steps can be obtained as an interpretation of different "action-proposition" pairs taken as arguments of operators O and F. Moreover, in the case of O we can have the same set of steps for pairs α, φ and β, φ even if $\mathcal{I}^a(\alpha) \neq \mathcal{I}^a(\beta)$ (the same holds for pairs α, φ and α, ψ). That is because we postulate that binary obligations emerge as results of trimming obligatory actions by means of their forbidden results or selecting from permitted actions those which lead to obligatory results. On the other hand a two-argument prohibition holds for any combination of action and proposition which interpretation is a subset of the interpretation of a forbidden pair.

The above remarks can be formalised with the use of the following satisfaction conditions:

$$\mathcal{M}, w \models \mathsf{O}(\alpha, \varphi) \iff (\mathcal{I}^a(w, \alpha) \in \mathcal{REQ}^a(w) \text{ or } \|\varphi\|^{\mathcal{M}} \in \mathcal{REQ}_s(w)) \ \& $$
$$\mathcal{I}^a(w, \alpha) \cap means(w, \|\varphi\|^{\mathcal{M}}) \in \mathcal{REQ}_s^a(w)$$

$$\mathcal{M}, w \models \mathsf{F}(\alpha, \varphi) \iff (\mathcal{I}^a(w, \alpha) \cup means(w, \|\varphi\|^{\mathcal{M}})) \in \mathcal{ILL}_s^a(w)$$

Our definitions of the operators make the conditions fulfilled.

3   Logics for Deontic Actions, Deontic States and a Bridge between Them

3.1   Logics for Deontic Actions and Deontic States

Deontic action logic is expressed in the language defined by conditions (13) and (15) without O^s and F^s operators. Its axiomatisation presented below corresponding with the $\mathcal{DAF}$ (see section 1.1) comes from [13].

$$\text{Boolean algebra for actions from } Act. \tag{28}$$

$$\mathsf{F}^a(\alpha \sqcup \beta) \equiv \mathsf{F}^a(\alpha) \wedge \mathsf{F}^a(\beta) \tag{29}$$

$$\mathsf{F}^a(\mathbf{0}) \tag{30}$$

$$\mathsf{O}^a(\alpha) \wedge \mathsf{O}^a(\beta) \to \mathsf{O}^a(\alpha \sqcap \beta) \tag{31}$$

$$\mathsf{O}^a(\alpha) \to \mathsf{F}^a(\overline{\alpha}) \tag{32}$$

$$\mathsf{O}^a(\alpha) \wedge \mathsf{F}^a(\beta) \to \mathsf{O}^a(\alpha \sqcap \overline{\beta}) \tag{33}$$

Deontic state logic is expressed in the language defined by conditions (14) and (15) without O^a and F^a operators. Its axiomatisation corresponding with the $\mathcal{DSF}$ (see section 1.1) is analogous to the axioms above (of course Boolean algebra is substituted by classical propositional calculus).

It is worth mentioning that we assume that deontic operators for actions and states have the same logical characterisation. But generally their axiomatisations are independent and may differ (see for instance a deontic state logic similar to ours presented in [8]).

3.2 Tautologies for Binary Deontic Operators

The formulas below are theses of the combined deontic logic. They correspond
to the axioms of deontic action (state) logic presented above.

$$F(\alpha \sqcup \beta, \varphi) \equiv F(\alpha, \varphi) \wedge F(\beta, \varphi) \tag{34}$$

$$F(\mathbf{0}, \varphi) \tag{35}$$

$$F(\alpha, \varphi \vee \psi) \equiv F(\alpha, \varphi) \wedge F(\alpha, \psi) \tag{36}$$

$$F(\alpha, \perp) \tag{37}$$

$$O(\alpha, \varphi) \wedge O(\beta, \varphi) \rightarrow O(\alpha \sqcap \beta, \varphi) \tag{38}$$

$$O(\alpha, \varphi) \wedge O(\alpha, \psi) \rightarrow O(\alpha, \varphi \wedge \psi) \tag{39}$$

$$O(\alpha, \varphi) \wedge F(\beta, \varphi) \rightarrow O(\alpha \sqcap \overline{\beta}, \varphi) \tag{40}$$

$$O(\alpha, \varphi) \wedge F(\alpha, \psi) \rightarrow O(\alpha, \varphi \wedge \neg \psi) \tag{41}$$

$$O(\alpha, \varphi) \rightarrow F(\overline{\alpha}, \neg \varphi) \tag{42}$$

$$P_{weak}(\alpha \sqcup \beta, \varphi) \equiv P_{weak}(\alpha, \varphi) \vee P_{weak}(\beta, \varphi) \tag{43}$$

$$P_{weak}(\alpha, \varphi \vee \psi) \equiv P_{weak}(\alpha, \varphi) \vee P_{weak}(\alpha, \psi) \tag{44}$$

Formulas (43) and (44) correspond to van der Meyden axioms ◇3. and ◇5. for
weak permission presented in [14]. Our forbiddance operator is strong; as such
it is formally similar to van der Meyden's strong permission π. So, by analogy,
our F operator satisfies the same axiom schemas (see π3. and π5. in [14]).

3.3 Bridging Formulas

Some bridging formulas, connecting defined operators with the primitive ones,
follow immediately from definitions (24) and (25).

$$O(\alpha, \varphi) \rightarrow O^a(\alpha) \vee O^s(\varphi) \tag{45}$$

$$O^a(\alpha) \wedge O^s(\varphi) \rightarrow O(\alpha, \varphi) \tag{46}$$

More interesting relations between binary and unary deontic operators can be
formulated for specific systems of norms. Let us, for instance, consider a system
in which there are no a-norms. No action should then be obligatory or forbidden.
However, we need to take into account that $\mathbf{0}$ is always forbidden by axiom (30).
Thus in the normative system of our interest we have the following formula:

$$F^a(\alpha) \rightarrow \alpha = \mathbf{0} \tag{47}$$

and its equivalent:

$$F^a(\alpha) \equiv \alpha = \mathbf{0} \tag{48}$$

It follows from (47) that only the universal action $\mathbf{1}$ can be obligatory in such a system, formally

$$O^a(\alpha) \to \alpha = \mathbf{1} \tag{49}$$

However, we may also require the stronger version of (49):

$$\neg O^a(\alpha) \tag{50}$$

If a-norms are absent, then all norms are s-norms. We can express this idea by the following formula:

$$((F^a(\beta_1) \to \beta_1 = \mathbf{0}) \wedge \neg O^a(\beta_2)) \to (F(\alpha, \varphi) \equiv F^s(\varphi)) \wedge (O(\alpha, \varphi) \equiv O^s(\varphi)) \tag{51}$$

In the weaker version we have:

$$(F^a(\beta) \to \beta = \mathbf{0}) \to (F(\alpha, \varphi) \equiv F^s(\varphi)) \tag{52}$$

Similarly, we can define a system in which there are no s-norms.

3.4   Individualised Norms in the Logic

To introduce individualised norms into the logic we shall use an additional operator of the most general forbiddance "$F^\#$". It is not possible to define it within the language, so we use the following metalanguage definition:

$$F^\#(\alpha, \varphi) \triangleq F(\alpha, \varphi) \ \& \ \forall \beta, \psi \ (F^\#(\beta, \psi) \Longrightarrow \beta \sqsubseteq \alpha \ \& \ \varphi \to \psi) \tag{53}$$

$F^\#(\alpha, \varphi)$ indicates a unique pair of arguments defining the space of forbidden actions and propositions. Everything that is not described by that pair is permitted.

Now we can introduce the operator of strong permission (for which we shall use the symbol P) using the following postulates:

$$P(\alpha \sqcup \beta, \varphi) \equiv P(\alpha, \varphi) \wedge P(\beta, \varphi) \tag{54}$$

$$P(\mathbf{0}, \varphi) \tag{55}$$

$$P(\alpha, \varphi \vee \psi) \equiv P(\alpha, \varphi) \wedge P(\alpha, \psi) \tag{56}$$

$$P(\alpha, \bot) \tag{57}$$

$$P(\alpha, \varphi) \wedge F(\alpha, \varphi) \to \alpha = \mathbf{0} \vee (\varphi \equiv \bot) \tag{58}$$

Moreover

$$F^\#(\alpha, \varphi) \to P(\overline{\alpha}, \neg\varphi) \tag{59}$$

Formulas (54) – (57) are analogous to formulas (34) and (37) for prohibition. Formula (58) states that no action or state should be at the same time forbidden and strongly permitted. Formulas (54) – (58) correspond to axioms defining strong permission in [9] and [13].

Let us notice that strong permission has some properties that are regarded as permission paradoxes. However, they are paradoxical only when we want to connect them with weak permission (lack of prohibition) usually used in natural language. In the context of strong permission (free choice permission) they are quite natural.

As we noticed in the considerations on the level of models if there exists any obligation in the normative system we can identify the weakest permission with the strongest obligation provided obligations are consistent, i.e., it is true that $\neg O(\mathbf{0})$:

$$O(\beta, \psi) \rightarrow (\mathsf{F}^{\#}(\alpha, \varphi) \rightarrow O(\overline{\alpha}, \neg\varphi)) \tag{60}$$

$$O(\alpha, \varphi) \wedge P(\alpha, \varphi) \rightarrow \mathsf{F}^{\#}(\overline{\alpha}, \neg\varphi) \tag{61}$$

A similar intuition about the relation between strong permission and obligation understood as the most specific norm was presented in [8].

4 Comparison with Dov Gabbay, Loïc Gammaitoni and Xin Sun's Paper

In this section we would like to refer to Dov Gabbay, Loïc Gammaitoni and Xin Sun's paper [4] and compare the theory contained therein with our work.

4.1 Actions

In the Gabbay at al. paper we find an ontology of action similar to ours. The counterparts of our action steps are present there as unlabelled transitions between states. Among actions—sets of transitions—the authors distinguish:

- particular actions being singletons
- atomic actions being unions of particular actions which share the same precondition (starting point)
- molecular actions being unions of atomic actions bearing different preconditions

The interpretation of union and intersection of actions is similar to ours, whereas the interpretation of action's negation is local and precondition sensitive, i.e., the authors make sure that the transitions belonging to a negation of a given action A have the same precondition as the ones in A. Postconditions of $-A$ are members of a set-theoretical difference of $\mathcal{W}$ and postconditions of A. Because any pair $\{\langle w, w' \rangle\}$, where $\langle w, w' \rangle \in \mathcal{W} \times \mathcal{W}$ is a particular action, the negation of an action may create unintuitive results, namely, every state which cannot be achieved by action A from w can be brought about by its negation.

In our model we define the interpretation of a negation of action A in a similar way, however the procedure of obtaining $-A$ is different. We first look for a set-theoretical difference between the set of possible actions $\mathcal{S}tep$ and the interpretation of A (carrying out $-A$ means the execution of an action from $\mathcal{S}tep$ which is not in A), so our interpretation of A's negation is essentially global (see condition (21)). Then we find those executions of the action $-A$ which can be carried out in a chosen state w (by referring to function $exe(w)$; see definition (22)).

The main difference between the two approaches to action's negation lies in the fact that we define it by referring to $Step$ which contains actions which can be executed. So we can be sure that our negation of an action does not go beyond $Step$.

4.2   Norms

The authors of [4] consider an action to be obligatory "if there is a legal document approving it to be obligatory". It corresponds to our $\mathcal{REQ}^a$. In fact we share the same intuitions. They, however, did not introduce a set of illegal/forbidden actions. Thus, there is no counterpart of $\mathcal{ILL}^a$ in the studied theory. Semantical counterpart of obligation in [4] behaves similarly to our $\mathcal{REQ}^a$. We share axioms (in fact just one axiom) for the operator of obligation and its satisfaction condition.

A state of affairs is obligatory "if it is true in those ideal worlds or outcomes defined through obligatory actions." F_t from [4] is a set of obligatory propositions[4]. F_t does not really correspond with our $\mathcal{REQ}_s$ for the reasons described below. The set F_t contains sets of ideal outcomes of actions when they are executed at t. $\mu(t, X)$ is a meaning assignment function that, for a given state t and action X, is responsible for selecting a set of ideal outcomes from all the results of the action. All sets selected in that way go to F_t. In our theory we do not assume any dependency between obligatory actions and obligatory states.

The range of function μ for a fixed t is (just) a subset of F_t. Thus, the authors of [4] do not exclude propositions which are not ideal results of any action from F_t. We also claim that a set of obligatory propositions $\mathcal{REQ}_s$ may contain propositions which can be unobtainable.

We understand that obligatory propositions do come from legal documents and we do not refer to the quite enigmatic concept of "ideal outcome".

The authors of [4] also assume that if propositions X and Y are obligatory, then their intersection is obligatory too. At the same time their satisfaction condition for state obligation excludes the empty set to be obligatory. The two conditions together exclude a propositional counterpart of (31) to be a tautology. Thus we can say that our obligations for propositions differ.

Conclusion and Future Perspectives

In the paper the idea of binding norms concerning actions (a-norms) and states (s-norms) is presented. We assume that a-norms and s-norms come from normative sources and are general norms which can be applied to different situations and agents. An agent, in order to comply with the norms in a certain situation, has to

---

[4] We would like to warn the reader that the notation in [4] may be confusing for the reader of the present paper. There F_t does not stand for actions forbidden in t.

take all of them into account, individualise them and find the actions which can be carried out and results which can be brought about without breaking the law.

We realise our idea by introducing two deontic logics accompanied by models— one for actions and one for states. We use them to define deontic notions binding the two. We extensively discuss formal properties of the obtained framework.

Since in our theory we take into account only one-step actions (transitions), we are interested in extending it to sequences of actions. As we have shown in [5] the move from one-step actions to their sequences is far from being trivial.

Acknowledgements. The authors would like to thank Marek Piechowiak for his remarks on general and individualised norms in the legal context.

References

1. Atienza, M., Manero, J.R.: A theory of legal sentences. Kluwer (1998)
2. Broersen, J., Gabbay, D., van der Torre, L.: Discussion Paper: Changing Norms Is Changing Obligation Change. In: Ågotnes, T., Broersen, J., Elgesem, D. (eds.) DEON 2012. LNCS, vol. 7393, pp. 199–214. Springer, Heidelberg (2012)
3. d'Altan, P., Meyer, J.-J.C., Wieringa, R.J.: An integrated framework for ought-to-be and ought-to-do constraints. Artificial Intelligence and Law 4(2), 77–111 (1996)
4. Gabbay, D., Gammaitoni, L., Sun, X.: The paradoxes of permission an action based solution. Journal of Applied Logic 12(2), 179–191 (2014)
5. Kulicki, P., Trypuz, R.: A deontic action logic with sequential composition of actions. In: Ågotnes, T., Broersen, J., Elgesem, D. (eds.) DEON 2012. LNCS, vol. 7393, pp. 184–198. Springer, Heidelberg (2012)
6. Kulicki, P., Trypuz, R.: Two faces of obligation. In: Brożek, A., Jadacki, J., Zarnic, B. (eds.) Theory of Imperatives from Different Points of View, Logic, Methodology and Philosophy of Science at Warsaw University 7. Wydawnictwo Naukowe, Semper (2013)
7. Pattaro, E.: The Law and The Right. A Treatise of Legal Philosophy and General Jurisprudence, vol. 1. Springer (2012)
8. Roy, O., Anglberger, A.J.J., Gratzl, N.: The Logic of Obligation as Weakest Permission. In: Ågotnes, T., Broersen, J., Elgesem, D. (eds.) DEON 2012. LNCS, vol. 7393, pp. 139–150. Springer, Heidelberg (2012)
9. Segerberg, K.: A deontic logic of action. Studia Logica 41, 269–282 (1982)
10. Stolpe, A.: A theory of permission based on the notion of derogation. J. Applied Logic 8(1), 97–113 (2010)
11. Trypuz, R., Kulicki, P.: A systematics of deontic action logics based on boolean algebra. Logic and Logical Philosophy 18, 263–279 (2009)
12. Trypuz, R., Kulicki, P.: Towards metalogical systematisation of deontic action logics based on boolean algebra. In: Governatori, G., Sartor, G. (eds.) DEON 2010. LNCS, vol. 6181, pp. 132–147. Springer, Heidelberg (2010)
13. Trypuz, R., Kulicki, P.: On deontic action logics based on Boolean algebra. Journal of Logic and Computation (2013)
14. van der Meyden, R.: The Dynamic Logic of Permission. Journal of Logic and Computation 6(3), 465–479 (1996)
15. von Wright, G.H.: Norm and Action: A Logical Inquiry. Routledge & Kegan Paul (1963)
16. Ziembinski, Z.: Practical Logic. Dordrecht, D. Reidel Pub. Co. (1976)

Author Index